RECEIVED
13 JUL 2004

NONEQUILIBRIUM THERMODYNAMICS

Transport and Rate Processes in Physical and Biological Systems

To Zuhal Selcuk and Can

NONEQUILIBRIUM THERMODYNAMICS

Transport and Rate Processes in Physical and Biological Systems

Yaşar Demirel

Department of Chemical Engineering
Virginia Tech
Blacksburg, VA, U.S.A.

2002
ELSEVIER
Amsterdam – Boston – London – New York – Oxford – Paris – San Diego –
San Francisco – Singapore – Sydney – Tokyo

ELSEVIER SCIENCE B.V.
Sara Burgerhartstraat 25
P.O. Box 211, 1000 AE Amsterdam, The Netherlands

First edition 2002

Library of Congress Cataloging in Publication Data
A catalog record from the Library of Congress has been applied for.

British Library Cataloguing in Publication Data
A catalogue record from the British Library has been applied for.

ISBN: 0-444-50886-4

♾ The paper used in this publication meets the requirements of ANSI/NISO Z39.48-1992 (Permanence of Paper).
Printed in The Netherlands.

Contents

Preface

Classical thermodynamics is based on a limited number of natural laws, which have led to a vast number of equations describing macroscopic behavior of various types of systems. However classical thermodynamics is mainly limited to energy conversion in equilibrium, and particularly applied to reversible and closed systems. Beside the equilibrium, there are instabilities, fluctuations, and evolutionary processes.

The objective of this book is to bring out and emphasize the unifying role of thermodynamics in transport phenomena, chemical reactions, and coupled processes in physical and biological systems by using the nonequilibrium thermodynamic approach. The development of nonequilibrium thermodynamics is based on the entropy generating character of irreversible processes to provide a link between the classical thermodynamics, and the transport and rate processes. In 1850, Clausius introduced the concept of noncompensated heat as a measure of irreversibility. In 1911, Jaumann introduced the concepts of the entropy production and the entropy flux. Donnan and Guggenheim in 1934 related the coupled natural processes to the second law of thermodynamics stated as "A finite amount of organization may be purchased at the expense of a greater amount of disorganization in a series of interrelated spontaneous action." After the publication of Onsager's reciprocal relations in 1931, Casimir, Meixner, Prigogine and De Groot made early attempts of a macroscopic and general theory for irreversible processes.

Irreversible processes cause entropy generation because of net thermodynamic forces and flows within the system. Nonequilibrium thermodynamics is mainly concerned with the analysis of entropy generation and the study of the relations between the conjugate flows and the forces. Using the general balance equations of mass, momentum, energy, entropy, and the Gibbs relation, entropy generation or the dissipation function can be derived. The rate of entropy generation or the dissipation function identifies the flows and forces that are related by the phenomenological equations. These equations contain the proportionality constants called the phenomenological coefficients. This coefficient matrix is symmetric according the Onsager's reciprocal relations. Such an analysis is necessary to understand the complex, coupled transport and the global behavior of physical and biological systems.

The theory treating near-equilibrium phenomena is called the linear nonequilibrium thermodynamics. It is based on the local equilibrium assumption in the system and phenomenological equations that linearly relate forces and flows of the processes of interest. Application of classical thermodynamics to nonequilibrium systems is valid for systems not too far from equilibrium. This condition does not prove excessively restrictive as many systems and phenomena can be found within the vicinity of equilibrium. Therefore equations for property changes between equilibrium states, such as the Gibbs relationship, can be utilized to express the entropy generation in nonequilibrium systems in terms of variables that are used in the transport and rate processes. The second law analysis determines the thermodynamic optimality of a physical process by determining the rate of entropy generation due to the irreversible process in the system for a required task.

Some processes may have forces operating far away from equilibrium where the linear phenomenological equations are no longer applicable. Such a domain of irreversible phenomena like some chemical reactions, periodic oscillations, and bifurcation, is examined by extended nonequilibrium thermodynamics. Extending the methods of thermodynamics to treat the linear and nonlinear phenomena, and such dissipative structures are attracting scientists from various disciplines.

This book introduces the theory of nonequilibrium thermodynamics and its use in transport and rate processes of physical and biological systems. The first chapter briefly presents the equilibrium thermodynamics. In the second chapter, the transport and rate processes have been summarized. The rest of the book covers the theory of nonequilibrium thermodynamics, dissipation function and various applications based on linear nonequilibrium thermodynamics. Extended nonequilibrium thermodynamics is briefly covered. All the parts of the book can be used in senior and graduate teaching in engineering and science.

All through this book, the work of many people who contributed to both theory and the applications of nonequilibrium thermodynamics has been visited and revisited. All those whose work has contributed in preparing this book are acknowledged, and greatly appreciated.

Y. Demirel, 2002

Chapter 1

Equilibrium Thermodynamics

1. BASIC DEFINITIONS

The name thermodynamics stems from the Greek word therme (heat) and dynamis (power), and can be defined as science of energy. Energy can cause a physical and chemical change. A process takes place in a *system*, which is formed by a geometrical volume of macroscopic dimensions subjected to controlled experimental conditions. A system contains a substance, which has a large amount of molecules. The properties of a system depend on the behavior of these molecules called the *microscopic state*, which is the main concern of *statistical thermodynamics*. On the other hand *Classical thermodynamics* is concerned with the *macroscopic state*, which is the average behavior of large group of molecules such as temperature or pressure.

Part of the physical universe with a specified boundary for observation is a thermodynamic system. An ideal thermodynamic system is a model system with simplifications to represent a real system that can be described by the theoretical thermodynamics approach; a simple thermodynamic system is a single state system wit no internal boundaries, and is not acted upon by external force fields or inertial forces. A composite system, however, has at least two simple systems separated by a barrier restrictive to one form of energy or matter. *Boundary* of the volume separates the system from its *surroundings*. Through the boundary of an *open system* energy and matter can be exchanged between the system and its surroundings. In an open system we may consider the two changes to the mass of component j that are the internal change taking place inside the system $d_{\mathrm{i}}m_j$, and the external change supplied from the surrounding $d_{\mathrm{e}}m_j$

$$dm_j = d_{\mathrm{e}}m_j + d_{\mathrm{i}}m_j \qquad (1)$$

At stationary state, system exchanges energy or matter at constant rate. The *isolated systems* cannot exchange energy and matter, while the *closed systems* exchange only energy with the surroundings. In a closed system, mass changes only due to the chemical reaction. *Adiabatic systems* are surrounded by walls, which prevent the exchange of matter and thermal energy. A system and its surrounding are considered the *universe*.

Statistical averaging over the microscopic coordinates of motion are observable and called *thermodynamic properties*, coordinates, variables, or parameters. If a thermodynamic variable is *a state function*, its change is independent of the *path* between the initial and final states, and depends on only the properties of initial and final states of the system. Infinitesimal change of a state function is an exact differential. Properties like mass m and volume V are defined by the system as a whole. Such properties are additive, and are called the *extensive properties*. Separation of the total change for a component into the external and internal parts may be generalized to any extensive property. All extensive properties are homogeneous functions of the first order in the mass of the system. For example, doubling the mass of a system at constant composition doubles the internal energy. The pressure P and temperature T define the values at each point of the system and are therefore called the *intensive properties*, which can be expressed as derivatives of extensive properties, such as the temperature $T = (\partial U / \partial S)_{V,N_i}$.

Energy may be transferred in the forms of heat or work through the boundary of a system. A mechanical work of expansion or compression proceeds with the observable motion of the coordinates of the particles of matter. Chemical work, on the other hand, proceeds with changes in internal energy due to changes in the chemical composition (mass action). Potential energy is a capacity for mechanical work related to position of a body, while the kinetic energy is the capacity for mechanical work related to motion of a body. Potential and kinetic energies are external, while the sensible heat and latent heat are internal energies.

Conservation of mass in an open system states that the change in the total mass is equal to the mass exchanged with the surroundings. During a process, energy can be transferred and converted from one form to another, while the total energy remains constant. This is known as the *conservation of energy* principle.

2. REVERSIBLE AND IRREVERSIBLE PROCESSES

Consider the equations that describe time dependent physical processes, if these equations are invariant with regard to the algebraic sign of the time, the process is called a *reversible process*; otherwise it is called an *irreversible process*. Reversible process refers to macroscopic processes that occur while the system remains near equilibrium. The process is called reversible because it can be reversed at any stage by a slight change in external parameter. In a reversible process, it would be possible to perform a second process in at least one way to restore the system and of its environment to their respective original states, except for differential changes higher than the first order. A reversible process proceeds with infinitesimal driving forces (e.g. gradients) within the system. Hence for a linear transport system, reversible change occurs slowly on the scale of macroscopic relaxation times, and dissipative effects must not be present. Time appears only through its arithmetic value in the equations for reversible processes. For example, the wave equation describing the propagation of waves in a non-absorbing medium is given by

$$\frac{1}{c_o^2}\frac{\partial^2 u}{\partial t^2} = \nabla^2 u \tag{2}$$

where c_o is the velocity of propagation and u is the amplitude of the wave. Eq. (2) is invariant for the substitution t or $(-t)$, hence the propagation of waves is a reversible process. For a simple reversible reaction, if one path is preferred for the backward reaction, the same path must also be preferred for the reverse reaction. This is called the principle of *microscopic reversibility*. Time can be measured by reversible, periodic phenomena, such as the oscillations of a pendulum. However, the direction of time cannot be determined by such phenomena; it is related to the unidirectional increase of entropy in all natural processes. Some ideal processes may be reversible and proceed in the forward and backward directions.

The Fourier equation

$$\frac{1}{\alpha}\frac{\partial T}{\partial t} = \nabla^2 T \tag{3}$$

is not invariant with respect to time, and it describes an irreversible process. In Eq. (3), α shows the thermal diffusivity. Irreversibility is a consequence of collisional dynamics in which transfer of mass, energy, and momentum take place. As a typical example of irreversibility, hydrodynamics specifies a number of nonequilibrium states by the mass density, velocity, and energy density of the fluid. The hydrodynamic equations then comprise a wide range of relaxation processes, such as heat flow, diffusion or viscous dissipation, which are all known as the irreversible changes. All natural processes proceed, or decay toward an equilibrium state; they can be expressed by phenomenological relations, such as Fourier's law of heat conduction, and dissipate their driving power.

The term irreversibilty has two different uses and has been applied to different 'arrows of time'. These arrows are not interrelated, however they seem to be connected to intuitive notion of causality. Mostly the word irreversibility refers to the direction of time evolution of a system. Irreversibility is also used to describe non-invariance of the changes with respect to the nonlinear time reversal transformation. For changes that generate the space-time symmetry transformations, irreversibility implies the impossibility to create a state, which evolves backward in time, and therefore irreversibility is a time asymmetry due to a preferred direction of time evolution.

The most prominent arrow of time is the thermodynamic arrow of time; the entropy S in an isolated system increases $dS/dt > 0$, until it reaches equilibrium $dS/dt = 0$, and hence displays a direction of change. Another arrow of time of classical physics is the radiation arrow. The phenomenological approach favoring the retarded potential over the solution of the Maxwell field equation is called the time arrow of radiation. These two arrows of time lead to Einstein-Ritz controversy: Einstein believed that irreversibility is based on probability consideration, while Ritz believed that an initial condition and thus causality is the basis of irreversibility. Causality and probability may be two aspects of the same principle since the arrow of time has a global nature [3].

Real physical processes progress with *dissipative phenomena*, such as mechanical or electrical friction, viscosity, and turbulence. These dissipative phenomena internally generate heat, and decrease the amount of energy available for work. In an isolated composite system, the change in internal energy of the subsystem equals the change in heat

$$dU = \delta Q = \frac{\partial U}{\partial S} dS \tag{4}$$

This phenomenon is associated with the level of entropy production due to irreversibility in the process. Entropy is not conserved in an isolated system; it is the extensive parameter of heat, and cannot be measured directly.

3. EQUILIBRIUM

If a physical system is isolated, its state changes irreversibly to a time-invariant state in which no physical or chemical change occurs, and a state of equilibrium is reached in a finite time. At equilibrium, all the irreversible processes vanish, and uniform temperature, pressure and chemical potentials are attained; this means that no thermodynamic force is present in the system. A thermodynamic system may be in *stable, metastable, unstable*, or in *neutral equilibrium*. In a stable system, a perturbation causes small departures from the original conditions, which are restorable. In an unstable equilibrium, even a small perturbation causes large irreversible changes. A metastable system may be stable and unstable according to the level and direction of perturbation. Any perturbation will not cause any change in neutral equilibrium. The emergence of macroscopic reversibility from microscopic irreversibilities is referred to as *dynamic equilibrium* with the mechanisms of the cancellation of opposite molecular processes.

3.1 The fundamental equations

A *fundamental equation* relates all extensive properties of a thermodynamic system, and hence contains all the thermodynamic information on the system. For example, the fundamental equation in terms of entropy S is given by

$$S = S(U,..,X_j,..) \tag{5}$$

The extensive properties of U and X are the canonical variables. The fundamental equation in terms of internal energy U is given by

$$U = U(S,..,X_j,..) \tag{6}$$

For the fundamental equations of entropy and internal energy the canonical variables consists of extensive parameters. These extensive properties for a simple system are S, U and V, and the fundamental equations defines a fundamental surface of entropy $S = S(U,V)$ in the Gibbs space of S, U, V.

Differential form of the fundamental equation contains the intensive thermodynamic properties. For example dS and dU are expressed by

$$dS = \left(\frac{\partial S}{\partial U}\right)_X dU + \sum_i \left(\frac{\partial S}{\partial X_i}\right)_{U,X \neq X_i} dX_i \tag{7}$$

$$dU = \left(\frac{\partial U}{\partial S}\right)_X dS + \sum_i \left(\frac{\partial U}{\partial X_i}\right)_{S,X \neq X_i} dX_i \tag{8}$$

Here the first-order partial derivatives are the intensive properties T, I, and Y. In terms of intensive properties, Eqs. (7) and (8) become

$$dS = \frac{1}{T} dU + \sum_i I_i dX_i \tag{9}$$

$$dU = TdS + \sum_i Y_i dX_i \tag{10}$$

The first term on the right-hand side of Eq. (9) or Eq. (10) represents the thermal energy (heat) terms associated with the thermodynamic temperature T, and remaining terms are the work terms. The pairs of intensive and extensive properties, for example $1/T$ and U, or I_i and X_i in Eq. (9) are called the conjugate properties.

3.2. Thermodynamic equilibrium

An extremum principle minimizes or maximizes the fundamental equation subject to certain constraints. For example, the principle of maximum entropy $(dS)_{\mathrm{U}} = 0$ and $(d^2S)_{\mathrm{U}} < 0$, and the principle of minimum internal energy $(dU)_{\mathrm{S}} = 0$ and $((d^2U)_{\mathrm{S}} > 0$ are the fundamental principles of thermodynamic equilibrium, and can be associated with the thermodynamic stability. The conditions of thermodynamic equilibrium can be established in terms of the extensive parameters *U* and *S*, or in terms of intensive parameters. For that we consider a composite system with two simple subsystems of *A* and *B* having a single component. Then the condition of equilibrium is expressed as

$$dU = (T_A - T_B)dS_A - (P_A - P_B)dV_A + (\mu_A - \mu_B)dN_A = 0 \tag{11}$$

Hence the thermal, mechanical, and chemical equilibrium conditions in terms of the intensive properties are

$$T_A = T_B;\ P_A = P_B;\ \mu_A = \mu_B \tag{12}$$

since the dS_A, dV_A, and $d\mu_A$ are the infinitesimal changes in independent variables. Similarly the equilibrium conditions are expressed in terms of entropy

$$dS = \left(\frac{1}{T_A} - \frac{1}{T_B}\right)dU_A + \left(\frac{P_A}{T_A} - \frac{P_B}{T_B}\right)dV_A - \left(\frac{\mu_A}{T_A} - \frac{\mu_B}{T_B}\right)dN_A = 0 \tag{13}$$

and the equilibrium conditions become

$$\left(\frac{1}{T_A} - \frac{1}{T_B}\right) = 0\ ;\ \left(\frac{P_A}{T_A} - \frac{P_B}{T_B}\right) = 0\ ;\ \left(\frac{\mu_A}{T_A} - \frac{\mu_B}{T_B}\right) = 0 \tag{14}$$

4. THERMODYNAMIC LAWS

4.1. The zeroth law of thermodynamics

Two systems in thermal contact eventually arrive at a state of thermal equilibrium. This state is uniquely defined by temperature, which is a universal function of the state properties and the internal energy. If a system 1 is in equilibrium with system 2, and if system 2 is in equilibrium with system 3, then system 1 is in equilibrium with system 3. This is called the *zeroth law of thermodynamics* and implies the construction of a universal temperature scale (stated first by Joseph Black in the 18th century, and named much later by Guggenheim). If a system is in thermal equilibrium, it is assumed that the energy is distributed uniquely over the volume. Once the energy of the system increases, the temperature of the system also increases ($dU/dT > 0$).

4.2. The first law of thermodynamics

Any function that may be expressed in terms of the volume V, pressure P, and the number of moles N_i, is called the *state function* of the system. A state function is a total derivative of the thermodynamic variables. The change in a state function accompanying the transition of the system from one state to another depends only on the initial and final states and not on the path between these states. If the system returns to its original state, the integral of the change is zero

$$\oint dU = 0 \tag{15}$$

Such systems are called the *cyclic processes*. The Poincare statement of the first law states that in a cyclic process the work done by a system equals the heat received by it.

In the first law of thermodynamics, the state function of internal energy U in a closed system is defined as the energy supplied by the surrounding that is equal to the sum of the heat δQ and the mechanical work $-\delta W$ performed at the boundaries of the system

$$dU = \delta Q - \delta W \tag{16}$$

The quantities δQ and δW are not independent of path. Therefore, it is not possible to define a function Q or W that depends only the initial and final states, and hence Q and W are not the state functions. In Eq. (16), we assume that the rate of heat flow δQ and work δW in a time dt is well defined. When we consider an open system, we have a flow of energy due to heat transfer and due to exchange of matter. The conservation of energy implies that the change in the internal energy of the system is equal to the energy exchanged with surroundings, and can be stated and utilized in terms of macroscopic properties.

In general, the value of δW shows all different forms of work, which are a product of an intensive variable and a differential of an extensive variable. For example, if the system is displaced by a distance dl under a force F, it performs a work of $-Fdl$, or if a quantity of electricity $-de$ is given off at an electric potential ψ, an electrical work of $-\psi de$ is obtained. If $-dN_i$ moles of substance i with the chemical potential μ_i is transported from the system to its surroundings, a chemical work of $-\mu dN_i$ is performed. With these considerations work becomes

$$\delta W = -PdV + Fdl + \psi de + \sum_{i=1}^{n} \mu_i dN_i + ... \tag{17}$$

For an open system, an additional contribution to the energy due to exchange of matter dU_m is added

$$dU = \delta Q - \delta W + dU_m \tag{18}$$

For systems with chemical reactions, the total energy may be considered as a function of T, V and N_i: $U = U(T, V, N_i)$. The total differential of U is given by

$$\begin{aligned} dU &= \left(\frac{\partial U}{\partial T}\right)_{V,N_i} dT + \left(\frac{\partial U}{\partial V}\right)_{T,N_i} dV + \sum_i^n \left(\frac{\partial U}{\partial N_i}\right)_{V,T,N_{i\neq k}} dN_i \\ &= \delta Q - \delta W + dU_m \end{aligned} \tag{19}$$

The exact form of the function $U(T, V, N_i)$ for a certain system is obtained empirically.

4.3. The second law of thermodynamics

If a system is in equilibrium then all the forces X_i are fully known from external parameters a_i, so that the first law can be written as

$$\delta Q = dU + \delta W = dU + \sum X_i(a_i) da_i \tag{20}$$

Eq. (20) is a Pfaffian equation and a_i is an independent variable. Caratheodory's theory states that starting from a known original state, there may be other states that cannot be reached by an adiabatic process along the path $\delta Q = 0$. This shows the existence of an integrating factor for δQ, hence we have

$$\lambda dS = d\eta \tag{21}$$

where $d\eta$ is a total differential of the variables a_i. Therefore, η must be a state function known as the entropy S, and the integrating factor is the reciprocal of absolute temperature T, so that we obtain

$$dS = \frac{\delta Q}{T} \tag{22}$$

Eq. (22) is a mathematical statement of the second law of thermodynamics for irreversible processes. The introduction of the integrating factor for δQ causes the thermal energy to be split into an extensive factor S and an intensive factor T.

Introducing Eq. (22) into Eq. (16) yields the combined first and second laws

$$dU = TdS - \delta W \tag{23}$$

Every system is associated with an energy and entropy. When the system changes from one state to another, the total energy is conserved. However the total entropy is not conserved and increases in irreversible processes, while it remains unchanged in reversible processes. The notion of entropy is not a direct intuitive concept. Boltzmann showed that entropy is proportional to the number of configurations Ω, a system can have

$$S = k \ln \Omega \tag{24}$$

where k is the Boltzmann constant. Eq. (24) shows that disorganization and randomness increase entropy, while organization and ordering decrease it. The concept of entropy as a measure of organized structure is attracting scientist from diverse fields such as physics, biology, communication and information. For example, the information entropy maximization method has been applied to nonequilibrium phenomena within the perspective of thermodynamic laws.

Entropy of an irreversible process can be measured and expressed as a function of the external and internal properties, regardless the energy content of the system. We can attain the same distribution of internal parameters imposed by a set of external parameters both reversibly and irreversibly. These different paths result in different work and energy changes in the system. However, it is assumed that a set of local parameters determines the entropy, and hence we can device an ideal process that would bring the system reversibly to any configuration of the irreversible process. For example, diffusion of a substance is a nonequilibrium process, and the local concentration profile is necessary to define the system. We may apply reversibly a centrifugal field to the system to maintain the same concentration profile in a state of equilibrium. The energy applied reversibly to the centrifugal field is different from those of the system in an irreversible diffusion process. Thus, the thermodynamic states of an irreversible diffusion process and the corresponding equilibrium system are different. However, the entropy will be the same in both systems, and is defined by the properties of the equilibrium state. Entropy may be computed as the corresponding entropy of the real system. The first law expresses the qualitative equivalence of the heat and work as well as the conservation of energy. The second law is a qualitative statement on the accessibility and direction of progress of real processes. For example, the efficiency of a reversible engine is a function of temperature only, and it cannot exceed unity. These statements are the results of the first and second laws, and can be used to define an absolute scale of temperature that is independent of any material properties used to measure it. A quantitative description of the second law emerges by using entropy and entropy generation.

5. ENTROPY AND ENTROPY GENERATION

By using Eq. (23) with a pressure-volume work pdV, we have

$$dS = \frac{dU + PdV}{T} \tag{25}$$

The entropy of a system is an extensive property, and it changes through the exchange of mass and energy. If a system consists of several processes, the total entropy is equal to the sum of the entropies produced by each process. The total change of the entropy dS results from the flow of entropy due to exchanges with surroundings d_eS, and from the changes inside the system d_iS

$$dS = d_\mathrm{e}S + d_\mathrm{i}S \tag{26}$$

The value of d_iS is zero when the change inside the system is reversible, and it is positive when the change is irreversible

$$d_\mathrm{i}S = 0 \quad \text{(Reversible change)};\ \ d_\mathrm{i}S > 0 \quad \text{(Irreversible change)} \tag{27}$$

For an isolated system there is no interaction with the surroundings so that

$$dS = d_\mathrm{i}S > 0 \tag{28}$$

The rate of entropy generation P is expressed by

$$P = \frac{d_\mathrm{i}S}{dt} \geq 0 \tag{29}$$

The entropy source strength Φ is the rate of entropy generation per unit volume

$$\Phi = \frac{dP}{dV} \geq 0 \tag{30}$$

The product of the entropy source strength and the absolute temperature of a reference system is called the dissipation function Ψ

$$\Psi = T\Phi \geq 0 \tag{31}$$

When interfacial phenomenon is considered, the entropy generation is based per unit surface area. The entropy source strength and the dissipation function are not state functions, and they depend on the path between the given states.

The second law of thermodynamics states that the sum of the entropy generations of all processes for any system and its environment is positive. The entropy of system may decrease only if an irreversible process in the environment causes a sufficiently large entropy generation to yield a positive total change.

In every macroscopic region of the system, the entropy generation due to irreversible processes is positive. A macroscopic region contains enough number of molecules for microscopic fluctuations to be negligible. Interference of the irreversible processes is possible when they occur in the same region of the system. The thermodynamic time, the natural direction of time, implies that irreversible processes produce entropy in any system, isolated, open or closed, and Eq. (28) holds for $dt > 0$. Nonequilibrium processes evolve in time in accordance to the second law of thermodynamics.

Eq. (26) can be applied to various irreversible processes. Let us consider a system consisting of two closed subsystems of I and II, and maintained at uniform temperatures of T^{I} and T^{II}, respectively. The total entropy dS is expressed as

$$dS = dS^{\mathrm{I}} + dS^{\mathrm{II}} = \frac{\delta^{\mathrm{I}}Q}{T^{\mathrm{I}}} + \frac{\delta^{\mathrm{II}}Q}{T^{\mathrm{II}}} \tag{32}$$

The interactions of heat in each subsystem are given by

$$\delta^{\mathrm{I}}Q = \delta_{\mathrm{i}}^{\mathrm{I}}Q + \delta_{\mathrm{e}}^{\mathrm{I}}Q; \quad \delta^{\mathrm{II}}Q = \delta_{\mathrm{i}}^{\mathrm{II}}Q + \delta_{\mathrm{e}}^{\mathrm{II}}Q \tag{33}$$

Using Eq. (33) and the conservation of energy $\delta_{\mathrm{i}}^{\mathrm{II}}Q + \delta_{\mathrm{i}}^{\mathrm{I}}Q = 0$, Eq. (32) yields

$$dS = \frac{\delta_{\mathrm{e}}^{\mathrm{I}}Q}{T^{\mathrm{I}}} + \frac{\delta_{\mathrm{e}}^{\mathrm{II}}Q}{T^{\mathrm{II}}} + \delta_{\mathrm{i}}^{\mathrm{I}}Q\left(\frac{1}{T^{\mathrm{I}}} - \frac{1}{T^{\mathrm{II}}}\right) = d_{\mathrm{e}}S + d_{\mathrm{i}}S \tag{34}$$

The entropy production per unit time is

$$\frac{d_{\mathrm{i}}S}{dt} = \frac{\delta_{\mathrm{i}}^{\mathrm{I}}Q}{dt}\left(\frac{1}{T^{\mathrm{I}}} - \frac{1}{T^{\mathrm{II}}}\right) > 0 \tag{35}$$

Eq. (35) shows that the rate of entropy generation is the product of flow (heat flux) $(\delta_{\mathrm{i}}^{\mathrm{I}}Q/dt)$ and the thermodynamic force $(1/T^{\mathrm{I}} - 1/T^{\mathrm{II}})$.

Eq. (25) is the total differential of the entropy as function of the variables U and V only. To generalize this relation, we should also consider the changes in the amounts of components. Using the mole amounts for the components, we have a general expression for the change of entropy from the Gibbs relation

$$dS = \frac{dU}{T} + \frac{P}{T}dV - \sum_j \frac{\mu_j}{T} dN_j \tag{36}$$

Eq. (36) is the main expression in the calculation of entropy generation.

For ideal systems the chemical potential is expressed by

$$\mu_j = \mu_j^o(T,P) + RT\ln x_j \tag{37}$$

where x_j is the mole fraction of component j. The chemical potential can also be defined in terms of the concentration of component j, $c_j = N_j/V$

$$\mu_j = \mu_j^o(T,V) + RT\ln c_j \tag{38}$$

For nonideal systems, we use the activity coefficient γ_j, and Eq. (37) becomes

$$\mu_j = \mu_j^o(T,P) + RT\ln \gamma_j x_j \tag{39}$$

In the presence of external fields, the potential energy is included in the chemical potential. When the external field is an electric field, we get

$$\mu_j = \mu_j^o(T,P) + RT\ln \gamma_j x_j + Fz_j\psi \tag{40}$$

where F is the Faraday, or electric charge per mole ($F = 96500$ coulombs/mole), z_j is the valence of the component j, and ψ is the electric potential.

Entropy production due to a chemical reaction in a closed is given by

$$dS = \frac{\delta Q}{T} + \frac{A}{T}d\xi \tag{41}$$

where A is the affinity of the chemical reaction, $A = -\sum \nu_j \mu_j$, ξ is the extent of the reaction, and ν is the stoichiometric coefficient. Eq. (41) shows that the entropy change contains two contributions: one is due to interactions with the surroundings $d_e S = \delta Q / T$, and the other is due to change in the system $d_i S = A d\xi / T$, which can be expressed in terms of the scalar rate of reaction J_r

$$\frac{d_i S}{dt} = \frac{1}{T} A J_r > 0 \tag{42}$$

where

$$J_r = \frac{d\xi}{dt}$$

Eq. (42) is similar to Eq. (35) in relating the rate of entropy generation to the product of the flows (here the rate of reaction) and the scalar thermodynamic force that is *A/T*. Eq. (42) can be readily extended to several chemical reactions taking place inside the system, and we obtain

$$\frac{d_i S}{dt} = \frac{1}{T} \sum_{k=1}^{l} A_k J_{r,k} \tag{43}$$

When the chemical reaction reaches the equilibrium affinity vanishes $A = -\sum \nu_j \mu_j = 0$. The entropy generated per unit time and unit volume is called the rate of volumetric entropy generation or the entropy source of density Φ

$$\Phi = \frac{A}{VT} J_r \geq 0 \tag{44}$$

The second law requires the total entropy production resulting from all the simultaneous reactions to be positive. This has been verified experimentally.

Sometimes a system has two simultaneous coupled reactions, such that

$$A_1 J_{r,1} < 0, \quad A_2 J_{r,2} > 0 \tag{45}$$

although, the sum of the entropy generation strengths is always positive

$$(A_1 J_{r,1} + A_2 J_{r,2}) / VT \geq 0 \tag{46}$$

Thermodynamic coupling allows one of the processes to progress in a direction opposite to that imposed by its own thermodynamic force. For example, in thermodiffusion a component diffuses against its concentration gradient resulting a negative entropy generation, which must be compensated by a spontaneous process with a positive and larger entropy production due to the heat flow. Such coupled processes are of great importance in physical and biological systems, such as the active transport of sodium or potassium.

6. THE GIBBS EQUATION

By introducing Eq. (17) into Eq. (23), we have

$$dU = TdS - PdV + Fdl + \psi de + \sum_{i=1}^{n} \mu_i dN_i + ... \tag{47}$$

Eq. (47) is a general expression; it relates the total change in internal energy to the sum of products of intensive variables of T, P, F, μ_i, ψ, and all possible changes in extensive properties (capacities) of dS, dV, dl, dN_i, and de. Brϕnsted work principle states that the overall work ΔW performed by a system is the sum of contributions due to the difference of extensive properties ΔK across a difference of conjugated potentials $(X_{i,1} - X_{i,2})$

$$\Delta W = \sum_{i=1}^{n} (X_{i,1} - X_{i,2}) \Delta K_i \tag{48}$$

Eq. (47) is more useful if it is integrated on the Pfaffian form, however this is not a straightforward step, since intensive properties are functions of all the independent variables of the system. The Euler relations for $U(S, V, l, e, N_i)$ is

$$U = S\left(\frac{\partial U}{dS}\right)_{V,l,e,N_i} + V\left(\frac{\partial U}{dV}\right)_{S,l,e,N_i} + l\left(\frac{\partial U}{dl}\right)_{V,S,e,N_i} + e\left(\frac{\partial U}{de}\right)_{V,l,S,N_i} + \sum_{i=1}^{m} N_i \left(\frac{\partial U}{\partial N_i}\right)_{S,V,l,e,N_j} + ... \tag{49}$$

Comparing Eq. (49) with Eq. (47) yields the definitions of intensive properties for the partial differentials

$$\left(\frac{\partial U}{dS}\right)_{V,l,e,N_i} = T; \quad \left(\frac{\partial U}{dV}\right)_{S,l,e,N_i} = -P; \quad \left(\frac{\partial U}{dl}\right)_{V,S,e,N_i} = F$$
$$\left(\frac{\partial U}{de}\right)_{V,l,S,N_i} = \psi; \quad \left(\frac{\partial U}{\partial N_i}\right)_{S,V,l,e,N_j} = \mu_i \tag{50}$$

The chemical potential μ indicates that the internal energy is a potential for chemical work (or mass action) $\mu_i dN_i$, and it is the driving force for change in the chemical composition of matter as a result of chemical reactions. The chemical potential cannot be measured directly, and the absolute values are expressed in terms of a reference state. However the change of chemical potential is of common interest.

By introducing the definitions given in Eq. (50) into Eq. (47), we obtain the integrated form of the Gibbs equation

$$U = TS - PV + Fl + \psi e + \sum_{i=1}^{n} \mu_i N_i \qquad (51)$$

Differentiation of Eq. (51) gives

$$\begin{aligned} dU = {} & TdS + SdT - PdV - VdP + Fdl + ldF + \psi de + ed\psi \\ & + \sum \mu_i dN_i + \sum N_i d\mu_i \end{aligned} \qquad (52)$$

Comparison of Eq. (52) with Eq. (47) indicates that the following relation must be satisfied

$$SdT - VdP + ldF + ed\psi + \sum N_i d\mu_i = 0 \qquad (53)$$

Eq. (53) is called the *Gibbs-Duhem relation*, which becomes particularly useful at isobaric, isothermal conditions, and when the force and electrical work are neglected

$$\sum_{i=1}^{n} N_i d\mu_i = \sum_{j=1}^{n} N_i \left(\frac{\partial \mu_i}{\partial N_j} \right)_l = 0 \qquad j = 1,2,..,n; \; i \neq j \qquad (54)$$

Eq. (54) determines the changes in chemical potential with the addition of any substance into the system.

From the Gibbs fundamental equation $f(U,S,V,\mathrm{N})$ we have the three functions of S, V and N, the respective differential relations, and the *Euler equations* given by

$$\begin{aligned} & S = S(U,V,N) \\ & dS = \frac{dU}{T} + P\frac{dV}{T} - \mu\frac{dN}{T}; \quad S = U\frac{1}{T} + P\frac{V}{T} - \mu\frac{N}{T} \end{aligned} \qquad (55)$$

$$\begin{aligned} & V = V(U,S,N) \\ & dV = \frac{dU}{P} + T\frac{dS}{P} + \mu\frac{dN}{P}; \quad V = U\frac{1}{P} + S\frac{T}{P} + N\frac{\mu}{P} \end{aligned} \qquad (56)$$

and

$$N = N(U,S,V)$$

$$dN = \frac{dU}{\mu} - T\frac{dS}{\mu} + P\frac{dV}{\mu}; \ N = U\frac{1}{\mu} - S\frac{T}{\mu} + V\frac{P}{\mu} \tag{57}$$

Using the molar specific volume $v = V/N$ and molar specific entropy $s = S/N$, a simplified version of the Gibbs-Duhem relation results by

$$\mu(T,P) = d\mu = -sdT + vdP \tag{58}$$

By partial differentiation, Eq. (58) can be transformed to a form called the *thermal equation*

$$\left(\frac{\partial \mu}{\partial P}\right)_T = v = v(T,P) \tag{59}$$

and the corresponding *caloric equation* is expressed by

$$\left(\frac{\partial \mu}{\partial T}\right)_P = s = s(T,P) \tag{60}$$

In Eq. (20), if X_k are the external variables to maintain the nonequilibrium distribution of internal parameters of ξ_k in a state of equilibrium, we have a potential energy of $-\sum \xi_k X_k$. This potential energy is the additional work of the external parameters to maintain the distribution of internal parameters. The internal energy of equilibrium system U_{eq} is related to the internal energy of the nonequilibrium system U by

$$-U_{\text{eq}} = U - \sum \xi_k X_k \tag{61}$$

The irreversible work δW_{eq} is related to the work necessary in reaching the same conditions irreversibly as

$$\delta W_{\text{eq}} = \delta W + \sum \xi_k X_k \tag{62}$$

The entropy change in the corresponding reversible process is given by

$$TdS = dU_{\text{eq}} + \delta W_{\text{eq}} \tag{63}$$

By inserting Eqs. (61) and (62) into Eq. (63), we have

$$\begin{aligned} TdS &= dU - \sum X_k d\xi_k - \sum \xi_k dX_k + \delta W + \sum \xi_k dX_k \\ &= dU + \delta W - \sum X_k d\xi_k \end{aligned} \tag{64}$$

The entropy term *TdS* in Eq. (64) is the same for the irreversible process and the corresponding reversible process. Therefore, Eq. (64) represents the Gibbs equation for irreversible process. With the first law of thermodynamics

$$dU + \delta W = \delta Q,$$

Eq. (64) becomes

$$TdS = \delta Q - \sum X_k d\xi_k \tag{65}$$

For an adiabatic process $\delta Q = 0$, and we have

$$TdS = -\sum X_k d\xi_k \tag{66}$$

Eq. (66) represents the change of entropy for an irreversible process in an adiabatic system as function of the internal and external parameters that make entropy measurable. This may be an important property to quantify the level of irreversibility of a change, and hence yields (i) a starting point to relate the economic implications of irreversibility in real process, and (ii) an insight into the interference between two processes in the system.

7. EQUATIONS OF STATE

An equation of state expresses an intensive property in terms of the extensive properties, and is obtained as the partial derivatives of the Euler equation. In the entropy representation we have the following equations of state

$$\frac{1}{T} = \left(\frac{\partial S}{dU}\right)_{V,l,e,N_i} \tag{67}$$

$$\frac{P}{T} = \left(\frac{\partial S}{dV}\right)_{V,l,e,N_i} \tag{68}$$

$$\frac{\mu}{T} = \left(\frac{\partial S}{dN}\right)_{V,l,e,N_i} \tag{69}$$

In contrast to a fundamental equation, an equation of state does not contain all the information on the system, since the intensive properties are partial derivatives of the extensive ones. To recover all the information, all the equations of state are inserted into the Euler equation.

8. THERMODYNAMIC POTENTIALS

At equilibrium, the extensive properties U, S V, N_i, and the linear combination of them are functions of state. Such combinations are the Helmholtz free energy, the Gibbs free energy and enthalpy, and called the *thermodynamic potentials*. Table 1 provides a summary of the thermodynamic potentials and the change of them. The thermodynamic potentials are extensive properties, while the *ordinary potentials* are the derivative of the thermodynamic potentials and are intensive properties.

A thermodynamic potential reaches an extremum value towards equilibrium under various conditions. The Helmholtz free energy A is particularly useful for systems at constant volume and temperature. Combining Eq. (23) and Eq. (71) at constant temperature yields

$$-(dA)_T = \delta W \tag{82}$$

Eq. (82) shows that the total reversible work performed by a system is equal to the decrease in the Helmholtz free energy. The Gibbs free energy is especially suitable for isothermal and isobaric system, and from Eqs. (23) and (75), we have

$$-(dG)_{T,P} = \delta W - PdV \tag{83}$$

Thus the decrease in the Gibss free energy is the useful work that is equal to the total work minus the pressure-volume work. Since PdV is usually negligible for condensed phase and living tissues, the use of thermodynamic potential G is common in such systems. For a closed system under isobaric conditions, and using Eq. (23) we have

$$(dH)_P = (\delta Q)_P \tag{84}$$

Eq. (84) shows that the enthalpy is the same with the heat exchanged with surrounding. The Gibbs free energy can be related to enthalpy

$$G = H - TS \tag{85}$$

Hess's law referring to the heat evolved in a chemical reaction is conveniently formulated in terms of enthalpy.

Table 1
Thermodynamic potentials and their changes

	Thermodynamic potentials	
Definition	1. Helmholtz free energy, A $A = U - TS$	(70)
Change	$dA = dU - TdS - SdT$	(71)
With Eq. (47)	$dA = -SdT - PdV + Fdl + \psi de + \sum \mu_i dN_i$	(72)
Chemical potential	$\mu_i = \left(\frac{\partial A}{\partial N_i}\right)_{T,V,l,e,N_j}$	(73)
Definition	2. Gibbs free energy, G $G = U - TS + PV$	(74)
Change	$dG = dU - TdS - SdT + PdV + VdP$	(75)
With Eq. (47)	$dG = -SdT + VdP + Fdl + \psi de + \sum \mu_i dN_i$	(76)
Chemical potential	$\mu_i = \left(\frac{\partial G}{\partial N_i}\right)_{T,P,l,e,N_j}$	(77)
Definition	3. Enthalpy, H $H = U + PV$	(78)
Change	$dH = dU + PdV + VdP$	(79)
With Eq. (47)	$dH = TdS + VdP + Fdl + \psi de + \sum \mu_i dN_i$	(80)
Chemical potential	$\mu_i = \left(\frac{\partial H}{\partial N_i}\right)_{S,P,l,e,N_j}$	(81)

8.1. Cross Relations

Maxwell first noted the cross relations based on the property of the total differentials of the state functions. The cross differentiations of a total differential of the state function are equal to each other. Table 2 summarizes the total differentials and the corresponding *Maxwell relations*. The Maxwell relations may be used to construct important thermodynamic equations of states.

Table 2
The total differentials of the state functions and the Maxwell relations for closed systems

The total differentials / The Maxwell relations

The internal energy: $U = U(S,V)$

$$dU = \left(\frac{\partial U}{\partial S}\right)_V dS + \left(\frac{\partial U}{\partial V}\right)_S dV = TdS - PdV \tag{86}$$

$$\left(\frac{\partial (T)_V}{\partial V}\right)_S = -\left(\frac{\partial (P)_S}{\partial S}\right)_V \tag{87}$$

The Hemlholtz free energy: $A = A(T,V)$

$$dA = \left(\frac{\partial A}{\partial T}\right)_V dT + \left(\frac{\partial A}{\partial V}\right)_T dV = -SdT - PdV \tag{88}$$

$$\left(\frac{\partial (S)_V}{\partial V}\right)_T = \left(\frac{\partial (P)_T}{\partial T}\right)_V \tag{89}$$

The Gibbs free energy: $G = G(T,P)$

$$dG = \left(\frac{\partial G}{\partial T}\right)_P dT + \left(\frac{\partial G}{\partial P}\right)_T dP = -SdT + VdP \tag{90}$$

$$-\left(\frac{\partial (S)_P}{\partial P}\right)_T = \left(\frac{\partial (V)_T}{\partial T}\right)_P \tag{91}$$

Enthalpy: $H = H(S,P)$

$$dH = \left(\frac{\partial H}{\partial S}\right)_P dS + \left(\frac{\partial H}{\partial P}\right)_S dP = TdS + VdP \tag{92}$$

$$\left(\frac{\partial (T)_P}{\partial P}\right)_S = \left(\frac{\partial (V)_S}{\partial S}\right)_P \tag{93}$$

The cross relations can be seen in a reversible change of a rectangular rubber sheet subjected to two perpendicular forces F_x and F_y under isothermal conditions. If the extent of stretching in both the directions of x and y are Δx and Δy, we have

$$F_x = M_{11}\Delta x + M_{12}\Delta y \tag{94}$$

$$F_y = M_{21}\Delta x + M_{22}\Delta y \tag{95}$$

The ordinary elastic moduli in the x and y directions are denoted by M_{11} and M_{22}, and are given by

$$M_{11} = \left(\frac{\partial F_x}{\partial x}\right)_y ; \quad M_{22} = \left(\frac{\partial F_y}{\partial y}\right)_x \tag{96}$$

However the coupling moduli M_{12} and M_{21} relate the force in one direction to the stretch in the other direction, and we have

$$M_{12} = \left(\frac{\partial F_x}{\partial y}\right)_x ; \quad M_{21} = \left(\frac{\partial F_y}{\partial x}\right)_y \tag{97}$$

From Eq. (50) we can express F_x and F_y as

$$F_x = \left(\frac{\partial U}{\partial x}\right)_y ; \quad F_y = \left(\frac{\partial U}{\partial y}\right)_{\mathrm{x}} \tag{98}$$

The cross relations impose that

$$\left(\frac{\partial F_x}{\partial y}\right)_x = \left(\frac{\partial F_y}{\partial x}\right)_y \tag{99}$$

Therefore the matrix of the moduli becomes symmetric

$$\begin{vmatrix} M_{11} & M_{12} \\ M_{21} & M_{22} \end{vmatrix} \tag{100}$$

where $M_{12} = M_{21}$. The similar matrices may occur in thermodynamic description of irreversible processes.

The second partial derivatives of the state functions at constant volume and entropy are

$$\left(\frac{\partial U}{\partial T}\right)_V = C_V ; \quad \left(\frac{\partial S}{\partial T}\right)_{\mathrm{V}} = \frac{C_{\mathrm{V}}}{T} ; \quad \left(\frac{\partial V}{\partial T}\right)_S = \alpha_S V ; \quad -\left(\frac{\partial V}{\partial P}\right)_S = \kappa_S V \tag{101}$$

where C_V is the heat capacity at constant volume, α_S is the adiabatic expansivity, and κ_S is the compressibility. The second partial derivatives of the state functions at constant pressure and temperature are

$$\left(\frac{\partial H}{\partial T}\right)_P = C_P; \left(\frac{\partial S}{\partial T}\right)_\mathrm{P} = \frac{C_\mathrm{P}}{T}; \left(\frac{\partial V}{\partial T}\right)_P = \alpha_P V; -\left(\frac{\partial V}{\partial P}\right)_T = \kappa_T V \tag{102}$$

Here C_P is the heat capacity at constant pressure, α_P is the isobaric expansivity, and κ_P is the isothermal compressibility.

8.2. Extremum Principles

Equilibrium thermodynamics has various extremum principles. At various conditions a thermodynamic potential will approach an extremum value as the system reaches to equilibrium state. For an isolated or closed system we may consider the following extremum principles:

- The entropy of an isolated system reaches the maximum possible value at equilibrium

$$d_i S > 0 \quad \text{at constant } U \text{ and } V \tag{103}$$

- For a closed system $dU = \delta Q - PdV = Td_\mathrm{e}S - PdV$.

Since the total entropy change $dS = d_\mathrm{e}S + d_\mathrm{i}S$, we have $dU = TdS - PdV - Td_\mathrm{i}S$

$$dU = -Td_\mathrm{i}S \le 0 \quad \text{at constant } S \text{ and } V \tag{104}$$

For having constant entropy, we keep T, V and dN_i constant, and the entropy produced $d_\mathrm{i}S$ has to be removed from the system. The decrease in energy is generally due to the conversion of mechanical energy into heat.

- The Helmholtz free energy reaches a minimum possible value at equilibrium

$$dA = -Td_\mathrm{i}S \le 0 \quad \text{at constant } T \text{ and } V \tag{105}$$

- The Gibbs free energy reaches a minimum possible value at equilibrium

$$dG = -Td_\mathrm{i}S \le 0 \quad \text{at constant } P \text{ and } T \tag{106}$$

- The enthalpy reaches a minimum possible value at equilibrium

$$dH = -Td_iS \leq 0 \quad \text{at constant } S \text{ and } P \tag{107}$$

REFERENCES

[1] I. Prigogine, Introduction to Thermodynamics of Irreversible Processes, Wiley, New York, 1967.
[2] D. Kondepudi and I. Prigogine, Modern Thermodynamics, From Heat Engines to Dissipative Structures, Wiley, New York, 1999.
[3] A. Bohm, H-D. Doebner, P. Kielanowski (eds.), Irreversibility and Causality, Springer-Verlag, Berlin, 1998, pp 181,283.
[4] A. Katchalsky and P.F. Curran, Nonequilibrium Thermodynamics in Biophysics, Harvard University Press, Cambridge, 1967.
[5] N.W. Tschoegl, Fundamentals of equilibrium and Steady-State Thermodynamics, Elsevier, Amsterdam, 2000.

Chapter 2

Transport and rate processes

INTRODUCTION

Thermodynamics first emerged as a science after the construction and operation of steam engines in 1697 by Thomas Savery and in 1712 by Thomas Newcomen in England. In the 1850s William Rankine, Rudolph Clausius, and William Thomson (later Lord Kelvin) developed the first and the second laws of thermodynamics.

Using a limited number of principles, *classical thermodynamics* describes systems in mechanical, thermal and chemical equilibrium; it provides a set of extremum principles (isolated system entropy attains a maximum value, while the free energy reaches a minimum at equilibrium), and mainly targets closed systems.

The classical thermodynamics has played an important role in physical chemistry and biochemistry. However, many physical and biological processes are nonequilibrium, open systems with irreversible changes. For such irreversible changes, the laws of classical thermodynamics provide a set of inequalities describing only the direction of change. Consequently, we can use kinetic equations and statistical models to describe the transport of matter, energy, electricity, as well as biological activities, such as nerve conduction, muscle contractions, and complex coupled phenomena. Kinetic and statistical models often require more detailed information than is available or sometimes readily obtainable. Therefore, it may be advantageous to have a phenomenological approach with thermodynamic principles to describe the natural processes. An approach towards this end is the formalism of the nonequilibrium thermodynamics to the investigation of physical and biological nonequilibrium systems with irreversible processes. In the formalism, the Gibbs equation is a key relation since it combines the first and second laws of thermodynamics. The Gibbs relation is combined with the general balance equations based on the local thermodynamic equilibrium to determine the rate of entropy generation.

There is a close connection among molecular mass, momentum, and energy transport, which can be explained in terms of a molecular theory for low-density monatomic gases; equations of continuity, motion, and energy can all be derived from the Boltzmann equation, and the expressions for the fluxes and transport properties are generated. Similar kinetic theories are also available for polyatomic gases, monatomic liquids, and polymeric liquids. In this chapter, we briefly summarize the kinetic theory, transport phenomena, and chemical reactions.

1. NONEQUILIBRIUM SYSTEMS

A system reaches the thermodynamic *equilibrium state* when it is left for a long time with no external disturbances, as shown in Fig 1. At equilibrium the internal properties are fully determined by the external properties. This leads to easy description of such systems; for example, if the temperature is not uniform within the system, heat is exchanged with the immediate surrounding until the system reaches a thermal equilibrium, at which the total internal energy U entropy S are completely specified by the temperature, volume and number of moles. The equilibrium state is stable, and the range of correlations is based on short-range intermolecular interactions; for small deviations the system can spontaneously return to the state of equilibrium. The existence of extremum principles is a characteristic property of equilibrium thermodynamics.

However, the natural systems consist of flows caused by the unbalanced driving forces, and hence the description of such systems requires a larger number of properties in space and in time. Such systems are away from the equilibrium state, and called the *nonequilibrium systems*; they can exchange energy and matter with the environment, and have finite driving forces. The formalism of *nonequilibrium thermodynamics* can describe such systems in qualitative and quantitative manner by replacing the inequalities of classical thermodynamics with equalities.

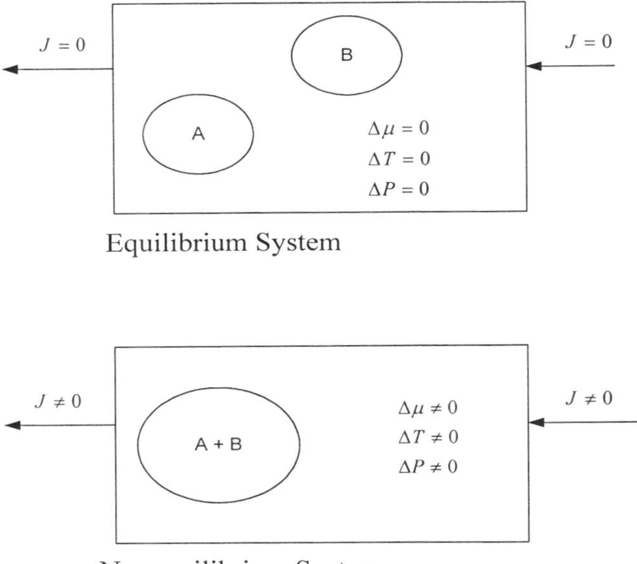

Fig. 1. Equilibrium and nonequilibrium systems.

Constant driving forces cause steady flows, and a stationary state is attained. For example, a constant temperature difference applied to a metal bar will induce a heat flow that will cause a change in all local temperatures. After a while, a constant distribution of temperature will be attained and the heat flow will become steady. The steady state flow and constant distribution of forces characterizing a system is the ultimate state of irreversible systems corresponding to the states of equilibrium in classical thermodynamics.

Steady states are typical of living systems. Since all the variables are constant during a stationary state, the entropy is constant that is the entropy generated by steady-state flow is equal to the entropy given off to the surroundings. Only the systems open to exchange of entropy with their environment can reach a steady state; all living systems are open systems capable of stable existence in a stationary state. Prigogine pointed out that steady nonequilibrium systems produce entropy at a minimal rate. The synthesis of large molecules and information, and developing and sustaining structural states require anti-entropic systems. Biological systems progress towards increasing in size and complexity, and they do not decay towards an equilibrium state. Still all living organisms obey the second law of thermodynamics and choose to produce less entropy by maintaining a steady state.

For systems not far away from equilibrium, a system is assumed to be an extrapolation of the equilibrium state. The states away from equilibrium are called the *thermodynamic branch* (Fig. 2). For the near equilibrium systems transport and rate processes are expressed in linear phenomenological equations and with the Onsager relations. The *linear nonequilibrium thermodynamics* theory determines the dissipation function or the rate of entropy generation to describe such systems. This theory is particularly useful to define and quantify the coupling between flows in physical and biological processes.

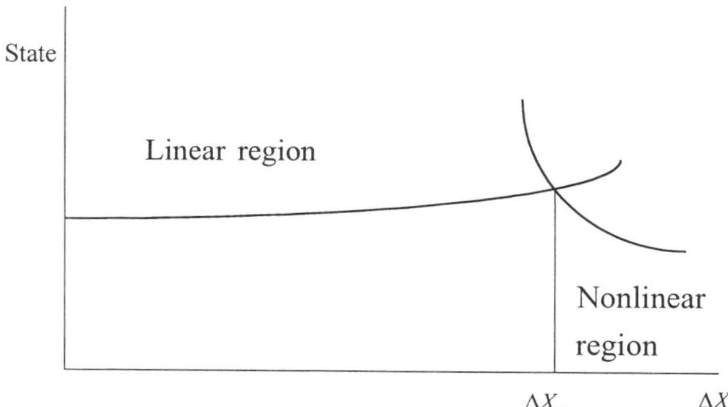

Fig. 2. Thermodynamic branch. ΔX_c indicates the critical distance from equilibrium state.

In nonequilibrium systems, the intensive properties of temperature, pressure and chemical potential are not uniform. However they all are defined locally. For example in a region k, we can define the densities of thermodynamic properties such as energy and entropy at local temperature. The energy density, the entropy density, and the amount of matter are expressed by $U_k(T, N_k)$, $S_k(T, N_k)$ and N_k, respectively. The total energy U, the total entropy S, and the total number of moles N of the system are determined by

$$U = \int_V U_k dV \tag{1}$$

$$S = \int_V S_k dV \tag{2}$$

$$N = \int_V N_k dV \tag{3}$$

Since we cannot define a single temperature for the whole system, the total entropy is not a function of the other extensive properties of U, V and N. When we define the local temperature, the entropy of a nonequilibrium system is defined in terms of an entropy density S_k, as given in Eq. (2).

In some systems the distance from equilibrium reaches a critical point, after which the states in the thermodynamic branch become unstable. Prigogine called these states as the *dissipative structures*; they are organized and ordered states, and need constant supply of energy. This part of the thermodynamic branch corresponds to the nonlinear region of the branch, and is studied by the *extended nonequilibrium thermodynamics* approach. In the nonlinear region chemical reactions and transport processes are characterized by the local potentials in which each macroscopic variable is described by an average quantity and fluctuating quantity. The compensation function and the generalized hydrodynamics are an integral part of the extended theory of irreversible processes.

The linear nonequilibrium thermodynamics approach can be used to design thermodynamically optimum processes. This trend is called the *second law analysis*, in which the rate of entropy generation, dissipation loss, or exergy destruction is calculated in the system. The entropy generation approach is especially important in terms of the process optimality as the each process contributing to the entropy generation can be identified and evaluated separately. Through the minimization of the excessive irreversibilities by the modified operating conditions or the modified design, a thermodynamic optimum can be attained for the process with a required task. The information on the trade-off between the various contributions of entropy generation, and equipartition of entropy generation or uniform driving force within the system may be used as thermodynamic optimum criteria for the system. The map of volumetric entropy generation rate and the distribution of irreversibility ratios identify the regions within the system, where excessive entropy generation occurs due to the irreversible processes. The second law analysis has been applied extensively in the thermal engineering field, separation by distillation, and chemical reactor design.

2. KINETIC APPROACH

Statistical mechanics can provide phenomenological description of the nonequilibrium processes. An alternative approach based on kinetic theory is favorable especially to describe the transport and rate phenomena. A kinetic theory of nonequilibrium systems has been developed for dilute monatomic gases, however a substantial progress has also been achieved in extending the theory to dense and real gases, and liquids.

A rigorous kinetic theory of monatomic gases at low density was developed early in the twentieth century by Chapman in England and independently by Enskog in Sweden. Initially the kinetic theory has been limited to low density, nonreacting systems of simple, spherical molecules with no internal degrees of freedom. Typically kinetic approaches start with the Boltzmann equation for the velocity distribution function for each component in a multicomponent system, and the time evolution of the distribution function is obtained by solving the governing kinetic equations with given initial conditions. Evolution of the velocity distribution function with time is calculated with an external force acting on a molecule and using an intermolecular potential energy function, such as the Lennard-Jones potential. The conservation laws appear in kinetic theory as the result of collisional phenomena for mass, momentum, and kinetic energy of molecules. The conservation relations together with the equations of mass, heat and momentum yield the equations of change describing the hydrodynamic fields of velocity, temperature, and concentration for reacting and nonreacting systems. The kinetic theory also provides us with the expressions for transport coefficients of momentum, energy, and mass.

If a gas mixture is at rest, the velocity distribution function is given by the Maxwell-Boltzmann distribution function obtained from equilibrium statistical mechanism. For nonequilibrium systems with near equilibrium forces of concentration, velocity, and temperature gradients, the Maxwell-Boltzman distribution function is multiplied by a correction factor, and the transport equations are represented as a linear function of forces. The equations yield the fluxes representing the molecular transport of mass, momentum, and energy. In these equations, the transport coefficients appear as the kinematic viscosity ν, the thermal diffusivity α, and the Fick diffusivity D_{ij}.

The kinetic theory leads to definitions of the temperature, pressure, internal energy, heat flux density, diffusion fluxes, entropy flux, and entropy source in terms of definite integrals of the distribution function with respect to the molecular velocities. The classical phenomenological expressions for the entropy flux and entropy source (the product of fluxes and forces) follows from the approximate solution of the Boltzmann kinetic equation. This corresponds to the linear nonequilibrium thermodynamics approach of irreversible processes, and to Onsager's symmetry relations with the assumption of local equilibrium.

If the collisions produce a chemical reaction, the Boltzmann equation is modified in obtaining the equations of change; these problems are addressed and analyzed in the context of quantum theory, reaction paths, saddle points and chemical kinetics. Mass, momentum and energy are conserved even in collisions, which produced a chemical reaction.

We can understand the irreversibility better by using the kinetic theory relationships in a maximum entropy formalism, and obtain kinetic equations for both dilute and dense fluids. A derivation of the second law, which states that the entropy generation must be positive in any irreversible process, appears within the framework of the kinetic theory. This is known as Boltzmann's H-theorem. Both conservation laws and transport coefficient expressions can be obtained via the generalized maximum entropy approach. Thermodynamic and kinetic approaches can be used to determine the values of transport coefficients in mixtures, such as the validation of Onsager's reciprocal relations experimentally [7,8].

The kinetic equations serve as a bridge between the microscopic domain and the behavior of macroscopic irreversible processes through the description of hydrodynamics in terms of interparticle collisions. Hydrodynamics can specify a large number of nonequilibrium states by a small number of reproducible properties such as the mass, density, velocity, and energy density of the fluid, same as the quantities conserved during collisions of the fluid particles. Therefore the hydrodynamic equations are capable of describing a wide range of relaxation processes of nonequilibrium states to equilibrium that are called decay processes expressed by phenomenological equations, such as Fourier's law of heat conduction. The decay rates are determined by the transport coefficients. Thermodynamics and kinetic principles can be used to determine such coefficients in mixtures. The reliable coefficients provide microscopic and macroscopic information, and validate the results of molecular dynamics.

3. TRANSPORT PHENOMENA

The majority of systems studied in physics, chemistry and biology contain irreversible processes. Beside the equilibrium states, the stationary states, not varying in time, are also of great interest to engineering. In the stationary states, the flows of mass and energy between a system and its environment do not change with time, allowing technological processes to be carried out in a continuous base.

There exist a large number of linear phenomenological laws describing irreversible processes in the form of proportionalities between the flow J_i and the conjugate driving force X_i

$$J_i = L_{ii} X_i \tag{4}$$

where L_{ii} is the constant called the phenomenological coefficient and can be related to the transport coefficients such as the thermal conductivity k, or the rate constant for a chemical reaction. The validity of Eq. (4) should be determined experimentally for a particular process. For example it will be valid for a conductor if the conductor obeys Ohm's law. This linear representation also implies that the driving force that indicates the distance from the equilibrium should not be too large for the system to be characterized in the nonlinear region of thermodynamic branch (Fig. 2).

3.1 Momentum transfer

In Fig. 3a, we have a fluid between two large parallel plates separated by a distance H. This system is initially at rest; however, at time $t = 0$ the lower plate is set to in motion by a constant force F in the positive x direction at a constant velocity $\mathbf{v}$. As the time proceeds, the fluid gains momentum, and the linear steady-state velocity profile is established. *Newton's law of viscosity* relates shear stress τ_{yx}, which is the force in the x direction on a unit area perpendicular to the y direction, and the velocity gradient in a Newtonian fluid, and is given for a one-dimensional flow as

$$\tau_{yx} = -\mu \frac{dv_x}{dy} \tag{5}$$

where μ is the viscosity. The shear stress is a tensor with magnitude, direction and orientation. The viscosity of a fluid is a measure of its resistance to the deformation rate. The concept of viscosity requires a study of the microscopic motion and collisions of the fluid molecules. Eq. (5) can describe the resistance to flow of all gases and all liquids of the *Newtonian fluids*. Most common fluids, such as water, air and kerosene are Newtonian fluids at normal temperature and pressure. Eq. (5) however, cannot describe the polymeric liquids, suspensions, pastes, slurries and other complex fluids, which are referred to as the *non-Newtonian fluids*. In non-Newtonian fluids the shear stress is not directly proportional to rate of deformation or shear rate, and may be represented by the power law model in terms of the flow behavior index. Some non-Newtonian fluids are classified as *thixotropic* fluids, which show decrease in viscosity with time, and *rheopectic* fluids, which show an increase in viscosity with time.

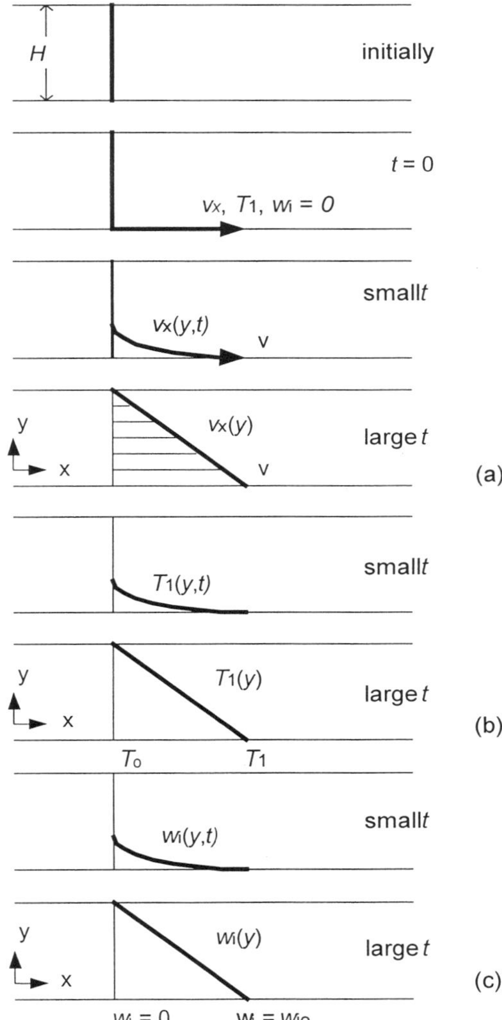

Fig.3. Steady state transport of (a) momentum, (b) heat, and (c) mass.

In the close vicinity of the moving solid surface, the fluid gains a certain amount of x-momentum, and transfers it to the adjacent layer of fluid, so that the fluid develops a motion in the x direction. Hence x-momentum is being transmitted through the fluid in the y direction, and τ_{yx} represents the flux of x-momentum; this interpretation is consistent with the molecular phenomena of momentum transport and the kinetic theory of fluids, and with heat and mass transport. In gases, the momentum is transported by the motion and collisions of molecules; in liquids the transport is mainly due to motion of pairs of molecules bonded with intermolecular forces. Momentum flows 'downhill' from a region of high velocity to a region of low velocity, and the velocity gradient therefore is a driving force.

In fluid dynamics, often, *kinematic viscosity* ν that is the viscosity divided by the density of the fluid is used

$$= \frac{\mu}{\rho} \tag{6}$$

Table 1 shows the units of quantities associated with the momentum flow.

The viscosities of fluids change over many orders of magnitude. The viscosity is a strong function of temperature and increases with temperature for gaseous systems at low density, while the viscosity usually decreases with increasing temperature in liquids. Extensive data on viscosities of pure gases and liquids are available, and Table 2 shows some experimental values of viscosities for fluids.

In Eq. (5), we have simple steady-state shearing flow with velocity function of y alone. In more complicated flows the velocity components fields may be expressed in three directions and with time, and in Cartesian coordinates we have

$$v_x = v_x(x, y, z, t); \quad v_y = v_y(x, y, z, t); \quad v_z = v_z(x, y, z, t) \tag{7}$$

In such a situation, there will be nine-stress components τ_{ij}. The viscous forces appear only when there are velocity gradients within the fluid. The forces per unit area (molecular stresses) acting on the body π, both by the thermodynamic pressure and by the viscous stresses are given by

$$\boldsymbol{\pi}_x = P\boldsymbol{\delta}_x + \tau_x; \quad \boldsymbol{\pi}_y = P\boldsymbol{\delta}_y + \tau_y; \quad \boldsymbol{\pi}_z = P\boldsymbol{\delta}_z + \tau_z \tag{8}$$

Here δ_x is the unit vector in x direction. The components of the forces are scalars and are defined as

$$\pi_{ij} = P\delta_{ij} + \tau_{ij} \tag{9}$$

Table 1
Units of quantities related to Eq. (5)

Quantity	SI	c.g.s.	British
τ_{yx}	Pa = N/ m^2	dyn/cm^2	lb$_f$/ft^2
v_x	m/s	cm/s	ft/s
μ	Pa· s	poise = g/cm· s	lb$_m$/ft· s
ν	m^2/s	cm^2/s	ft^2/s

where i and j may be x, y, or z, and δ_{ij} is the *Kronecker delta*, which is 1 if $i = j$, and zero if $i \neq j$. The stresses

$$\pi_{xx} = P + \tau_{xx}; \quad \pi_{yy} = P + \tau_{yy}; \quad \pi_{zz} = P + \tau_{zz} \tag{10}$$

are called *normal stresses*, where as the remaining six quantities are called *shear stresses*. They have two subscripts associated with the coordinates, and are referred to as the components of molecular momentum flux tensors, or the components of molecular stress tensor, as they are associated with molecular motion. Usually, the viscous stress tensor τ, and the molecular stress tensor π are simply referred to as *stress tensor*. For a Newtonian fluid the stresses may be expressed in terms of velocity gradients and viscosities in rectangular coordinates as follows

Table 2
Viscosities of gases and liquids at 1 atm pressure

Substance	T, K	Gases μ, mPa· s	Liquids μ, mPa· s
Water	273		1.787
	288		1.140
	298		0.890
	313		0.653
	333		0.463
	353		0.3548
	373	0.0121	0.2821
Air	273	0.01716	
	293	0.0181	
	313	0.01908	
	353	0.02087	
	373	0.02173	
i-Butane, i-C_4H_{10}	296	0.0076	
Methane, CH_4	293	0.0109	
Carbon dioxide, CO_2	293	0.0146	
Nitrogen, N_2	293	0.0175	
Oxygen, O_2	293	0.0204	
Hydrogen, H_2	300	0.0089	
Mercury, Hg	293		1.552
Ethanol, C_2H_5OH	273		1.786
	323		0.694
Acetone, $(CH_3)_2CO$	273		0.283
	298		0.224
Benzene, C_6H_6	293		0.649
Glycerol	298		934.0

$$\tau_{xy} = \tau_{yx} = \mu\left(\frac{\partial v_y}{\partial x} + \frac{\partial v_x}{\partial y}\right) \tag{11}$$

$$\tau_{yz} = \tau_{zy} = \mu\left(\frac{\partial v_z}{\partial y} + \frac{\partial v_y}{\partial z}\right) \tag{12}$$

$$\tau_{zx} = \tau_{xz} = \mu\left(\frac{\partial v_x}{\partial z} + \frac{\partial v_z}{\partial x}\right) \tag{13}$$

$$\pi_{xx} = -P - \frac{2}{3}\mu\nabla\cdot\mathbf{v} + 2\mu\frac{\partial v_x}{\partial x} \tag{14}$$

$$\pi_{yy} = -P - \frac{2}{3}\mu\nabla\cdot\mathbf{v} + 2\mu\frac{\partial v_y}{\partial y} \tag{15}$$

$$\pi_{zz} = -P - \frac{2}{3}\mu\nabla\cdot\mathbf{v} + 2\mu\frac{\partial v_z}{\partial z} \tag{16}$$

Eqs. (11)-(13) can be written in vector-tensor notation

$$\tau = -\mu(\nabla\mathbf{v} + (\nabla\mathbf{v})^T + (\frac{2}{3}\mu - \kappa)(\nabla\cdot\mathbf{v})\boldsymbol{\delta} \tag{17}$$

where δ is the unit tensor with components δ_{ij}, $\nabla\mathbf{v}$ is the *velocity gradient tensor* with components $(\partial/\partial x_i)v_j$, $(\nabla\mathbf{v})^{\mathrm{T}}$ is the transpose of the velocity gradient tensor with components $(\partial/\partial x_j)v_i$, and $(\nabla\cdot\mathbf{v})$ is the divergence of the velocity vector.

The generalization in Eq. (17) involves the viscosity μ and the dilatational viscosity κ to characterize the fluid. Usually it is not necessary to know the κ in fluid mechanic problems. For gases we often assume it to be close to ideal monatomic gas, for which κ is practically zero. We also assume that liquids are incompressible $(\nabla\cdot\mathbf{v}) = 0$, and therefore the term κ can be neglected.

When experimental data are not available, the viscosity can be estimated for gases at low temperature using the molecular theory

$$\mu = \frac{2}{3\pi}\frac{\sqrt{\pi m\kappa T}}{\pi d^2} \tag{18a}$$

where d and m are respectively the diameter and mass of the spherical molecules with negligible attraction between them, κ is the Boltzmann constant, and πd^2 is called the collision cross section. Eq. (18a) predicts the viscosity satisfactorily as function of temperature without the effect of pressure up to about 10 atm. However, the predicted temperature dependence is less satisfactory. Experimental data shows that viscosity increases more rapidly than $T^{1/2}$. To describe the temperature dependency, it is necessary to replace the rigid-sphere model by one that can represent the attractive and repulsive forces more accurately. It is also necessary to consider dense gases and use the Boltzmann equation to obtain the molecular velocity distribution in nonequilibrium systems more accurately.

The viscosity of a pure monatomic gas of molecular weight M may be expressed in terms of the Lennard-Jones parameters as

$$\mu = 2.6693\ 10^{-5} \frac{\sqrt{MT}}{\sigma^2 \Omega_\mu} \tag{18c}$$

where T is in K, σ is the collision diameter in Å, and μ is in g/cm s. The dimensionless quantity Ω_μ is called the *collision integral* for viscosity; it describes the deviation from rigid sphere behavior, and varies slowly with the dimensionless temperature $\kappa T/\varepsilon$, of the order of magnitude of unity.

Another approach is the *corresponding-state correlation,* which is based on the principle of corresponding states widely used for correlating thermodynamic data. This approach is shown in Fig. 4, where the *reduced viscosity* $\mu_r = \mu/\mu_c$ is plotted versus the *reduced temperature* $T_r = T/T_c$ and the *reduced pressure* $P_r = P/P_c$, and shows a global view of the pressure and temperature dependence of viscosity. The viscosity of a gas approaches a low-density limit at about 1 atm pressure, and increases with increasing temperature, whereas the viscosity of a liquid decreases with increasing temperature. The critical viscosity can be estimated if critical P-V-T data are available using the following empirical relations

$$\mu_c = 61.6(MT_c)^{1/2}(V_c)^{-2/3} \quad \text{and} \quad \mu_c = 7.70M^{1/2}(P_c)^{2/3}(T_c)^{-1/6} \tag{18d}$$

Here μ_c is in micropoises, P in atm, T_c in K, and V_c in cm^3/g-mole. Fig 4 can also be used to approximate viscosities of mixtures with the following definitions of the *pseudocritical properties*

$$P_c' = \sum_{i=1}^{n} x_i P_{ci} \quad T_c' = \sum_{i=1}^{n} x_i T_{ci} \quad \mu_c' = \sum_{i=1}^{n} x_i \mu_{ci} \tag{19}$$

Here x_i are the mole fraction of the mixtures. We may use Fig. 4 same as for pure fluids, but with the pseudocritical properties instead of the critical properties. This procedure yields reasonable accuracy for mixtures of chemically similar substances.

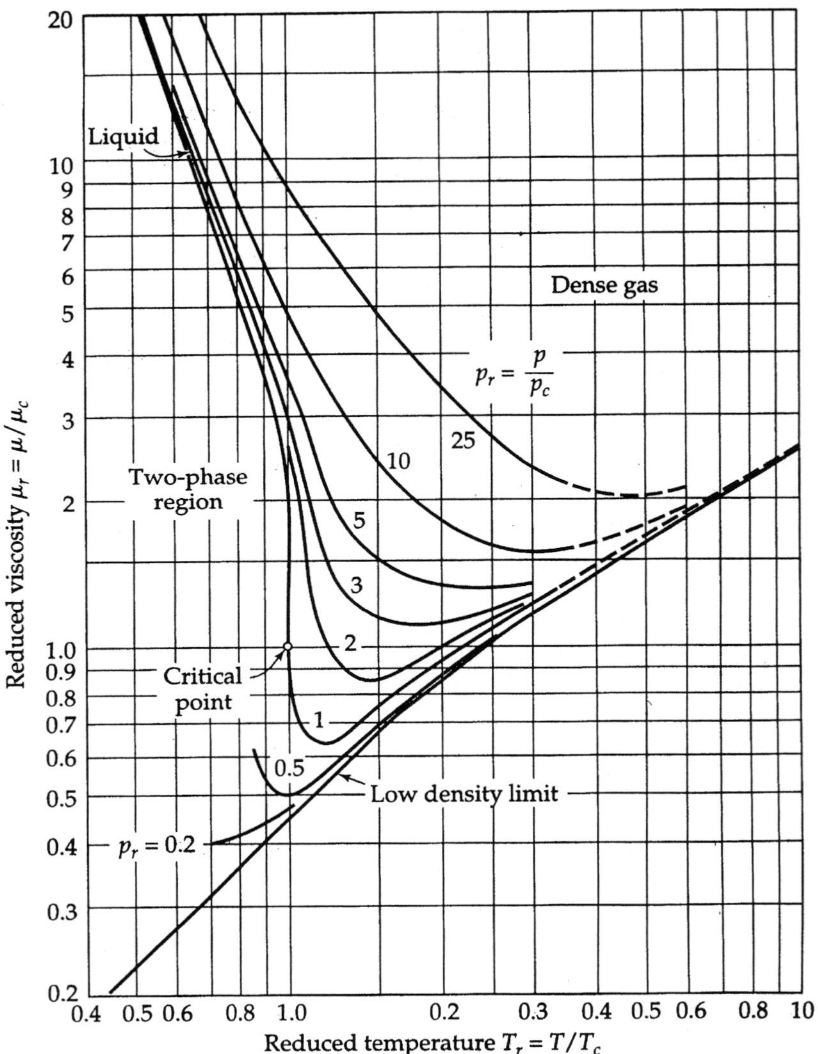

Fig. 4. Change of reduced viscosity as a function of reduced temperature and reduced pressure. [O. A. Hougen, K.M. Watson, R.A Ragatz, C. P. P. Charts, Wiley, New York, 2nd edition (1960)].

The following semiempirical relation also predicts the viscosity of a gas mixture within an average deviation of about 2%

$$\mu_{\text{mix}} = \sum_{i=1}^{n} \frac{x_i \mu_i}{\sum_j x_j \Phi_{ij}} \tag{20}$$

where the dimensionless quantities Φ_{ij} are given by

$$\Phi_{ij} = \frac{1}{\sqrt{8}} \left(1 + \frac{M_i}{M_j}\right)^{-1/2} \left[1 + \left(\frac{\mu_i}{\mu_j}\right)^{1/2} \left(\frac{M_j}{M_i}\right)^{1/4}\right]^2 \tag{21}$$

Here n is the number of chemical species in the mixture, x_i is the mole fraction of species of i, μ_i is the viscosity of species i at the system temperature and pressure, and M_i is the molecular weight of species i. The dependence of viscosities on composition is nonlinear for some mixtures of light and heavy gases, and many additional empirical equations are available for estimating viscosities of gases and gas mixtures [5].

3.2. Heat transfer

Consider a solid slab of area A located between two large parallel plates with a distance H apart, as shown in Fig. 3b. Initially the solid material is at a temperature T_o. At time $t = 0$ the lower plate is brought to a higher temperature T_1, and maintained at this temperature. A linear steady-state temperature distribution is developed due to a constant rate of heat flow q through the slab. It is found that for small values of temperature difference $\Delta T = T_1 - T_o$, the heat flux that is the rate of heat flow per unit area is proportional to the temperature decrease over distance H. When the slab thickness approaches zero, the one-dimensional form of Fourier's law relates the heat flow (flux) q and temperature gradient, and we have in the direction y

$$q_y = -k \frac{dT}{dy} \tag{22}$$

where k is the thermal conductivity, which is assumed as independent of direction. If the temperature varies in all three directions, we get the three-dimensional form of Fourier's law in vector form

$$\mathbf{q} = -k \nabla T \tag{23}$$

Eq. (23) is applicable to an isotropic medium only, so the heat is conducted with the same thermal conductivity k in all the directions. The thermal conductivity k is a property of a conducting medium, and is mainly a function of temperature. High pressure affects the thermal conductivity in a gas medium.

Eq. (22) for heat conduction and Eq. (5) for momentum transfer are similar, and the flux is proportional to the negative of the gradient of a macroscopic variable; the coefficient of proportionality is a physical property characteristic of the medium and dependent on the temperature and pressure. In a three-dimensional transport, Eq. (23) and Eq. (17) differ because the heat flux is a vector with three components, and the momentum flux τ is a second-order tensor with nine components.

Beside the thermal conductivity k, the thermal diffusivity α is also widely used, and α is defined as

$$\alpha = \frac{k}{\rho C_p} \tag{24}$$

Here C_p is the heat capacity at constant pressure. The thermal diffusivity α has the same units as the kinematic viscosity ν, and they play similar roles in the equations of change for momentum and energy. The dimensionless ratio

$$Pr = \frac{\nu}{\alpha} = \frac{C_p \mu}{k} \tag{25}$$

is called the Prandtl number Pr, which shows the relative ease of molecular momentum and energy transport in the hydrodynamic and thermal boundary layers, respectively. The Prandtl number for gases is near unity, and hence the magnitudes of energy and momentum transfer by diffusion are comparable. For liquid metals, $Pr \ll 1$ and thermal diffusion is much larger than momentum diffusion; this is opposite for liquids for which $Pr \gg 1$. The units that are commonly used for thermal conductivity and heat transport are given in Table 3.

Table 3
Units of quantities related to heat flux.

Quantity	SI	c.g.s.	British
q_y	W/m^2	cal/cm^2·s	Btu/ft^2·h
k	W/m·K	cal/cm· s· °C	Btu/h·ft· °F
C_p	J/kg·K	cal/g· °C	Btu/lb$_m$· °F
α	m^2/s	cm^2/s	ft^2/s

Thermal conductivity can vary from about 0.01 W/m K for gases to about 1000 W/m K for pure metals. Some experimental values of the thermal conductivities are given in Table 4. When available, experimental values should be used in calculations, otherwise several empirical relations may provide satisfactory predictions [5]. In the corresponding state approach, the reduced thermal conductivity $k_r = k/k_c$ is plotted as a function of the reduced temperature and the reduced pressure, as shown in Fig. 5, which is based on a limited amount of experimental data for monatomic substances, and may be used for rough estimates for polyatomic substances. Fig. 5 shows that the thermal conductivity of a gas approaches a limiting function of T at about 1 atm pressure. The thermal conductivities of gases at low density increase with increasing temperature, while for most liquids the thermal conductivity decreases with increasing temperature. This correlation may change in polar and associated liquid region, for example water exhibits a maximum in the curve of k versus T. The corresponding-states provide a global view of the behavior of the thermal conductivity of fluids.

Table 4
Thermal conductivities of gases and liquids.

Substance Gas	T, K	k, W/m·K	C_p, J/kg·K	Pr
		Gases at 1 atm pressure		
Hydrogen, H_2	100	0.06799	11,192	0.682
	300	0.1779	14,316	0.720
Oxygen, O_2	100	0.00904	910	0.764
	300	0.02657	920	0.716
Carbon dioxide, CO_2	200	0.0095	734	0.783
	300	0.01665	846	0.758
Methane, CH_4	200	0.02184	2087	0.721
	300	0.03427	2227	0.701
Liquids at their saturation pressures				
Water, H_2O	300	0.6089	4.183	6.02
	350	0.6622	4.193	2.35
	400	0.6848	4.262	1.35
Ethanol, C_2H_5OH	250	0.1808	2.120	35.8
	300	0.1676	2.454	15.2
	350	0.1544	2.984	8.67
Carbon tetrachloride, CCl_4	250	0.1092	0.8617	16.0
	350	0.0893	0.9518	5.13
Acetone, $(CH_3)_2CO$	250	0.1478	2.197	5.68
	300	0.1274	2.379	4.13
	350	0.1071	2.721	3.53
1-Pentene, C_5H_{10}	200	0.1461	1.948	8.26
	300	0.1153	1.907	3.72

The thermal conductivities of dilute monatomic gases are well understood. The thermal conductivity of a dilute gas of rigid spheres of diameter d is expressed as

$$k = \frac{2}{3\pi}\frac{\sqrt{\pi m \kappa T}}{\pi d^2}C_v \tag{26a}$$

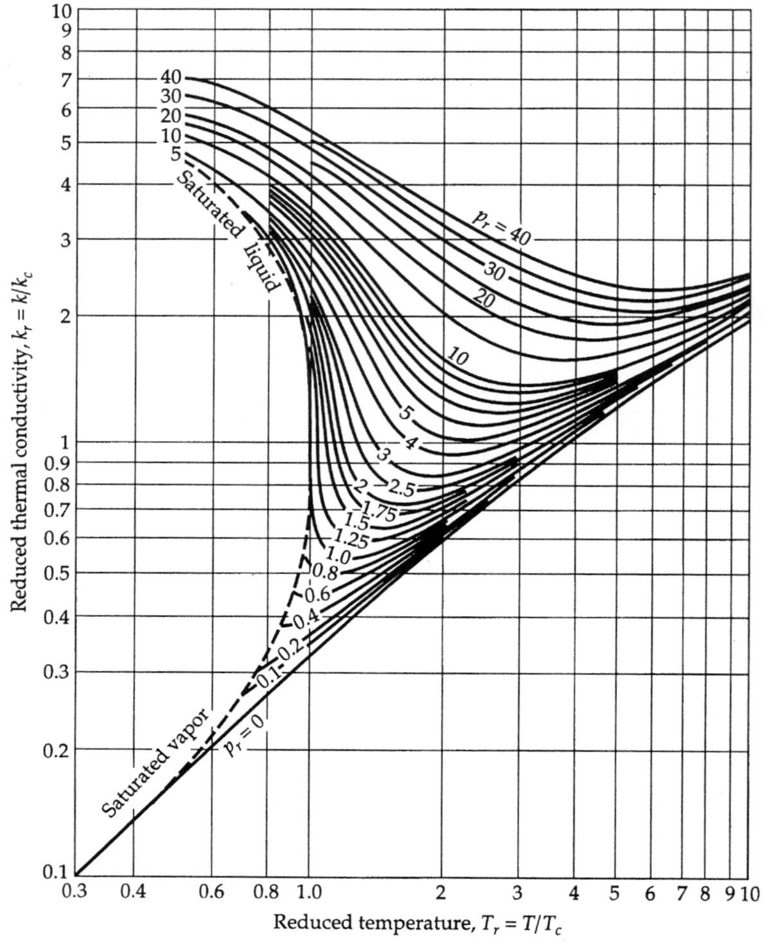

Fig.5. Change of reduced thermal conductivity with reduced temperature and reduced pressure for monatomic substances [O. A. Hougen, K. M. Watson, and R. A. Ragatz, Chemical Process Principles Charts, 2nd edition, Wiley, New York (1960)].

where m is the mass of a molecule, and κ is the Boltzmann constant. Eq. (26a) predicts that k is independent of pressure, and the prediction is satisfactory up to about 10 atm for most gases. The predicted temperature dependence is weak. The Chapman-Enskog formula at low density and temperature produce better predictions, and is given by

$$k = \frac{25}{32}\frac{\sqrt{\pi m \kappa T}}{\pi \sigma^2 \Omega_k} C_v \quad \text{or} \quad k = 1.9891x10^{-4}\frac{\sqrt{T/M}}{\sigma^2 \Omega_k} \tag{26b}$$

In the second form of this equation, k is in cal/cm s K, T in K, σ in Angstrom, and the collision integral for thermal conductivity Ω_k is identical to that for viscosity Ω_μ. The values of collision integrals are given for the Lennard-Jones intermolecular potential as a function of the dimensionless temperature $\kappa T/\varepsilon$.

A simple semiempirical equation for polyatomic gases at low densities is given by

$$k = \left(C_p + \frac{5}{4}\frac{R}{M} \right)\mu \tag{27}$$

This equation is called the *Eucken formula*, and it can provide a simple method of estimating the Prandtl number for nonpolar polyatomic gases at low density

$$\Pr = \frac{C_p \mu}{k} = \frac{C_p}{C_p + (5/4)R} \tag{28}$$

The thermal conductivities for gas mixtures k_{mix} at low densities may be estimated as

$$k_{\text{mix}} = \sum_{i=1}^{n} \frac{x_i k_i}{\sum_j x_j \Phi_{ij}} \tag{29}$$

The x_i are the mole fractions, and the k_i are the thermal conductivities of the pure gases. The coefficients Φ_{ij} are identical to those appearing in Eq. (20) in the viscosity equation.

3.3. Mass transfer

We consider a thin, horizontal, fused-silica plate of area A and thickness H. initially both horizontal surfaces of the plate are in contact with air (Fig. 3c). We assume that the air is completely soluble in silica. At time $t = 0$, the air below the plate is replaced by pure helium, which is appreciably soluble in silica. The helium slowly penetrates into the plate by molecular motion, and eventually appears in the air above the plate. This molecular transport of one substance relative to another is known as *diffusion*. The air above the plate is removed rapidly, so that there is no measurable helium concentration there. In this system, the indices i shows helium, and j shows silica, and the concentrations are given by the mass fractions w_i and w_j respectively. Eventually, the concentration profile tends towards a straight line with increasing t, and we have $w_i = w_{io}$ at the bottom surface, and $w_i = 0$ at the top surface of the plate. At steady-state diffusion, it is found that mass flux vector that is the flow rate of helium per unit area $\mathbf{j}_i$ is proportional to the concentration gradient, and we have

$$\mathbf{j}_i = -\rho D_{ij} \frac{dw_i}{dy} \tag{30}$$

where ρ and D_{ij} are the density and the diffusivity coefficient of the silica-helium system respectively. Eq. (30) is called the one-dimensional form of Fick's first law of diffusion, and indicates that mass flows from high to low concentration region. It is valid for any binary fluid, or solid solution, provided that $\mathbf{j}_i$ is defined as the mass flux relative to the mixture mass average velocity $\mathbf{v}$, and in general we have

$$\mathbf{j}_i = \rho w_i (\mathbf{v}_i - \mathbf{v}) \tag{31}$$

here the term $(\mathbf{v}_i - \mathbf{v})$ is called the diffusion velocity. The mass flux $\mathbf{j}_j$ is defined analogously. As the two chemical species interdiffuse there is a shifting of the center of mass in the y direction if the molecular weights of components i and j differ. The fluxes $\mathbf{j}_i$ and $\mathbf{j}_j$ are measured with respect to the motion of the center of mass, and $\mathbf{j}_i + \mathbf{j}_j = 0$. If we express the diffusion in three directions, we have the vector form of Fick's law

$$\mathbf{j}_i = -\rho D_{ij} \nabla w_i \tag{32}$$

where ∇w_i is the concentration gradient in terms of the mass fraction.

We can define the molecular mass flux vector $\mathbf{j}_i$ and the convective mass flux vector $\rho\mathbf{v}$ to get the combined mass flux vector, and similarly for the combined molar flux vector

$$\mathbf{n}_i = \mathbf{j}_i + \rho_i \mathbf{v} \tag{33}$$

Eq. (33) can be interpreted as the mass flux with respect to mass average velocity $\mathbf{v}$, and using the diffusion velocity $(\mathbf{v}_i - \mathbf{v})$, we obtain

$$\mathbf{j}_i = \rho_i(\mathbf{v}_i - \mathbf{v}) = \mathbf{n}_i - \rho_i\mathbf{v} = -\rho D_{ij}\nabla w_i \tag{34}$$

For a pair *i-j* there is just one diffusivity coefficient $D_{ij} = D_{ji}$, and in general diffusivity is a function of pressure, temperature and composition. With respect to mobility of the particles, the diffusion coefficients are generally higher for gases and lower for solids. Diffusivities of gases at low densities are almost independent of concentration, increase with temperature, and vary inversely with pressure. Liquid and solid diffusivities are strongly concentration-dependent and generally increase with temperature. Table 5 shows some of experimental binary diffusivities for gases and liquid systems.

The molecular theory yields the self-diffusivity of component *i* at low density

$$D_{ii} = \frac{2}{3\pi}\frac{\sqrt{\pi m_i \kappa T}}{\pi\, d_i^2}\frac{1}{\rho} \tag{35a}$$

Eq. (35a) can be compared with Eq. (26a) for the thermal conductivity, and with Eq. (18a) for the viscosity. For binary gas mixtures at low pressure, D_{ij} is inversely proportional to the pressure, increases with increasing temperature, and is almost independent of the composition for given gas pair. For an ideal gas law $P = cRT$, the Chapman-Enskog kinetic theory yields the binary diffusivity at low density

$$D_{ij} = 0.0018583\sqrt{T^3\left(\frac{1}{M_i}+\frac{1}{M_j}\right)}\frac{1}{P\sigma_{ij}^2\Omega_{D.ij}} \tag{35b}$$

Here D_{ij} in cm^2/s, σ_{ij} is in Angstrom, T is in K, and P is in atm. The dimensionless quantity $\Omega_{D,ij}$, is the collisional integral for diffusion, and is a function of the dimensionless temperature $\kappa T/\varepsilon_{ij}$. The parameters σ_{ij} and ε_{ij} are those appearing in the Lennard-Jones potential between molecules of *i* and *j*.

At low pressures the following expression has been developed from a combination of kinetic theory and corresponding-states approach

$$D_{ij} = a\left(\frac{T}{\sqrt{T_{ci}T_{cj}}}\right)^b\left(\frac{1}{M_i}+\frac{1}{M_j}\right)\frac{(P_{ci}P_{cj})^{1/3}}{P}(T_{ci}T_{cj})^{5/12} \tag{35c}$$

Here D_{ij} is in cm^2/s, P is in atm, and T is in K. Analysis of experimental data gives the dimensionless constants $a = 2.745\ 10^{-4}$ and $b = 1.823$ for nonpolar gas pairs, excluding helium and hydrogen, and $a = 3.64\ 10^{-4}$ and $b = 2.334$ for pairs of water and a nonpolar gas. Eq. (35b) predicts the data at atmospheric pressure within an average deviation of 6 to 8%.

A corresponding states plot of the self-diffusivity D_{AA^*} that is the interdiffusion of labeled molecules of A at the low-pressure limit is shown in Fig. 6. The reduced self-diffusivity that is cD_{AA^*} at pressure P and T divided by cD_{AA^*} at the critical point is plotted as a function of the reduced pressure and reduced temperature. Fig. 6 shows that cD_{AA^*} increases strongly with increasing temperature, especially for liquids. At each temperature the value of cD_{AA^*} decreases towards a low-pressure limit.

Table 5
Molecular diffusivities of gases and liquids.

Gas pair (*i-j*)	*T*, K	D_{ij}, cm^2/s
Air-Ammonia	273	0.198
Air-Carbon dioxide	273	0.136
Air-Ethanol	298	0.132
Air-n-Octane	298	0.0602
Air-Water	298	0.260
Air-Acetone	273	0.110
Air-Benzene	298	0.880
Air-Naphtelene	300	0.620
Air-Hydrogen	298	0.410
Carbon dioxide-Benzene	318	0.0715
Carbon dioxide-Ethanol	273	0.0693
Carbon dioxide-Methanol	298.6	0.105
Carbon dioxide-Water	298	0.164
Hydrogen-Ammonia	293	0.849
Hydrogen-Benzene	273	0.317
Hydrogen-Ethane	273	0.439
Hydrogen-Water	293	0.850
Oxygen-Ammonia	293	0.253
Oxygen-Benzene	296	0.094
Oxygen-Ethylene	293	0.182
Liquid pair; Solute *i* (concentration in gmole/liter)- solute *j*		
Ammonia (3.5)-Water	278	$1.24\ 10^{-5}$
Ammonia (1)-Water	288	$1.24\ 10^{-5}$
Ethanol (3.75)-Water	283	$0.50\ 10^{-5}$
Ethanol (2.0)-Water	289	$0.90\ 10^{-5}$
Ethanol (0.05)- Water	283	$0.83\ 10^{-5}$
Chloroform (2.0)-Ethanol	293	$1.25\ 10^{-5}$
Sodium chloride (0.05)-Water	291	$1.26\ 10^{-5}$
Sodium chloride (0.05)-Water	250	$1.24\ 10^{-5}$
Sodium chloride (0.05)-Water	350	$1.24\ 10^{-5}$

At critical point, the following the empirical relation can be used for estimating self-diffusivity between i and labeled species i^* $(cD_{ii^*})_c$

$$(cD_{ii^*})_c = 2.96\ 10^{-6}\left(\frac{1}{M_i}-\frac{1}{M_{i^*}}\right)^{1/2}\frac{P_{ci}^{2/3}}{T_{ci}^{1/6}} \tag{36}$$

Where c is in g mole/cm^3, D_{ii} is in cm^2/s, T_c is in K, and P_c is in atm. For binary diffusion of chemically dissimilar species the following relation can be used at low pressure

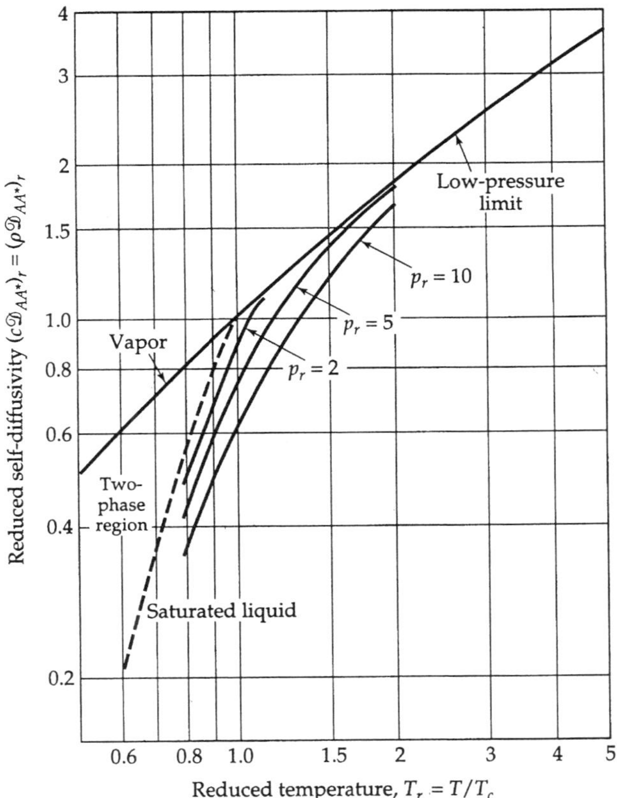

Fig. 6. Change of reduced self-diffusivity with reduced temperature and reduced pressure [J. J. van Loef and E.G.D. Cohen, Physica A, 156, 522-533 (1989), B.I.Lee and M.G. Kesler, AIChE J., 23, 510-527 (1975)].

$$(cD_{ij})_c = 2.96\ 10^{-6}\left(\frac{1}{M_i} - \frac{1}{M_j}\right)^{1/2} \frac{(P_{ci}P_{cj})^{1/3}}{(T_{ci}T_{cj})^{1/12}} \tag{37}$$

In Fig. 6, and Eqs. (36) and (37) c-multiplied diffusion coefficients are used because their dependence on pressure and temperature is simpler, and they are frequently used in mass transfer calculations.

For diffusion in liquids, it is necessary to rely on empirical expressions; Wilke-Chang equation gives the diffusivity for small concentrations of i in j as

$$D_{ij} = 7.4 10^{-8} \frac{T\sqrt{\psi_j M_j}}{\mu V_i^{0.6}} \tag{38}$$

Here V_i is the molar volume of the solute i in cm^3/gmole as liquid at its normal boiling point, μ is the viscosity of the solution in centipoises, ψ_j is an association parameter for the solvent, and T is the absolute temperature in K. Recommended values of association parameters are 2.6 for water, 1.9 for methanol, 1.5 for ethanol, and 1.0 for benzene, ether, heptane, and other unassociated solvents. Eq. (38) should be used only for dilute nondissociating solutions.

Binary and multicomponent diffusion are different in nature; in binary diffusion flux of species i is always proportional to the negative of the concentration gradient of species i. In multicomponent diffusion, however, other interesting situations can arise: (i) in *reverse diffusion*, a substance diffuses in a direction opposite to the direction imposed by its own concentration gradient; (ii) in *osmotic diffusion*, a substance diffuses although its concentration gradient is zero; (iii) in *diffusion barrier*, a substance does not diffuse even though its concentration gradient is nonzero.

All these phenomenological systems show that in relatively slow processes the conjugate flow J is largely determined by frictional forces, and is linearly related to the conjugate force X

$$J = LX \tag{39}$$

where the coefficient L is a proportionality factor, which is not necessarily constant but independent of both J and X.

4. THE MAXWELL-STEFAN EQUATIONS

For multicomponent diffusion in gases at low density the following equation provides a satisfactory approximation

$$\nabla x_i = -\sum_{i=1}^{n} \frac{x_i x_j}{D_{ij}^{'}} (\mathbf{v}_i - \mathbf{v}_j) = -\sum_{i=1}^{n} \frac{1}{cD_{ij}^{'}} (x_j N_i - x_i N_j) \tag{40}$$

The $D_{ij}^{'}$ are the binary diffusivities, and there are (½)n(n-1) of them required for n-component system. These relations are called the *Maxwell-Stefan equations*. For gases at low density, dense gases, liquids, and polymers the Maxwell-Stefan equation can be used with the diffusivities called the Maxwell-Stefan diffusivities, which can be related to the Fick diffusivities through the thermodynamic correction factor for nonideal liquid systems.

5. TRANSPORT COEFFICIENTS

The fluxes of momentum, energy, and mass in one-dimensional and at constant densities are expressed as follows

$$\tau_{xy} = -\nu \frac{d}{dx}(\rho v_y) \tag{41}$$

$$q_x = -\alpha \frac{d}{dx}(\rho C_p T) \qquad (C_p = \text{constant}) \tag{42}$$

$$J_{Ax} = -D_{AB} \frac{d}{dx}(\rho_A) \tag{43}$$

Eqs. (41) to (43) state, respectively, that momentum transport occurs because of a gradient in momentum concentration, energy transport is due to a gradient in energy concentration, and mass transport is the result of a gradient in mass concentration; Therefore these three transport processes show analogies. However, these analogies do not apply in two- and three- dimensional transport processes, since τ is a tensor quantity with nine components, while J_A and q are vectors with three components.

The mass diffusivity D_{ij}, the thermal diffusivity $\alpha = k / \rho C_p$, and the momentum diffusivity or kinematic viscosity $= \mu / \rho$ all have dimensions of (length)^2/time, and are called the transport coefficients. The ratios of these quantities yield the following dimensionless groups of Prandtl number Pr, the Schmidt number Sc, and the Lewis number Le, which are expressed by

$$Pr = \frac{\nu}{\alpha} = \frac{C_p \mu}{\rho} \tag{44}$$

$$Sc = \frac{\nu}{D_{ij}} = \frac{\mu}{\rho D_{ij}} \tag{45}$$

$$Le = \frac{\alpha}{D_{ij}} = \frac{k}{\rho C_p D_{ij}} \tag{46}$$

These dimensionless groups of fluid properties play an important role in dimensionless equations, and for systems where simultaneous transport processes occur.

The close interrelations among mass, momentum, and energy transport can be explained in terms of a molecular theory of monatomic gases at low density. The continuum equations of continuity, motion, and energy can all be derived from one starting point that is the Boltzmann equation for the velocity distribution function, from which the molecular expressions for the fluxes and transport properties are produced. The discussion of the dependence of the fluxes on the driving forces is closely related to the nonequilibrium thermodynamic approach. Similar discussions are also available for polyatomic gases, monatomic liquids, and polymeric liquids. In kinetic theories for monatomic liquids, the expressions for the momentum and heat fluxes include contributions associated with forces between two molecules; for polymers, additional forces within the polymer chain is necessary.

6. ELECTRIC CHARGE FLOW

Ohm found that the flow of electricity that is the current I is directly proportional to electric potential difference or applied voltage $\Delta\psi$

$$I = \frac{\Delta\psi}{R} \tag{47}$$

where R is the resistance of the medium to the current. The value of resistance is influenced by the medium configuration, and for many materials is independent of current. When an electric field is applied, the free electrons experience the acceleration in a direction opposite to that of the field, and the flow of charge is called the electric current. Quantum mechanics predicts that there is no interaction between an accelerating electron and atoms in a perfect crystal lattice. Since the current reaches a constant value after the electric field is applied, there exist frictional forces, which counter the acceleration. The frictional forces are the result of the scattering electrons by imperfections in the crystal due to impurity atoms, dislocations, and vacancies. Thermal vibrations of the atoms may also cause frictional forces. The frictional forces cause the resistance which may be described by the drift velocity v_d and the mobility of an electron μ_e. The drift velocity represents the average electron velocity, and the electron mobility indicates the frequency of scattering phenomena, and has units of m^2/Volt s

$$v_d = \mu_e \varepsilon \tag{48}$$

where ε is the electric field intensity, and defined as the voltage difference between two points divided by the distance l separating them

$$\varepsilon = \frac{\Delta\psi}{l} \tag{49}$$

The conductivity of most materials may be expressed in terms of number of free electrons n per unit volume, absolute magnitude of the electrical charge of an electron ($|e|=1.6 \times 10^{-19}$ C) in Coulomb, and the mobility of electron

$$\sigma = n|e|\mu_e \tag{50}$$

The electrical conductivity specifies the electrical character of the material. The solid materials, in three groupings of conductors, semiconductors and insulators, exhibit a wide range of electrical conductivities. Metals have the conductivities on the order of 10^7 $(\Omega\ m)^{-1}$, insulators have the conductivities ranging between 10^{-10} and 10^{-20} $(\Omega\ m)^{-1}$, and the conductivities of semiconductors changes from 10^{-6} $(\Omega\ m)^{-1}$ to 10^4 $(\Omega\ m)^{-1}$.

The resistivity ρ is independent of specimen geometry and related to resistance R

$$\rho = \frac{RA}{l} \tag{51}$$

where A is the cross sectional area normal to the direction of the current. For most metals and their alloys resistivity increases with temperature due to the increase in thermal vibration and other irregularities, such as plastic deformations, which serve as electron-scattering centers. The resistivity also changes with composition for alloys. The electrical conductivity can be also defined as the reciprocal of the resistivity with units in Ω m

$$\sigma = 1/\rho \tag{52}$$

Using Eq. (51) and Eq. (52) Ohm's law may be expressed by

Table 6
Units and symbols of electrical parameters.

Quantity	Symbol	SI
Electrical charge, coulomb	Q	C
Electrical potential, volt	ψ	kg m^2/s^2 C
Electric field strength, volt/meter	ε	kg m/s^2 C
Electric current, ampere	I	C/s
Resistance, ohm	R	kg m^2/s C^2
Resistivity, ohm-meter	ρ	kg m^3/s C^2
Conductivity, (ohm-meter)$^{-1}$	σ	s C^2/kg m^3
Current flux (density)	J_e	C/s m^2
Electron mobility	μ_e	m^2/V s = s C/kg

$$J_e = \sigma \varepsilon \tag{53}$$

where J_e is the current flux (density) that is the current per unit of specimen area I/A. The units of electrical parameters are given in Table 6.

7. THE RELAXATION THEORY

For describing transport phenomena, we have used one of the constitutive equations of Newton's law of viscosity, Fourier's law, or Fick's law; each of them relates fluxes with conjugate thermodynamic driving forces. The conservation laws for the momentum, heat and mass transfer lead to parabolic equations of change, which suggest that the velocity of propagation of an external disturbance such as thermal at any point in the transfer medium is infinite. This can be seen in Fig (3), when the surface of the semi-infinite solid material suddenly is brought to T_1 from initial uniform temperature T_o. The solution of temperature profile shows that at time $t = 0$, temperature $T = T_o$, but for $t > 0$ temperature everywhere is expressed as $T(y,t)$, implying that change of surface temperature is felt everywhere in the material. This phenomena is explained by the hypothesis of heat flux relaxation, which states that Fourier's law is an approximation to a more exact equation called the *Maxwell-Cattaneo equation*

$$\mathbf{q} = -k\nabla T - \tau_q \frac{\partial \mathbf{q}}{\partial t} \tag{54}$$

where τ_q is the relaxation time of heat flux. The analogous equations for the irreversible fluxes of momentum and mass can also be expressed. For example, for the mass transfer an identical equation to Eq. (54) is obtained from the nonstationary version of the Maxwell-Stefan equation, and for the momentumdiffusion a similar equation is obtained from the Maxwell equation for viscoelastic fluids. The relaxation time for heat transfer is $\tau_q = 10^{-12}$ s for metals, $\tau_q = 10^{-9}$ s for gases at normal conditions, and $\tau_q = 10^{-11}$ - 10^{-13} s for typical liquids. Relaxation times can be greater in rarefied gases, viscoelastic liquids, capillary porous bodies ($\tau_q = 10^{-4}$ s), liquid helium ($\tau_q = 4.7\text{x}10^{-3}$ s), turbulent flows ($\tau_q = 10^{-3} - 10^{3}$ s), and dispersed systems

Combining the thermal energy conservation with Eq. (54) yields

$$\frac{\partial T}{\partial t} = \alpha\left(\nabla^2 T - \frac{\partial^2 T}{c_o^2 \partial t^2}\right) \tag{55}$$

where α is the thermal diffusivity, and c_o is the propagation speed of the internal wave, and given by

$$c_o = \sqrt{\alpha / \tau_q} \tag{56}$$

Eq. (55) is a hyperbolic type, and its solution for the semi-infinite solid medium suggests that two regions exist in a solid; the first region is called the disturbed region where the heat diffusion occurs, and the second region that is undisturbed. Fourier's law of heat transfer predicts the heat diffusion everywhere in the medium. However, as soon as the surface temperature changes, the wall heat flux **q**(0,0) does not start instantaneously, and rather grows gradually with the rate, which depends on the current relaxation time and not the relaxation in state. For example, chemical reaction phenomena may illustrate the state relaxation, and heat and viscous stress relaxations and also current relaxation in electric circuits associated with a change in the magnetic energy may illustrate the current relaxation. The wall heat flux reaches a maximum and decreases in time, since temperature gradient at the wall decreases. Therefore the Fourier and Fick laws are inappropriate for description of the short-time effects, which may be theoretically important although the relaxation times are typically very small.

No exact general criterion is available when it is necessary to include the relaxation terms in the equations of change, however relaxation terms are necessary for viscoelastic fluids, dispersed systems, rarefied gases, capillary porous mediums, and helium, in which the frequency of the fast variable transients may be comparable to the reciprocal of a longest relaxation time.

8. CHEMICAL REACTIONS

Chemical reaction rate is dependent on the collisions of components, and defined as the extent of this activity per second per unit volume (mole L^{-1} s^{-1}). Since the number of collisions of a substance is proportional to its concentration, the chemical reaction rate is proportional to the product of concentrations. Thus for a single homogeneous chemical reaction

$$\nu_1 A(g) + \nu_2 B(g) = \nu_3 C(g) + \nu_4 D(g) \tag{57}$$

the flow of reaction (velocity) J_r refers to the difference between the forward rate $r_f = k_f c_A c_B$ and backward rate $r_b = k_b c_C c_D$, and we have

$$J_r = r_f - r_b = k_f c_A c_B - k_b c_C c_D \tag{58}$$

where k_f and k_b are the forward and backward reactions rate constants respectively.

The ratio of mole changes of reacting components are

$$-\frac{dN_A}{\nu_1} = -\frac{dN_B}{\nu_2} = \frac{dN_C}{\nu_3} = \frac{dN_D}{\nu_4} \tag{59}$$

where ν_i are the stoichiometric coefficients, which are positive for products and negative for reactants. The extent of reaction ξ is defined as

$$d\xi = \frac{dN_i}{\nu_i} \tag{60}$$

For a single homogeneous reaction, a generalized reaction rate J_r is a scalar value, and can be expressed in terms of the extent of reaction

$$J_r = \frac{d\xi}{dt} \tag{61}$$

The affinity A is defined by

$$A = -\sum_i \nu_i \mu_i \tag{62}$$

where μ_i is the chemical potential of component i. For chemical reaction system given in Eq. (57), the affinity becomes

$$A = \nu_1 \mu_A + \nu_2 \mu_B - (\nu_3 \mu_C + \nu_4 \mu_D) \tag{63}$$

The affinity of chemical reaction can be represented by the negative of change of the Gibbs free energy. If A is greater than zero, the reaction is driven from left to right; if it is smaller than zero, the reaction proceeds from right to left; but if $A = 0$, no reaction takes place.

For a chemical reaction, the dissipation function Ψ is expressed by

$$\Psi = AJ_r \geq 0 \tag{64}$$

Although Gibbs free energy is used in equilibrium states, affinity relates irreversibility of chemical reaction to entropy production. The rate of entropy production is

$$\frac{d_i S}{dt} = \left(\frac{A}{T}\right)\frac{d\xi}{dt} \geq 0 \tag{65}$$

At constant temperature and pressure the Gibbs free energy is related to the affinity as follows

$$-A_k = \left(\frac{\partial G}{\partial \xi}\right)_{T,P} \tag{66}$$

Eq. (66) can be interpreted that the affinity is the Gibbs free energy of reaction.

Concentration oscillations can occur about a nonequilibrium value of ξ in systems that are far away from equilibrium.

Some of the historical developments in thermodynamics, and transport phenomena and rate process are given in Table 7.

9. COUPLED PROCESSES

When two or more processes occur simultaneously in a system, they couple (interfere) and cause new effects. Some of such cross-phenomena are as follows:
(i) In 1808, Rous, a colloid chemist, observed that imposing an electrical potential difference across the porous wet clay led not only to the expected flow of electricity but also to a flow of water. He later applied hydrostatic pressure to the clay and observed a flow of electricity. This *electrokinetic* experiment was the first to demonstrate the existence of coupled phenomena. This experiment proved that a flow may not only be driven by its directly conjugated force but may also be coupled to other, nonconjugated forces. Therefore the electric current is evidently caused by an electromotive force, but it may be coupled to a hydrostatic pressure. When two chambers containing electyrolytes are separated by a porous wall, an applied potential generates a pressure difference called the *electrosmotic pressure*. Also mass flow may generate an electric current called the *streaming current*.

(ii) There are two reciprocal phenomena of thermo-electricity arising from the interference of heat and electrical conductions: the first is called the *Peltier effect*. This effect is known as the evolution or the absorption of heat at junctions of metals resulting from the flow of an electrical current. The other is the *thermoelectric force* resulting from the maintenance of the junctions made of two different metals at different temperatures. This is called the *Seebeck effect*.
(iii) Another well-known example of interference is the coupling of diffusion and heat conduction causing the *thermal diffusion* (*Soret effect*) that is the occurrence of the concentration gradient as a result of the temperature gradient. *Dufour effect* is the temperature difference arising due to the concentration gradient.
(iv) The coupling between chemical reactions and transport in biological membranes, such as the sodium and potassium pumps, are known as *active transport*, in which the metabolic reactions cause the transport of a substance against the direction imposed by its thermodynamic force mainly electrochemical potential gradient.

Table 7
Excursion into History [2]

1687	Viscosity	Newton
1736	Mass point	
1749	Hydrodynamic pressure	
1750	Newton's basic law of motion	Euler
1755	Ideal equation of motion	
1736-1819	Steam Engine	James Watt
1811	Propogation of heat in solids	Baron-Jaseph Fourier
1821-1828	Viscous equation of motion	Cauchy
1822	Equation of heat conduction	Fourier
1822-1838	Navier-Stokes equation	Navier-Saint-Venant
1824	Carnot cycle, reversibility	Sadi Carnot
1842	Conservation of energy	Robert Mayer
1847	Conservation of force	Hermann von Helmholtz
1848	Absolute temperature	Kelvin
1850	First law of thermodynamics	Rudolf Clausius
1852	Second law of thermodynamics	Kelvin
1864	Electromagnetic field	Maxwell
1865	Entropy	Rudolf Clausius
1872	Entropy (Statistical)	Boltzmann
1873	Equation of state for real fluids	Van der Waals
1872-1957	Thermodynamic affinity	De Donder
1901	Statistical mechanics	Gibbs
1905	Mass-momentum-energy relations	Einstein
1931	Reciprocal relations	Lars Onsager
1941	Dissipative structures	Prigogine

(v) The *diffusion potential* is the interference between diffusion and electrical conduction in an anisotropic crystal where heat conduction occurs in one direction caused by a temperature gradient in a different direction.
(vi) The gradients of electrical potential and pressure govern the behavior of ionic systems, selective membranes, and ultracentrifuges. In electrokinetic phenomena, induced dipoles can cause separations, such as dielectrophoresis and magnetophoresis, which may be especially important in specialized separations.

In an isothermal system there are three mechanical driving forces that cause the transport of mass with respect to the mean fluid motion: (i) the concentration gradient, (ii) the pressure gradient, and (iii) external forces affecting the various chemical substances unequally. In a multi component fluid, we have fluxes of momentum, energy, and mass, each resulting from an associated thermodynamic driving force. There will be a contribution to each flux stemming from each driving force in the system. This is the result of coupling that can occur only between conjugate flux-force pairs, which are tensors of equal order or differ in order by two. The momentum flux is a tensor of order two, and it depends only on the velocity gradient. The energy flux, which is a vector, depends both on the temperature gradient and on the mechanical driving forces leading to the Dufour effect. The mass flux, which is also a vector, depends both on the mechanical driving forces (concentration gradient that causes ordinary diffusion, pressure and force diffusion) and on the temperature gradient leading to the Soret effect.

The cross phenomenological coefficients relate the two coupling effects of Dufour and Soret. In order to describe the coupling effects the thermal diffusion ratio is introduced beside the transport coefficients of viscosity, thermal conductivity, and diffusivity,

For mixtures energy flux contain the conductive flux q_c, and the contributions resulting from the interdiffusion q_d of various substances and the Dufour effect q_D; we therefore express the total energy flux relative to the mass average velocity

$$q = q_c + q_d + q_D \tag{67}$$

where $q_c = -k\nabla T$ and $q_d = \sum \bar{h}_i J_i$, and $\bar{h}_i$ is the partial molal enthalpy.

When we express the energy flux e with respect to fixed stationary coordinates by neglecting the Dufour effect, viscous effect, and kinetic energy, we have

$$e = -k\nabla T + \sum \bar{h}_i J_i + Pv + \rho uv \tag{68}$$

or

$$e = -k\nabla T + \sum h_i N_i \tag{69}$$

Eq. (69) is the usual starting point on simultaneous heat and mass transfer

The expression for mass flux is associated with the mechanical driving forces and thermal driving force

$$J_i = J_{i,x} + J_{i,P} + J_{i,g} + J_{i,T} \tag{70}$$

where $J_{i,x}$ is the ordinary diffusion, $J_{i,P}$ is the pressure diffusion, $J_{i,g}$ is the forced diffusion, and $J_{i,T}$ is the thermal diffusion, can be expressed by

$$J_{i,x} = \frac{c^2}{\rho RT} \sum_i^n M_i M_j D_{ij} \left[x_j \sum_{\substack{k=1 \\ k \neq j}}^n \left(\frac{\partial \overline{G}_j}{\partial x_k} \right)_{T,P,x_l} \nabla x_k \right] \qquad (l \neq j,k) \tag{75}$$

$$J_{i,P} = \frac{c^2}{\rho RT} \sum_i^n M_i M_j D_{ij} \left[x_j M_j \left(\frac{\overline{V}_i}{M_j} - \frac{1}{\rho} \right) \nabla P \right] \tag{76}$$

$$J_{i,g} = -\frac{c^2}{\rho RT} \sum_i^n M_i M_j D_{ij} \left[x_j M_j \left(g_j - \sum_{k=1}^n \frac{\rho_k}{\upsilon} g_k \right) \right] \tag{77}$$

$$J_{i,T} = -D_i^T \nabla \ln T \tag{78}$$

where $\overline{G}_j$ and $\overline{V}_j$ are the partial Gibbs free energy and partial volume, respectively, D_{ij} are multicomponent diffusion coefficients, and D_i^T are multicomponent thermal diffusion coefficients; these coefficients show the following properties

$$D_{ii} = 0 \text{ and } \sum_i D_i^T = 0 \tag{79}$$

and

$$\sum_{i=1}^n (M_i M_l D_{il} - M_i M_k D_{ik}) = 0 \tag{80}$$

The ordinary diffusion depends on the partial Gibbs free energy and concentration gradient. The pressure diffusion is considerable only for high-pressure gradient, such as centrifuge separation. The forced diffusion is mainly important in electrolytes and the local electric field strength. Each ionic substance may be under the influence of a different force. If the external force is the gravity then $J_{i,g}$ vanishes since all g_i are the same.

The thermal diffusion is the tendency for substances to diffuse because of a temperature gradient; under the very steep temperature gradients separation of mixtures is possible.

For a binary system mass flux by ordinary, pressure, forced, and thermal is expressed as follows

$$J_A = -J_B = -\left(\frac{c^2}{\rho}\right) M_A M_B D_{AB}$$

$$\left[\left(\frac{\partial \ln a_A}{\partial \ln x_A}\right)\nabla xA + \frac{MAxA}{RT}\left(\frac{\bar{V}_A}{M_A} - \frac{1}{\rho}\right) - \frac{\mathrm{M_A}\rho_\mathrm{B}\mathrm{x_A}}{\rho\mathrm{RT}}(g_A - g_B) + k\nabla \ln T\right] \tag{81}$$

where $(d\bar{G}_A)_{T,P} = RTd\ln a_A$, and k_T is the thermal diffusion ratio $k_T = \frac{\rho}{c^2 M_A M_B}\frac{D_A^T}{D_{AB}}$.

When k_T is positive component A moves to the colder region, otherwise it moves to the warmer region. Some typical values of k_T for binary fluid systems are given in Table 8

Table 8
Experimental values of thermal diffusion ratios for liquids and gases at low density and pressure

Components (Liquids) A-B	T, K	x_A	k_T
$C_2H_2Cl_4$-n-C_6H_{14}	298	0.5	1.08
$C_2H_4Br_2$-n-$C_2H_4Cl_2$	298	0.5	0.225
$C_2H_2Cl_4$-CCl_4	298	0.5	0.060
CBr_4-CCl_4	298	0.09	0.129
CCl_4-CH_3OH	313	0.5	1.23
CH_3OH-H_2O	313	0.5	-0.137
Component (Gases)			
Ne-He	330	0.20	0.0531
	330	0.60	0.1004
N_2-H_2	264	0.294	0.0548
	264	0.775	0.0633

Using the formulation of nonequilibrium thermodynamics, the cross-effects are mathematically described by the addition of new terms into the phenomenological laws. For thermal diffusion, for example, a term proportional to the temperature gradient is added to the right hand side of Fick's law. These relations need certain phenomenological coefficients, which can be related to the transport coefficients such as thermal conductivity, ordinary and thermal diffusion coefficients, and Dufour coefficient. Therefore, the measured transport coefficients will facilitate the determination of phenomenological coefficients. The coupled phenomena are experimentally verified, and they are part of a comprehensive theory of nonequilibrium thermodynamics and phenomenological approach. Studies on the cross-effects and the coupled processes in biological and physical systems are attracting scientists from various fields.

REFERENCES

[1] I. Prigogine, Introduction to Thermodynamics of Irreversible Processes, Wiley, New York, 1967.

[2] R.B. Bird. W.E. Stewart and E.N. Lightfoot, Transport Phenomena, 2nd ed., Wiley, New York, 2002.

[3] D. Straub, Alternative Mathematical Theory of Non-equilibrium Phenomena, Academic Press, New York, 1997.

[4] J.R. Welty, C.E. Wicks and R.R. Wilson, Fundamentals of Momentum, Heat and Mass Transfer (3rd ed.), Wiley, New York, 1984.

[5] R.C. Reid, J.M. Prausnitz and B.E. Poling, The Properties of Gases and Liquids, 4th ed., McGraw-Hill, New York, 1987.

[6] A.H.P. Skelland, Diffusional Mass Transfer, Krieger Publishing Company, Malabar, FL, 1985.

[7] D.G. Miller, The Onsager relation:experimental evidence. .In Foundation of Continuum Mechanics, Eds. J.J. Delgado Domingos, M.N. Nina, J.H. Whitelaw, Mac Millan, London, 1974.

[8] R.L. Rowley and F.H. Horne, J. Chem. Phys., 68 (1978) 325.

[9] S. Sieniutycz and P. Salamon (Eds), Flow, Diffusion and Rate Processes, Advances in Thermodynamics, Vol. 6, Taylor & Francis, New York, 1992.

[10] S. Sieniutycz, In Flow, Diffusion and Rate Processes, Advances in Thermodynamics, Vol. 6, Taylor & Francis, New York, 1992.

[11] W.D. Callister, Jr. Material Science and Engineering. An Introduction, 5th ed., Wiley, New York, 2000.

Chapter 3

Linear nonequilibrium thermodynamics

INTRODUCTION

The basic engineering of classical fluid mechanics is competing with theories relating to some branches of thermodynamics. These theories, specified as linear or extended irreversible thermodynamics or as rational thermodynamics have a place in modern physics. Essential ideas are related to keywords such as instability, nonequilibrium, motion, and irreversibility. Irreversible processes are controlled by the phenomenological laws with a broken symmetry in time; physical systems identified by permanently stable and reversible behavior are rare; unstable phenomena occurring macroscopically in the physical system are caused by inherent fluctuations of the respective state variables. Close to equilibrium the fluctuations do not disturb the equilibrium; the trend towards equilibrium is distinguished by asymptotically vanishing dissipative contributions. In contrast, nonequilibrium states can amplify the fluctuations, and any local disturbance can move even a whole system into an unstable state. Equilibrium states can be classifieds by correlations having an average range of action with typical values of about 10^{-10} m . Correlations existing far from equilibrium can extend to macroscopic distance in an order of one centimeter or more. This difference is an important indication of qualitative difference between the equilibrium and nonequilibrium states.

An irreversible phenomenological macro world, and a micro world determined by linear and reversible quantum laws should be related to each other; the extensive work of Prigogine and his co-workers has a success in unifying the basic micro- and macroscopic ideas of matter.

The first attempts to develop nonequilibrium thermodynamics are associated with certain thermal diffusion and thermoelectric phenomena. Later, Onsager initiated the basic equations of the theory, and Casimir, Meixner and Prigogine refined and developed the theory further. This chapter outlines the principles of nonequilibrium thermodynamics for systems not far away from the thermodynamic equilibrium. In this region the transport and rate equations are expressed in linear forms and the Onsager reciprocal relations are valid. Therefore this region is called the *linear* or *Onsager region* and the formulations are known as the *linear nonequilibrium thermodynamics* theory. In this region, instead of thermodynamic potentials and entropy, a new property called the entropy generation appears. The formulation of linear nonequilibrium thermodynamics is proved to be valid and useful for a large range of transport and rate processes of physical and biological systems. Some of the keywords and concepts are local thermodynamic equilibrium, microscopic reversibility, phenomenological equations and coefficients, Curie-Prigogine principles, entropy generation, and dissipation function.

1. LOCAL THERMODYNAMIC EQUILIBRIUM

A local thermodynamic state is determined as elementary volumes at individual points for a substance of the system that is not in equilibrium. These volumes are small that the substance in them can be treated as homogeneous with a sufficient number of molecules for the phenomenological laws to be applicable to them. This local state shows the *microscopic reversibility* that is the symmetry of all mechanical equations of motion of individual particles with respect to time; when there are two alternative paths for a simple reversible reaction, and one of these paths is preferred for the backward reaction, the same path must also be preferred for the forward reaction. Onsager's derivation is based on the assumption of microscopic reversibility. The reversibility of molecular behavior give rise to a kind of symmetry in the way the transport processes are coupled to each other. Although a thermodynamic system as a whole may not be in equilibrium, the local states may be in *local thermodynamic equilibrium*; all intensive thermodynamic variables become functions of position and time, $T = T(x,t)$, $P = P(x,t)$, $\mu = \mu(x,t)$; specific entropy and specific internal energy may be determined at every point in the same way as for substances in equilibrium, and the temperature and pressure are definable by $T = \left(\frac{\partial U}{\partial S}\right)_V$, $P = -\left(\frac{\partial U}{\partial V}\right)_S$. Also the classical-thermodynamic equations of state relating the thermodynamic functions to the state properties, such as the Gibbs relation

$$dU = TdS - Pd\left(\frac{1}{\rho}\right) + \sum_{i=1}^{n} \mu_i dx_i \tag{1}$$

and the Gibbs-Duhem equation

$$-SdT + \left(\frac{1}{\rho}\right)dP = \sum_{i=1}^{n} x_i d\mu_i \tag{2}$$

in a multicomponent medium are assumed to be valid.

Prigogine expanded the molecular distribution function in an infinite series around the equilibrium molecular distribution function f_o

$$f = f_o + f_1 + f_2 + \dots \tag{3}$$

Eq. (1) is valid not only for an equilibrium system, but also for a nonequilibrium system that is characterized by the equilibrium distribution function of f_o+f_1 representing a nonequilibrium system sufficiently close to equilibrium. Prigogine's analysis applies only to the mixtures of monoatomic gases, and is also dependent on the Chapman-Enskog model. The domain of validity of the local equilibrium is not known in general from a microscopic perspective. For a stationary system, in which thermodynamic quantities at an arbitrary point in space do not vary with time, the concept of local equilibrium holds. Except for turbulent systems and shock wave phenomena, the assumption of local thermodynamic equilibrium is widely applicable [3].

The range of validity of the local thermodynamic equilibrium is determined only by experiment. Experiments show that the postulate of local thermodynamic equilibrium is valid if the gradients of intensive thermodynamic functions are small, and their local values vary slowly in comparison with the local state of the system. The change in an intensive parameter is comparable to the molecular mean free path, and energy dissipation rapidly damps large deviations from the equilibrium.

2. SECOND LAW OF THERMODYNAMICS

Let us consider the system shown in Fig. 1. In the region denoted by I, irreversible processes occur. The region II is isolated, and is the local environment of I. Equilibrium everywhere is attained in the region II. In the first possibility, region I may be a closed system that can exchange only heat with II. The first law expresses the internal energy change dU^I as a consequence of its gaining a quantity of heat from II δQ^I, and performing a quantity of work δW^I on I, so that we have $dU^I = \delta Q^I - dW^I$. According to the second law, the region I obeys the general inequality

$$dS \equiv \frac{\delta Q_{\mathrm{rev}}}{T} > \frac{\delta Q}{T} \tag{4}$$

where dS, the entropy gain, is determined by δQ_{rev}, the heat that could be absorbed in a reversible change. Since we are considering real (irreversible) processes, TdS is always greater than δQ, the actual heat absorbed. It follows that

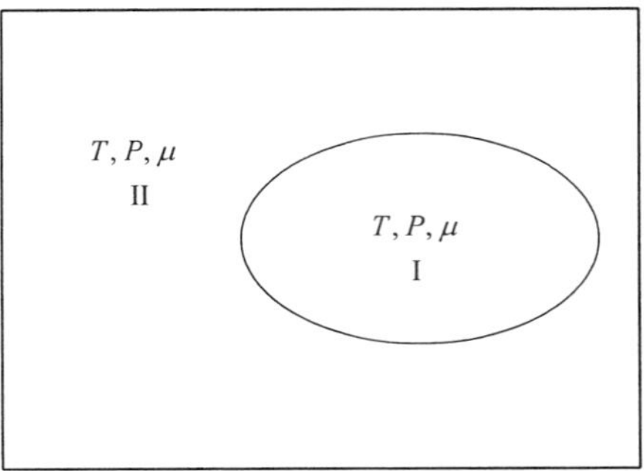

Fig. 1. A closed system with a subsystem I.

in practice, the region 1 fails to absorb the maximum amount of heat, which theoretically might be transformed into work. Then for real processes it is necessary to write

$$dS = \frac{\delta Q}{T} + \frac{\delta Q'}{T} \tag{5}$$

Here δQ is the actual uptake of heat and $\delta Q'$ is the additional heat that would have been absorbed from II if the change had been made reversibly, hence it is a positive quantity. In the actual change of state, the entropy increment dS is completed by the production of a quantity of entropy $\delta Q'/T$. This may be, for example, entropy of mixing or reaction within I. The relation of Carnot-Clausius gives the change of the entropy of a closed system

$$dS = d_\mathrm{e}S + d_\mathrm{i}S = \frac{\delta_e Q}{T} + d_\mathrm{i}S \tag{6}$$

Eq. (6) indicates two contributions: entropy inflow from the environment dS_e, and entropy production inside the volume under consideration $d_\mathrm{i}S$. The explicit calculation of $d_\mathrm{i}S$ as a function of the appropriate variables is essentially the basis of nonequilibrium thermodynamics. The state of equilibrium can be characterized either by minimum (zero) entropy production, or by maximum entropy. The $d_\mathrm{e}S$ is due to the conduction flow of internal energy, convection entropy flow transported along with the macroscopic flow of the substance as a whole, and the resultant entropy flux caused by diffusion of the individual components. The quantity, $d_\mathrm{e}S$ may be positive or negative, and zero at the special case. For a closed, thermally homogeneous system

$$d_{\mathrm{e}}S = \frac{\delta_{\mathrm{e}}Q}{T} = \frac{dU + pdV}{T} \text{ and } d_{\mathrm{i}}S > 0 \tag{7}$$

where $\delta_e Q$ is the elementary heat due to thermal interaction between the system and the environment. For a closed systems if the change of mole numbers dN_k were due to irreversible chemical reactions, the entropy production is expressed by

$$d_i S = -\frac{1}{T}\Sigma \mu_k dN_k > 0 \tag{8}$$

The rate of entropy production is

$$\frac{d_{\mathrm{i}}S}{dt} = -\frac{1}{T}\Sigma \mu_k \frac{dN_k}{dt} > 0 \tag{9}$$

where μ_k is the chemical potential that can be related to measurable quantities, such as P, T, and N_k. Eq. (9) can also be expressed in terms of the affinity A, and we get

$$\frac{d_{\mathrm{i}}S}{dt} = \left(\frac{A_k}{T}\right)\frac{d\xi_k}{dt} \geq 0 \tag{10}$$

For a reaction B = 2D the affinity is $A = \mu_A - 2\mu_D$, and $d\xi/dt$ is the velocity of reaction. At thermodynamic equilibrium the affinity A and the velocity of reaction vanish.

When a mass diffusion occurs in a closed system from higher chemical potential μ_2 to lower potential μ_1, we have the entropy generation expressed by

$$d_{\mathrm{i}}S = -\left(\frac{\mu_2 - \mu_1}{T}\right)d\xi > 0 \tag{11}$$

where $-dN_1 = dN_2 = d\xi$. Here the flow of mass from one region to another is associated with an extent of reaction, although no real chemical reaction takes place.

Entropy production due to electrical conduction is

$$\frac{d_{\mathrm{i}}S}{dt} = \frac{VI}{T} \tag{12}$$

where the product *VI* of potential difference and current is the heat generated, and called the Ohmic heat per unit time. Here the flow is the electric current and the corresponding force is *V/T*; the linear phenomenological equation is expressed by

$$I = L_e \frac{V}{T} \tag{13}$$

where L_e is the phenomenological coefficient. From the Ohm law $V = IR$, where R is the resistance; hence $L_e = T/R$, where R is the resistance.

For an isolated system $d_eS = 0$ and $d_iS > 0$. For an open system, we have

$$d_eS = \frac{dU + pdV}{T} + (d_eS)_{\text{matter}} \quad \text{and } d_iS > 0 \tag{14}$$

Systems that exchange entropy with their surrounding may undergo spontaneous transformation to dissipative structures, and self-organization. The forces exist in irreversible processes produce entropy and create these organized states, which range from convection patterns of Benard cells and biological activity.

In a stationary state the total entropy change of the system does not change with time, and we have

$$\frac{dS}{dt} = \frac{d_eS}{dt} + \frac{d_iS}{dt} = 0 \tag{15}$$

Eq. (15) shows that the entropy exchange with the surrounding must be negative

$$\frac{d_eS}{dt} = -\frac{d_iS}{\mathrm{dt}} < 0 \tag{16}$$

For the total entropy to be constant the entropy flowing out of the system is equal to the entropy entering the system plus the entropy generated within the system.

$$\frac{d_iS}{dt}(J_{s,\text{in}} - J_{s,\text{out}}) = 0 \tag{17}$$

From Eqs. (15) and (17), we have

$$\frac{d_eS}{dt} = (J_{s,\text{in}} - J_{s,\text{out}}) < 0 \tag{18}$$

So that, the stationary state is maintained through the decrease in exchanged entropy between the system and surrounding.

Entropy production inside the elementary volume considered caused by irreversible phenomena is the local value of the sum of entropy increments. By the second law of thermodynamics, the entropy generation dS_i is always positive for irreversible and zero for reversible phenomena.

The second possibility is that region I is an open system than can exchange both matter and energy with II. In place of the first law, this time we have to take into account of changes in compositions and consider the Gibss equation that plays a major role in the development of nonequilibrium thermodynamics. The total entropy change in the system is given by

$$dS_{\text{total}} = dS^{\text{I}} + dS^{\text{II}} \tag{19}$$

From Eq. (6) we have

$$dS^{\text{I}} = d_e S^{\text{I}} + d_i S^{\text{I}} \tag{20}$$

so that

$$dS^{\text{II}} = d_e S^{\text{II}} = -d_e S^{\text{I}} \tag{21}$$

since the entropy change in II is solely due to exchange with I. From Eqs. (19) to (21), it is clear that

$$dS_{\text{total}} = d_i S^{\text{I}} \tag{22}$$

Thus $d_i S^{\text{I}}$ represents total increase of entropy in the environment due to processes taking place in the system I. It can be shown that the entropy generation is associated with a loss of free energy, or capability for work. At constant temperature and pressure, the Gibbs free energy G measures the maximum work capability; the changes in each region are given by

$$dG^{\text{I}} = dU^{\text{I}} + PdV^{\text{I}} - TdS^{\text{I}} \tag{23}$$

$$dG^{\text{II}} = dU^{\text{II}} + PdV^{\text{II}} - TdS^{\text{II}} \tag{24}$$

Summing these two equations, and remembering that volume and internal energy are constant in an isolated system, we have

$$dG_{\text{total}} = dG^{\text{I}} + dG^{\text{II}} = -TdS_{\text{total}} \tag{25}$$

Since for any real process dS_{total} is necessarily positive, the free energy of the entire system decreases. We shall be concerned with the rate of decrease of free energy or the dissipation of free energy, which is given by the dissipation function Ψ, which will be elaborated later.

For a large class of systems that are not in thermodynamic equilibrium, the thermodynamic variables such as temperature, pressure, concentration, and internal energy are well-defined concepts locally; they are functions of time and position; this is the assumption of local equilibrium. In most hydrodynamic and chemical systems, local equilibrium assumption is a highly satisfactory approximation.

Definition of entropy in nonhomogeneous systems can be expressed in terms of entropy density $S(T,N_k)$, which is a function of the temperature field $T(x)$ and the mole number density $N_k(x)$; these densities can be measured, and internal energy and entropy can be expressed by these densities. The total energy and entropy of the system is obtained by the following integrations

$$S = \int_V \{S[T(x)],[N_k(x)]\}\, dV \tag{26}$$

$$U = \int_V \{U[T(x)],[N_k(x)]\}\, dV \tag{27}$$

The definitions in Eqs. (26) and (27) are dependent upon the well-defined temperature locally, in the nonequilibrium system, the total entropy S is generally not a function of the total entropy U and the total volume V.

3. PHENOMENOLOGICAL EQUATIONS

Spontaneous decaying phenomena towards equilibrium of a nonequilibrium system not far away from equilibrium can be expressed in linear relations between flows J_i and driving forces X_i.

$$J_i = L_{ii} X_i \tag{28}$$

where L_{ii} is called the *phenomenological coefficient*. For example Fourier's law relates heat flow to the force that is the temperature gradient, while Fick's law provides a relation between mass diffusion and concentration gradient. These laws present the relationships between the conjugate flows and forces, and consider a single force and a single flow, and are not capable of explaining simultaneous processes such as combined heat and mass flows. The validity of Eq. (28) must be determined experimentally for a certain type of process; for example, Eq. (28) holds only for a conductor, which obeys Ohm's law.

If a nonequilibrium system consists several flows caused by various forces Eq. (28) may be generalized in the linear region of thermodynamic branch, and we obtain

$$J_i = \sum_k L_{ik} X_k \tag{29}$$

These equations are called the *phenomenological equations*, which are capable of describing the multiflow systems and the effect of the nonconjugate forces on a flow. For example, ionic diffusion in an aqueous solution may be related to concentration, temperature as well as the imposed electromotive force. In the following section the relationships between flows and forces are summarized.

3.1. Flows and forces

Mass flow, heat flow, and chemical reaction rate are all called the 'flows' or fluxes J_i. They are caused by the chemical potential gradient $X_i = -grad\mu_i$, themperature gradients, and the chemical affinity respectively, and are called the thermodynamic 'forces', or just forces X_i. The affinity A is given by

$$A_j = -\sum_{i=1}^{n} \nu_{ij}\mu_i \qquad (j = 1,2,\ldots,l)$$

where ν_{ij} is the stoichiometric coefficient of the ith component in the jth reaction, n is the number of components in the reaction, and l is the number of reactions.

Generally any force X_i can produce any flow J_i. The flows may have vectorial or scalar characters. *Vectorial flows* are directed in space, such as mass, heat, and electric current. *Scalar flows* have no direction in space, such as those of chemical reactions. The other more complex flow is the viscous flow characterized by tensor properties. At an equilibrium state the flows and forces become zero

$$J_{i,\text{eq}}(X_i = 0) = 0 \tag{30}$$

As an example, for component i the diffusional flow vector $\mathbf{j}_i$ may be defined as the number of moles per unit area A per unit time t in a specified direction, and expressed by

$$\mathbf{j}_i = \frac{1}{A}\frac{dN_i}{dt} \tag{31}$$

Considering a small area dA at any point x, y, z perpendicular to average velocity $\mathbf{v}_i$, in which $\mathbf{v}_i$ is virtually constant, the volume occupied by the particles passing dA in unit time will be $\mathbf{v}_i dA$. If the concentration per unit volume is c_i then the total amount of substance is $c_i \mathbf{v}_i dA$, and the local flow is given by the amount of substance passing in unit area per unit time is

$$\mathbf{j}_i = c_i \mathbf{v}_i \tag{32}$$

Generally, the three quantities $\mathbf{j}_i$, c_i and $\mathbf{v}_i$ in Eq. (32) are functions of the time and space coordinates.

If the small area dA is not perpendicular to the flow vector, we consider a unit vector $\mathbf{i}$, perpendicular to dA, whose direction will specify the direction of the area

$$d\mathbf{A} = \mathbf{i}dA \tag{33}$$

The scalar product $\mathbf{v}_i \cdot d\mathbf{A}$ gives the volume dV, which is multiplied by the local concentration c_i to find differential flux dQ_i that is the amount of the substance passing an area making any angle with the velocity vector $\mathbf{v}_i$

$$dQ_i = \mathbf{j}_i \cdot d\mathbf{A} = c_i \mathbf{v}_i \cdot d\mathbf{A} \tag{34}$$

For a volume enclosed by a surface area A, the total amount of substance i leaving that volume is

$$Q_i = \int_A \mathbf{j}_i \cdot d\mathbf{A} \tag{35}$$

A negative value of Q indicates a net influx of the substance i in the volume. The divergence of the flow $\mathbf{j}_i$ is $\lim_{V\to 0}(Q_i / V)$, and is expressed as

$$\nabla \cdot \mathbf{j}_i = \frac{\partial \mathbf{j}_{i,x}}{\partial x} + \frac{\partial \mathbf{j}_{i,y}}{\partial y} + \frac{\partial \mathbf{j}_{i,z}}{\partial z} \tag{36}$$

Here $\mathbf{j}_{i,x}$, $\mathbf{j}_{i,y}$ and $\mathbf{j}_{i,z}$ are the Cartesian coordinates of the vector $\mathbf{j}_i$. As the volume V and the product $\mathbf{j}_i \cdot d\mathbf{A}$ are scalars, the divergence is also a scalar quantity. The point of positive divergence means a source of component i, while a negative divergence indicates a sink, and at points of div $\mathbf{j}_i = 0$, there is no accumulation and no removal of material. These lead to the transformation of the surface integral of a flow into a volume integral of a divergence

$$\int_V \nabla \cdot \mathbf{j}_i \cdot dV = \int_A \mathbf{j}_i \cdot d\mathbf{A} \tag{37}$$

The divergence of the mass flux vector $\rho\mathbf{v}$ is used in the continuity equation

$$\frac{\partial \rho}{\partial t} = -(\nabla \cdot \rho\mathbf{v}) \tag{38}$$

Eq. (38) states that the rate of increase of the density within a small volume element is equal to the net rate of mass influx to the element divided by its volume. Similarly the local equivalent of the law of conservation of mass for an individual component i is expressed as

$$\frac{\partial c_i}{\partial t} = -(\nabla \cdot \mathbf{j}_i) \tag{39}$$

Eq. (39) cannot describe a flow process for a reacting component.

Another conserved property is the total energy, and in terms of local energy density ρ_e to each point in the system, we have

$$\frac{\partial \rho_e}{\partial t} = -(\nabla \cdot \mathbf{j}_e) \tag{40}$$

where $\mathbf{j}_e$ is the energy flow.

The total entropy of a system is expressed in terms of the local entropy density s_v

$$S = \int_V s_v dV \tag{41}$$

The total entropy changes with time as follows

$$\frac{dS}{\partial t} = \int_V \frac{\partial s_v}{\partial t} dV \tag{42}$$

The entropy flow $\mathbf{j}_s$, on the other hand, is the result of the exchange of entropy with the surroundings, and may be expressed by

$$\frac{d_e S}{\partial t} = -\int_A \mathbf{j}_s \cdot d\mathbf{A} = -\int_V (\nabla \cdot \mathbf{j}_s)\, dV \tag{43}$$

In an irreversible process, the entropy is generated Φ in any local element of the system, and the total entropy production is

$$\frac{d_i S}{dt} = \int_V \Phi\, dV \tag{44}$$

From Eqs. (6) and (9), the total change in entropy becomes

$$\frac{dS}{dt} = \frac{d_e S}{dt} + \frac{d_i S}{dt} \tag{45}$$

By inserting Eqs. (42)-(44) into Eq. (45), we obtain

$$\int_V \frac{\partial s_v}{\partial t}\, dV = -\int_V (\nabla \cdot \mathbf{j}_s)\, dV + \int_V \Phi\, dV \tag{46}$$

Therefore, for any local change, an irreversible process in a continuous system is described as

$$\frac{\partial s_v}{\partial t} = -\nabla \cdot \mathbf{j}_s + \Phi \tag{47}$$

Eq. (47) shows the expression for a nonconservative change in local entropy density, and allows the isolation of the entropy generation from the total change in entropy and evaluation of the dependence of Φ on flows and forces.

Stationary or steady-state flow process resemble equilibria in their invariance with time; partial time differentials of density, concentration, or temperature will vanish, although flows continue to occur in the system, and entropy is being produced. If the property is conserved, the divergence of the corresponding flow must vanish, hence the steady flow of a conserved quantity is constant and source-free. At equilibrium all the constant values of steady state flows becomes zero.

At steady-state flow the local entropy density must remain constant because of the condition $\partial s_v / \partial t = 0$. However, the divergence of entropy flow does not vanish, and we obtain

$$\Phi = \nabla \cdot \mathbf{j}_s \tag{48}$$

Eq. (48) indicates that in a steady state the entropy generated at any point of a system must be removed by a flow of entropy at that point. In the state of equilibrium, all the flows including the flow of entropy vanishes, and we obtain the necessary and sufficient condition for equilibrium as

$$\Phi = 0 \tag{49}$$

Eq. (47) may be useful in describing the state of a system. For example, the state of equilibrium is the only state that can be achieved for an adiabatic system, since the entropy generated by irreversible processes cannot be exchanged with the surroundings.

For the thermodynamic forces, such as a difference in chemical potential of component *i*, μ_i, proper spatial characteristics must be assigned to vectorial forces for the description of local processes. For this purpose, we consider all points of equal μ_i as potential surface. For two neighboring equipotential surfaces with chemical potentials μ_i and $\mu_i + d\mu_i$, the change in μ_i with number of moles N is $\partial\mu_i / \partial N$, which is the measure of the local density of equipotential surfaces. At any point on the potential surface we construct a perpendicular unit vector with the direction corresponding to the direction of maximal change in μ_i. With the unit vectors in the direction *x*, *y*, and *z* coordinates denoted by **i**, **j**, and **k** respectively, the gradient of the field in Cartesian coordinates is defined by

$$\text{grad}\,\mu_i = \mathbf{i}\frac{\partial \mu_i}{\partial x} + \mathbf{j}\frac{\partial \mu_i}{\partial y} + \mathbf{k}\frac{\partial \mu_i}{\partial z} \tag{50}$$

A thermodynamic driving force occurs when a difference in potential exists, and its direction is the maximal decrease in μ_i. Consequently at the point *x, y, z* the local force **X** causing the flow of component *i* is expressed by

$$\mathbf{X}_i = -\text{grad}\,\mu_i \tag{51}$$

For a single dimensional flow, Eq. (51) becomes

$$\mathbf{X}_i = -\mathbf{i}\frac{\partial \mu_i}{\partial x} \tag{52}$$

From the definition of the chemical potential, we have

$$-\frac{\partial \mu_i}{\partial x} = \frac{\partial}{\partial x}\left(\frac{-\partial G}{\partial N_i}\right) = \frac{\partial}{\partial N_i}\left(\frac{-\partial G}{\partial x}\right) \tag{53}$$

where $-dG$ shows the free energy available to perform a useful work δW, and the differential of work with distance $\delta W/dx$, is a force. Therefore $\mathbf{X}_i$ is a force per mole of component i causing a flow in the direction of the unit vector. The overall thermodynamic force that is the difference in chemical potential for the transport of substance between regions 1 and 2 in discontinues systems is the integral of Eq. (38)

$$\int_1^2 \mathbf{X}_i dx = -\mathbf{i}\int_1^2 \frac{\partial \mu_i}{\partial x} dx = \mathbf{i}(\mu_{i,1} - \mu_{i,2}) = \Delta\mu_i \tag{54}$$

Here $\Delta\mu_i$ is a difference in potential, while $\mathbf{X}_i$ is a conventional force used in classical mechanics.

The same discussions on the electric potential ψ that causes a current at the point x, y, z lead to the definition of electric force $\mathbf{X}_e$

$$\mathbf{X}_e = -\text{grad}\,\psi \tag{55}$$

where $\mathbf{X}_e$ is the force per unit charge, or the local intensity of the electric field.

When we consider with the difference in electric potential between two points instead of local electric forces, the quantity of electromotive force $\Delta\psi$ is defined in a single direction by

$$\Delta\psi = -\mathbf{i}\int_1^2 \frac{d\psi}{dx} dx = \mathbf{i}(\psi_1 - \psi_2) \tag{56}$$

Other types of forces of irreversible processes may be derived similarly.

In general, the flows and forces are complicated nonlinear functions of one another. However, we can expand the nonlinear dependence of the flows J_i and the forces X_i in Taylor series about the equilibrium

$$J_i = J_{i,\text{eq}}(X_i = 0) + \sum_{j=1}^{n}\left(\frac{\partial J_i}{\partial X_j}\right)_{\text{eq}} X_j + \frac{1}{2!}\sum_{j=1}^{n}\left(\frac{\partial^2 J_i}{\partial X_j^2}\right)_{\text{eq}} X_j^2 + \dots. \tag{57}$$

$$X_i = X_{i,\text{eq}}(J_k = 0) + \sum_{k=1}^{n}\left(\frac{\partial X_i}{\partial J_k}\right)_{\text{eq}} J_k + \frac{1}{2!}\sum_{k=1}^{n}\left(\frac{\partial^2 X_i}{\partial J_k^2}\right)_{\text{eq}} J_k^2 + \dots \tag{58}$$

If we neglect the higher order terms, Eqs. (57) and (58) become linear relations, and we have the general type of *phenomenological equations* for the irreversible phenomena

$$J_i = \sum_{i=1}^{n} L_{ik} X_i \qquad (i,k = 1,2,\ldots,n) \tag{59}$$

$$X_i = \sum_{k=1}^{n} K_{ik} J_k \tag{60}$$

Eq. (59) shows that any flow is caused by contributions of all the forces present in the system, whereas Eq. (60) shows that any force is the result of all the flows. The coefficients L_{ik}, *conductance coefficients*, and K_{ik}, *resistance coefficients*, are called the *phenomenological coefficients*. The straight coefficients with the same indices relate the conjugated forces and flows. The cross coefficients with $i \neq k$ represent the coupling phenomena. According to the principle of Curie-Prigogine, vector and scalar flows are able to couple only in a nonisotropic medium. This theory has important consequences in activities taking place in the living cells.

Onsager's reciprocal relations state that, provided a proper choice is made for the flows and forces, the matrix of phenomenological coefficients is symmetrical. These relations are proved to be an implication of the property of 'microscopic reversibility', that is the symmetry of all mechanical equations of motion of individual particles with respect to time *t*. The Onsager reciprocal relations are the results of the global gauge symmetries of the Lagrangian, which is related to the entropy production of the system considered. This means that the results in general are valid for an arbitrary process.

The cross-phenomenological coefficients are defined as

$$L_{ik} = \left(\frac{\partial J_i}{\partial X_k}\right)_{X_j} = \left(\frac{J_i}{X_k}\right)_{X_j=0} \qquad (i \neq k) \tag{61}$$

$$K_{ik} = \left(\frac{\partial X_i}{\partial J_k}\right)_{J_i} = \left(\frac{X_i}{J_k}\right)_{J_i=0} \qquad (i \neq k) \tag{62}$$

The phenomenological coefficients are not the function of the thermodynamic forces and flows; on the other hand, they can be functions of the parameters of the local state as well as the nature of a substance. The coefficients L_{ik} and K_{ik} are the conductance and the resistance types, respectively. These phenomenological coefficients obey the Onsager relations, and the values of L_{ik} with the same indices are positive

$$L_{ii} > 0 \qquad (i=1,2,\ldots,n) \tag{63}$$

and all coefficients of different indices must satisfy the condition

$$L_{ii}L_{kk} > \frac{1}{4}(L_{ik} + L_{ki})^2 \quad (i \neq k;\ i,k = 1,2,..,n) \tag{64}$$

The matrix of phenomenological coefficients L_{ki} and K_{ki} are related by

$$\mathbf{K} = \mathbf{L}^{-1} \tag{65}$$

where $\mathbf{L}^{-1}$ is the inverse of the matrix $\mathbf{L}$.

For a two force-flow process, we have the phenomenological equations in terms of the flows

$$J_1 = L_{11}X_1 + L_{12}X_2 \tag{66}$$

$$J_2 = L_{21}X_1 + L_{22}X_2 \tag{67}$$

The forces are obtained by solving Eqs. (66) and (67)

$$X_1 = \frac{L_{22}}{L_{11}L_{22} - L_{12}L_{21}}J_1 - \frac{L_{12}}{L_{11}L_{22} - L_{12}L_{21}}J_2 \tag{68}$$

$$X_2 = -\frac{L_{21}}{L_{11}L_{22} - L_{12}L_{21}}J_1 + \frac{L_{11}}{L_{11}L_{22} - L_{12}L_{21}}J_2 \tag{69}$$

The phenomenological equations can also be expressed in terms of the forces

$$X_1 = K_{11}J_1 + K_{12}J_2 \tag{70}$$

$$X_2 = K_{21}J_1 + K_{22}J_2 \tag{71}$$

The following relations link the phenomenological coefficients

$$K_{11} = \frac{L_{22}}{|\mathbf{L}|};\ K_{12} = -\frac{L_{12}}{|\mathbf{L}|};\ K_{21} = -\frac{L_{21}}{|\mathbf{L}|};\ \mathrm{K}_{22} = \frac{L_{11}}{|\mathbf{L}|} \tag{72}$$

where $|\mathbf{L}|$ is the determinant of the matrix $\mathbf{L}$.

For an elementary step reaction, the flow J_r and the affinity A are expressed in terms of forward r_f and backward r_b reaction rates as follows

$$J_r = r_f - r_b \tag{73}$$

$$A = RT \ln \frac{r_f}{r_b} \tag{74}$$

These equations are solved together to express the flow

$$J_r = r_f (1 - e^{-A/RT}) \tag{75}$$

Close to the thermodynamic equilibrium, where $A/RT << 1$, we can expand Eq. (75) as

$$J_r = r_{f,\mathrm{eq}} \frac{A}{RT} \tag{76}$$

On the other hand we have the following linear phenomenological equations for chemical reaction taking place in the system

$$J_{ri} = \sum_{i,j=1}^{l} L_{ij} A_j \tag{77}$$

We can compare these linear phenomenological equations with Eq. (76) to obtain the phenomenological coefficient as

$$L_{ij} = \frac{r_{f,\mathrm{eq},ij}}{RT} \tag{78}$$

Here we have $r_{f,\mathrm{eq}} = r_{b,\mathrm{eq}}$.

For an overall reaction with l number of intermediate reactions, the linear phenomenological law is valid, if every elementary reaction satisfies the condition $A/RT << 1$, and the intermediate reactions are fast and hence a steady state is reached.

The formulation of linear nonequilibrium thermodynamics is based on the combination of the first and second laws of thermodynamics incorporating the balance equations including the entropy balance. Therefore any changes and developments in the process can be taken into account through these equations. The linear nonequilibrium thermodynamics approach is widely recognized as a useful phenomenological theory, which describes the coupled transport without the need for detailed mechanism of complex processes.

4. CURIE-PRIGOGINE PRINCIPLE

Consider a flow $\mathbf{j}_i$ with tensorial rank m. The value of m is zero for a scalar quantity, it is unity for vector quantity, and it is two for a dyadic. If a conjugate force X_j also has a tensorial rank m, than the coefficient L_{ij} is a scalar, and is consisted with the isotropic character of the system. The coefficients L_{ij} are determined by the isotropic medium; they need not to vanish, and hence the flow $\mathbf{j}_i$ and the force X_j can interact. If a force X_j has a tensorial rank different from m by an even integer k, than L_{ij} has a tensor of rank k. In this case $L_{ij}\,X_j$ is a tensor product. Since a tensor coefficient L_{ij} of even rank is also consistent with the isotropic character of the fluid system; the L_{ij} is not zero, and hence $\mathbf{j}_i$ and X_j can interact. However, for a force X_j whose tensorial rank differs from m by an odd integer k^*, L_{ij} has a tensorial rank of k^*. A tensor coefficient L_{ij} of odd rank implies a nonisotropic character for the system. Consequently, such a coefficient must vanish for an isotropic system, and $\mathbf{j}_i$ and X_j would not interact. For example, if k^* is unity, then L_{ij} would be vector. By definition, an isotropic system cannot have a vector quantity associated with it. Therefore, the vectorial flows $\mathbf{q}$ and $\mathbf{j}_i$ can only be related to the vector forces. The scalar reaction rates can be functions of the scalar forces and the trace of the dyadic, but not the vector forces. We observe, therefore coupling or cross effects between the diffusion flow and heat flow, and between the various chemical reactions. According to the *Curie-Prigogine principle*, vector and scalar quantities are able to interact only in a nonisotropic medium. This theory has important consequences in chemical reactions and transport processes taking place in the living cells.

5. DISSIPATION FUNCTION

The time derivative of the entropy is called the entropy generation rate, and can be calculated from the laws of conservation of mass, energy, momentum, and the second law of thermodynamics expressed as equality. If the *entropy generation* is taken per unit time and per unit volume, it is called the *volumetric rate of entropy generation* Φ

$$\Phi = \frac{d_i S}{dt dV} \tag{79}$$

The product of the entropy generation rate and the absolute temperature is called the *dissipation function* Ψ, which is also a positive quantity

$$\Psi = T\Phi = \sum X_i J_i \tag{80}$$

When phenomena at the interface between two phases are considered, the amount of entropy produced is taken per unit surface area.

The nonequilibrium thermodynamics approach calculates the entropy generation rate for a specified task of a process. This calculation is based on the hypotheses of positive and definite entropy production due to irreversible processes in the system and of Gibbs' relation given by

$$TdS = dU + PdV - \sum \mu_i dN_i \tag{81}$$

As the Gibbs relation is a fundamental relation and to be valid even outside thermostatic equilibrium, the entropy depends explicitly only on energy, volume and concentrations.

By introducing the linear phenomenological equations, [Eq. (59)], into the dissipation function, [Eq. (80)], we have

$$\Psi = \sum_{i,k=1}^{n} L_{ik} X_i X_k \geq 0 \tag{82}$$

This equation shows that the dissipation function is a quadratic form in all the forces. In continuous systems, the base of reference for diffusion flow affects the values of transport coefficients and the entropy production due to diffusion. Prigogine proved the invariance of the entropy production for an arbitrary base of reference if the system is in mechanical equilibrium (dv/dt = 0), and the divergence of viscous tensors vanishes. The same conclusion holds for the entropy generation

$$\Phi = \sum_{i,k=1}^{n} l_{i,k} X_i' X_k' \geq 0 \tag{83}$$

where for clarity, the phenomenological coefficients have been denoted by lowercase letters, and are associated by the following obvious relationship

$$l_{ik} = TL_{ik} \tag{84}$$

The dissipation function can be written in the following form

$$\Psi = \sum_{i=1}^{n} L_{ii} X_i^2 + \sum_{i,k=1}^{n} \frac{L_{ik} + L_{ki}}{2} X_i X_k \geq 0 \qquad (i \neq k) \tag{85}$$

The quadratic form of Eq. (82) may also be written in a matrix form

$$\Psi = \sum_{i,k=1}^{n} L_{ik} X_i X_k = [X_1\ X_2 ... X_n] \begin{bmatrix} L_{11}\ L_{12}....L_{1n} \\ L_{21}\ \ L_{22}...L_{2n} \\ .... \\ L_{m1}\ L_{m2}....L_{nn} \end{bmatrix} \begin{bmatrix} X_1 \\ X_2 \\ .... \\ X_n \end{bmatrix} \geq 0 \quad (86)$$

Eq. (68) can also be expressed in matrix form in terms of the conductance L and resistance K coefficients, and we obtain

$$\Psi = \mathbf{X}^T \mathbf{L} \mathbf{X} = \mathbf{J}^T \mathbf{K} \mathbf{J} \geq 0 \quad (87)$$

Eq. (87) shows that the dissipation function is a quadratic form in all forces and in all flows. A necessary and sufficient condition for $\Psi \geq 0$ is that all its principal minors be non-negative

$$\begin{vmatrix} L_{ii} & L_{ik} \\ L_{ki} & L_{kk} \end{vmatrix} = L_{ii} L_{kk} - L_{ik} L_{ki} \geq 0 \quad (88)$$

If only a single force occurs, Eq. (66) becomes

$$\Psi = L_{ii} X_i^2 \geq 0 \quad (89)$$

then the phenomenological coefficients cannot be negative $L_{ii} \geq 0$.

For a system with a metastable equilibrium, we have $J_i = 0$ and $\Psi = 0$; $X_k \neq 0$. If there is no metastable equilibrium and all forces and flows are independent, the inequality sign holds in Eq. (88).

The increment of dissipation function can be split into two contributions

$$d\Psi = d_{\mathrm{X}} \Psi + d_{\mathrm{J}} \Psi \quad (90)$$

where

$$d_{\mathrm{X}} \Psi = \sum_i J_i dX_i$$

$$d_{\mathrm{J}} \Psi = \sum_i X_i dJ_i$$

When the system is not far away from equilibrium, and the linear phenomenological equations are valid, we have $d_{\mathrm{X}} \Psi = d_{\mathrm{J}} \Psi = d\Psi / 2$ and a stationary state with $d\Psi \leq 0$.

The dissipation function for l chemical reactions is given in terms of the affinity and the velocity of reaction by

$$\Psi = \sum_{i=1}^{l} J_{ri} A_i \geq 0 \tag{91}$$

Classical thermodynamics states that the change of entropy generation as a result of the irreversible phenomena inside a closed adiabatic system is always positive. This principle admits a situation such that the entropy may decrease at some place in the system, providing that entropy production at another place compensates this loss.

The quantities Φ and Ψ are scalars, and hence they are the products of two scalars, the dot product of two vectors, or double dot products of two tensors of rank two. For an isotropic medium, the dissipation function or entropy production rate can be split into three nonnegative parts

$$\Psi = \Psi_o + \Psi_1 + \Psi_2 = \sum_{i=1}^{n_o} J_i X_i + \sum_{i=1}^{n_1} \mathbf{J_i} \cdot \mathbf{X_i} + \sum_{i=1}^{n_2} \mathbf{J_i} : \mathbf{X}_i \tag{92}$$

where n_o is the number of scalar, n_1 is the number of vectorial, and n_2 is the number of tensorial (rank two) thermodynamic forces. The choice of thermodynamic forces must be made, so that in the equilibrium state when the thermodynamic forces vanish ($X_i = 0$) the entropy generation must also be zero. It must be noted that in contradistinction to entropy, the entropy generation rate and the dissipation function are not state functions since they depend on the path of change between the given states.

The theory of nonequilibrium thermodynamics involves finding the conjugated flows and forces J_i and X_i, respectively, from the volumetric rate of entropy production Φ or from the dissipation function Ψ, and to establish the phenomenological equations. The Onsager reciprocal relations relate the phenomenological coefficients pertaining to interactions or coupling between the processes. The form of the expressions for the dissipation function does not uniquely determine the thermodynamic forces or generalized flows. For an open system, for example, we may define the energy flow in various ways. We may also define the diffusion in several alternative ways depending on the choice of reference velocity. Thus we may transform the flows and the forces in various ways. If such forces and the flows, which are related by the phenomenological coefficients obeying the Onsager relations, are subjected to a linear transformation then the dissipation function is not affected by that transformation.

6. VARIATION OF ENTROPY PRODUCTION

In the steady state, a system loses minimal amount of available energy. The concept of least dissipation is the physical principle underlying the evolution of life. The living systems are endowed with a series of regulating mechanisms that preserve the steady state and bring the organisms back to their unperturbated condition. The range of principle of the minimum entropy production is restricted to phenomena close to equilibrium and obeying the Onsager relations. The Onsager reciprocal relations are satisfied only when the flows are expressed as a function of their respective conjugate thermodynamics forces. For rapid metabolic processes the phenomenological equations may not hold, and the general laws applicable to all possible rate phenomena are still yet to be developed. Attempts are being made to extend the range of validity of the variational principles so as to include stationary states away from equilibrium. In equilibrium thermodynamics, the systems tend to maximize the entropy or to minimize the free energy or the other thermodynamic potentials according to the conditions compatible with that potential. Prigogine demonstrated that in linear nonequilibrium thermodynamics entropy generation in stationary states is minimum.

The dissipation function for a two-flow system can be represented by the conductance coefficients L_{ij}

$$\Psi = L_{11}X_1^2 + L_{22}X_2^2 + (L_{12} + L_{21})X_1X_2 \tag{93}$$

Eq. (93) produces a paraboloid like change of dissipation with respect to forces X_1 and X_2, as seen in Fig. 2. The system tends to minimize the entropy and eventually reaches to zero entropy generation if there are no restrictions on the forces. On the other hand, if we externally fix the value of one of the forces for example $X_2 = X_{20}$, then the system will tend toward the stationary state characterized by the minimum entropy generation at $X_2 = X_{20}$. The system will move along the parabola of Fig. 2 and stops at point Φ_o. At the minimum the derivation of Ψ with respect to X_2 is zero

$$\frac{d\Psi}{dX_2} = 2L_{22}X_2 + (L_{12} + L_{21})X_1 = 0 \tag{94}$$

If the Onsager relations are valid that is $L_{12} = L_{21}$, then Eq. (94) becomes

$$0 = 2(L_{22}X_2 + L_{12}X_1) = 2J_2 = 0 \tag{95}$$

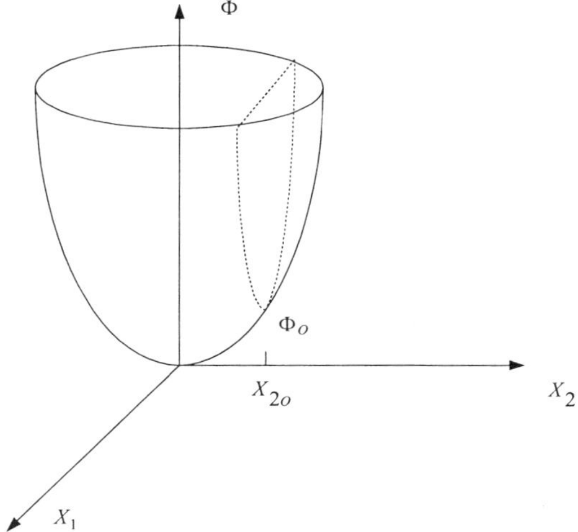

Fig. 2. Representation of entropy generation in terms of the forces.

since J_2 is the flow given by the phenomenological equations. Therefore a stationary point with respect to mass flux characterizes the state of minimum entropy generation, and the minimum dissipation. Therefore, if a system has n independent forces (X_1, X_2,..,X_n), and j of them are held constant (X_1, X_2,...,X_j = constant), then the flows with $J_i = J_{j+1}$, J_{j+2},..., J_n disappear at the stationary state with minimum entropy production. Such a state is called a stationary state of jth order.

For example, the total entropy production in a system of heat and mass reaches a minimum value at the nonequilibrium stationary state. For the following chemical reaction system

$$A \xleftrightarrow{1} B \xleftrightarrow{2} C$$

the entropy production is minimized at the nonequilibrium stationary state in the linear region, when the reaction velocities are equal to each other $J_{r1} = J_{r2}$.

As the system approaches the stationary state, the total entropy decreases to a minimum value; when the system reaches the state of equilibrium to total entropy production vanishes, while the entropy increases to a maximum value.

The study of the behavior of the stationary state of jth order after the disturbance may be helpful. For this purpose we designate the symbol s for all forces and flows in the stationary state. Assuming that the force X_k has been disturbed by δX_k, and keeping the remaining forces constant, we have

$$X_k = X_{s,k} + \delta X_k \qquad n \geq k \geq j+1 \tag{96}$$

The new value of the flow J_k becomes

$$J_k = J_{s,k} + L_{kk}\delta X_k \tag{97}$$

According to Prigogine theorem, $J_{s,k} = 0$, and we have

$$J_k = L_{kk}\delta X_k \tag{98}$$

After the disturbance, the rate of entropy generation will become

$$\Phi = \sum_{i=1}^{n} J_i X_i = \sum_{i,j=1}^{n} L_{ij} X_{s,i} X_{s,j} + [\sum_{i=1}^{n}(L_{ik} + L_{ki}) X_i \delta X_k] + L_{kk}(\delta X_k)^2 > 0 \tag{99}$$

In Eq. (85), the first summation term on the right-hand side is the minimum entropy production corresponding to the stationary state. The second sum is zero according to the Onsager reciprocal relations and the Prigogine principle. Therefore, we have

$$L_{kk}(\delta X_k)^2 > 0 \tag{100}$$

or from Eq. (84) we get

$$J_k \delta X_k > 0 \tag{101}$$

This inequality is identical to the Le Chatelier principle for nonequilibrium stationary states, since the disturbance δX_k has the same sign with the flow J_k, indicating the decrease in the effect of the disturbance. For example the increase in gradient of chemical potential will cause the mass flow, and diminish the gradient. Hence the stationary state will return to the original status.

Change of entropy generation can be measured by a sensitive calorimeter during the growth of bacterial colonies. The heat dissipated by the system is directly related to the entropy generation. The high precision calorimetric measurements indicate that the entropy increases sharply and reaches to a maximum, and finally decreases to a minimum. This is a stationary state, in which the colony no longer grows. The colony is limited to maintaining itself. If no grow occurs, the system can be described by means of linear phenomenological laws, and tends toward the minimum entropy generation.

Similarly the resting states of biological cells can often be interpreted as stationary states described by linear laws; the active phases or growth are mainly examples of nonlinear phenomena.

Application of minimum entropy generation can also be seen in the determination of certain geometric relations in the bronchial tree. The bronchial trees show a great variability of the dimension of the bronchioles (diameter, length), and yet high degree of organization and ordering principle exist. The closer observation of the structure brings about new levels of details called self-similarity.

REFERENCES

[1] S. Wisniewski, B. Staniszewski and R. Szymanik, Thermodynamics of Nonequilibrium Processes, D. Reidel Publishing Company, Dordrecht, 1976.
[2] I. Prigogine, Introduction to Thermodynamics of Irreversible Processes, Wiley, New York, 1967.
[3] D. Kondepudi, I. Prigogine, Modern Thermodynamics From Heat Engines to Dissipative Structures, Wiley, New York, 1999.
[4] S.R. De Groot, Thermodynamics of Irreversible Processes, North-Holland Publishing Company, Amsterdam, 1966.
[5] N.W. Tschoegl, Fundamentals of equilibrium and Steady-State Thermodynamics, Elsevier, Amsterdam, 2000.
[6] O. Levenspiel, Chemical Reaction Engineering, Wiley, New York, 1999.

Chapter 4

Balance equations and entropy generation

1. INTRODUCTION

A change in fluid media (except in rare gases and shock waves) is described by balance equations of extensive quantities. Intensive quantities specifying the local state of continuous medium appear in these balance equations. Intensive parameters described by the macroscopic properties of the medium are based on a large number of particles. Fluctuations occurring in turbulent flow deviate relatively little from the local equilibrium state, and in general the linear phenomenological equations can be applied to time averaged turbulent flows.

Mechanics of continuous medium is necessary to determine the thermodynamic state of a fluid. The properties of a fluid can be determined at points, which are at rest relative to a reference frame or moving along with the fluid. Every nonequilibrium intensive parameter in a fluid changes in time and in space. Consider temperature as function of time and space $T = T(t, x, y, z)$, the *total differential* of T is expressed as

$$dT = \frac{\partial T}{\partial t}dt + \frac{\partial T}{\partial x}dx + \frac{\partial T}{\partial y}dy + \frac{\partial T}{\partial z}dz \qquad (1)$$

Dividing the total differential by the time differential we obtain the *total time derivative* of T

$$\frac{dT}{dt} = \frac{\partial T}{\partial t} + \frac{\partial T}{\partial x}\frac{dx}{dt} + \frac{\partial T}{\partial y}\frac{dy}{dt} + \frac{\partial T}{\partial z}\frac{dz}{dt} \qquad (2)$$

The partial time derivative of T, $(\partial T/\partial t)$, shows the time rate of change of temperature of the fluid at a fixed position at constant x, y, and z

$$\frac{dT}{dt} = \frac{\partial T}{\partial t} \qquad (3)$$

If the derivative in Eq. (3) vanishes then temperature field becomes stationary. The terms dx/dt, dy/dt, and dz/dt are the components of the velocity of the observer different from the velocity of the fluid. If the velocity of observer is the same with the mass average velocity of the fluid $\mathbf{v}$ with components v_x, v_y, and v_z, than the rate of change of temperature is given by

$$\frac{DT}{Dt} = \frac{\partial T}{\partial t} + v_x \frac{\partial T}{\partial x} + v_y \frac{\partial T}{\partial y} + v_z \frac{\partial T}{\partial z} \tag{4}$$

or

$$\frac{DT}{Dt} = \frac{\partial T}{\partial t} + \mathbf{v} \cdot \nabla T \tag{5}$$

The special operator in Eq. (5) is called the *substantial time derivative*, and means that the time rate of change if observer moves with the substance.

A scalar or a vector function expressed in terms of $\partial / \partial t$ can be converted into the substantial form; for a scalar function $T = T(x, y, z, t)$, we have

$$\begin{aligned} \rho \frac{DT}{Dt} &= \frac{\partial(\rho T)}{\partial t} + \left(\frac{\partial(\rho v_x T)}{\partial x}\right) + \left(\frac{\partial(\rho v_y T)}{\partial y}\right) + \left(\frac{\partial(\rho v_z T)}{\partial z}\right) \\ &= \rho\left(\frac{\partial T}{\partial t} + v_x \frac{\partial T}{\partial x} + v_y \frac{\partial T}{\partial y} + v_z \frac{\partial T}{\partial z}\right) + T\left(\frac{\partial \rho}{\partial t} + \frac{\partial(\rho v_x)}{\partial x} + \frac{\partial(\rho v_y)}{\partial y} + \frac{\partial(\rho v_z)}{\partial z}\right) \end{aligned} \tag{6}$$

The second term in the second line of Eq. (6) is the equation of continuity and vanishes, so that in vector form we have

$$\rho \frac{DT}{Dt} = \frac{\partial(\rho T)}{\partial t} + (\nabla \cdot \rho\, \mathbf{v} T) \tag{7}$$

This equation is valid for every local quantity, which may be a scalar, an element of a vector, or an element of a tensor.

An extensive quantity E for a fluid in volume V can be expressed in terms of the specific quantity e, and we have

$$E = \int_V \rho\, e\, dV \tag{8}$$

The partial time derivative of E pertaining to the entire body is equal to the total differential

$$\frac{\partial E}{\partial t} = \frac{dE}{dt} = \frac{d}{dt} \int_V \rho\, e dV + \int_V \frac{\partial(\rho\, e)}{\partial t} dV \tag{9}$$

In Eq. (9) the quantity ρe is determined per unit volume when observer at rest. The amount of substance entering through an elementary surface area dA per unit time is $\rho \mathbf{v} \cdot d\mathbf{A}$, where $d\mathbf{A}$ is a vector with magnitude dA and direction taken as the normal to the surface. Along with the substance there is a convection flux $\rho \mathbf{v}e$, and the amount transported per unit time is $-\int(\rho \mathbf{v}e) \cdot d\mathbf{A}$. The conduction flux $\mathbf{J}_e$ is a vector with the same direction as the flow, and the amount transported per unit time by means of conduction without a flow of substance is $-\int \mathbf{J}_e \cdot d\mathbf{A}$.

The rate of generation of E inside the elementary volume of substance at a given point is expressed as

$$\Phi_e = \frac{dE}{dVdt} \tag{10}$$

For the entire volume at rest relative to coordinate system, the balance equation per unit time is expressed as

$$\frac{dE}{dt} = \int_V \frac{\partial(\rho e)}{\partial t} dV = -\int_A (\rho e\mathbf{v}) \cdot d\mathbf{A} - \int_A \mathbf{J}_e \cdot d\mathbf{A} + \int_V \Phi_e dV \tag{11}$$

Using the Gauss-Ostrogradsky theorem, Eq. (11) can be written over the entire volume as follows

$$\frac{dE}{dt} = \int_V \frac{\partial(\rho e)}{\partial t} dV = -\int_V [\nabla \cdot (\rho e\mathbf{v})]\, dV - \int_V (\nabla \cdot \mathbf{J}_e)\, dV + \int_V \Phi_e dV \tag{12}$$

From Eq. (12) the local balance equation for a fixed observer becomes

$$\frac{\partial(\rho e)}{\partial t} = -\nabla \cdot (\rho e\mathbf{v}) - \nabla \cdot \mathbf{J}_e + \Phi_e \tag{13}$$

The local balance equation for properties subject to a conservation law is called the *conservation equation,* which is given for e as follows

$$\frac{\partial(\rho e)}{\partial t} = -\nabla \cdot (\rho e\mathbf{v}) - \nabla \cdot \mathbf{J}_e \tag{14}$$

If the system is in a stationary state the extensive property E does not change with time $dE/dt = 0$, and we have

$$\nabla \cdot (\mathbf{J}_e + \rho e\mathbf{v}) = 0 \tag{15}$$

Eq. (15) shows that the net amount of E exchanged through the boundary must be zero, and the divergence of the sum of the conduction and convection flows governed by a *conservation law* is equal to zero in the stationary state. For the amounts of components $e = 1$, $\mathbf{J}_e = 0$, and $\Phi_e = 0$, and Eq. (13) becomes

$$\frac{\partial \rho}{\partial t} = -\nabla \cdot (\rho \mathbf{v}) = -\rho(\nabla \cdot \mathbf{v}) - \mathbf{v} \cdot \nabla \rho \tag{16}$$

The local balance equations for an observer moving along with the fluid are expressed in substantial time derivative form. From Eq. (13), we can express the substantial time derivative of e as

$$\rho \frac{De}{Dt} = -\nabla \cdot \mathbf{J}_e + \Phi_e \tag{17}$$

On the right-hand side of this equation, the divergence of the convection flux of e, $-\nabla \cdot (\rho e \mathbf{v})$, vanishes since the observer (coordinate system) is moving along with the fluid. In terms of the conservation law, where the source term vanishes, and Eq. (17) becomes

$$\rho \frac{De}{Dt} = -\nabla \cdot \mathbf{J}_e \tag{18}$$

Engineering systems mainly involve a single-phase fluid mixture with n components, subject to fluid friction, heat transfer, mass transfer, and involving with a number of l chemical reactions. A local thermodynamic state of the fluid is specified by two intensive parameters, for example, velocity of the fluid and the chemical composition in terms of component mass fractions w_i. For a unique description of the system, balance equations must be derived for the mass, momentum, energy and entropy. The balance equations, considered per unit volume basis, can be written in terms of the partial time derivative with an observer at rest, and in terms of the substantial derivative with an observer moving together with the fluid. Later the balance equations are used in the Gibbs relation to determine the rate of entropy generation. The balance equations allow us to clearly identify the importance of the local thermodynamic equilibrium postulate in deriving the expressions for the entropy generation.

1.1. The mass balance equations

The mass balance equation for component i is similar to the general form given in Eq. (13) after setting $e = w_i$, and $J_e = j_i$. Amount of component produced or consumed inside a unit volume per unit time is the result of chemical reactions. The mass balance equation is

$$\frac{\partial \rho_i}{\partial t} = -\nabla \cdot (\rho_i \mathbf{v}) - \nabla \cdot \mathbf{j}_i + M_i \sum_{j=1}^{l} \nu_{ij} J_{r,j} \tag{19}$$

Here $J_{r,j}$ is the chemical reaction rate per unit volume for reaction j and ν_{ij} the specific stoichiometric coefficient of species i in the chemical reaction j, and M_i the molecular mass of component i.

The mass flow of component i, $\rho_i \mathbf{v}_i$, is a vector showing the flow of a component relative to a motionless coordinate system. On the other hand, diffusion flow shows the transport of a component relative to a coordinate system moving at the reference velocity $\mathbf{v}_r$. The diffusion flow relative to the centre-of-mass velocity $\mathbf{v}$ (or mass average velocity) is given by

$$\mathbf{j}_i = \rho_i (\mathbf{v}_i - \mathbf{v}) \tag{20}$$

where $(\mathbf{v}_i - \mathbf{v})$ is the diffusion velocity. Mass average velocity is given by

$$\mathbf{v} = \frac{1}{\rho} \sum_{i=1}^{n} \rho_i \mathbf{v}_i = \sum_{i=1}^{n} w_i \mathbf{v}_i \tag{21}$$

Here w_i is the mass fraction of component i. We can express the molar diffusion flow $\mathbf{j}_{i,M}$ based on the molar average velocity $\mathbf{v}_M$

$$\mathbf{j}_{i,M} = c_i (\mathbf{v}_i - \mathbf{v}_M) \tag{22}$$

$$\mathbf{v}_M = \frac{1}{c} \sum_{i=1}^{n} c_i \mathbf{v}_i \tag{23}$$

or based on the volume average velocity $\mathbf{v}_v$

$$\mathbf{j}_{i,V} = c_i (\mathbf{v}_i - \mathbf{v}_v) \tag{24}$$

$$\mathbf{v}_v = \frac{1}{V} \sum_{i=1}^{n} \overline{V}_i \mathbf{v}_i \tag{25}$$

where $\overline{V}_i$ is the partial molar volume. The sum of diffusion flows of all components is zero

$$\sum_{i=1}^{n}\mathbf{j}_i = \sum_{i=1}^{n}\mathbf{j}_{i,M} = \sum_{i=1}^{n}\mathbf{j}_{i,V} = 0 \tag{26}$$

Of the n diffusion flows, only n-1 of them are independent.

From Eqs. (19) and (5), mass balance can be expressed in substantial time derivative, and we have

$$\frac{D\rho_i}{Dt} = \frac{\partial\rho_i}{\partial t} + \mathbf{v}\cdot\nabla\rho_i = -\nabla\cdot(\rho_i\mathbf{v}) - \nabla\cdot\mathbf{j}_i + M_i\sum_{j=1}^{l}\nu_{ij}J_{r,j} + \mathbf{v}\cdot\nabla\rho_i \tag{27}$$

Using Eq. (16), Eq. (27) is expressed by

$$\frac{D\rho_i}{Dt} = -\rho_i(\nabla\cdot\mathbf{v}) - \nabla\cdot\mathbf{j}_i + M_i\sum_{j=1}^{l}\nu_{ij}J_{r,j} \tag{28}$$

When an observer moves at the center-of-mass velocity of the fluid, the conservation equation is obtained from the substantial derivative of the density, and Eq. (16), and given by

$$\frac{D\rho}{Dt} = \frac{\partial\rho}{\partial t} + \mathbf{v}\cdot\nabla\rho = -\rho(\nabla\cdot\mathbf{v}) \tag{29}$$

The conservation of mass of substances is also obtained by replacing the density in Eq. (29) with the specific volume $v = 1/\rho$

$$\frac{Dv}{Dt} = v(\nabla\cdot\mathbf{v}) \tag{30}$$

Using Eq. (19), the balance equation for the amount of substance can also be written in terms of mass fraction w_i

$$\frac{\partial(\rho\, w_i)}{\partial t} = -\nabla\cdot(\rho\, w_i\mathbf{v}) - \nabla\cdot\mathbf{j}_i + M_i\sum_{j=1}^{l}\nu_{ij}J_{r,j} \tag{31}$$

With the substantial derivative Eq. (31) becomes

$$\rho\frac{D(w_i)}{Dt} = -\nabla\cdot\mathbf{j}_i + M_i\sum_{j=1}^{l}\nu_{ij}J_{r,j} \tag{32}$$

In stationary state $dm/dt = 0$, and from Eq. (15) we have

$$\nabla \cdot (\rho \mathbf{v}) = 0 \tag{33}$$

1.2. The momentum balance equations

Fluid motion may be described by applying Newton's second law to a particle. The momentum flow of a substance is equal to the product of the mass flow $\rho\mathbf{v}$ and the barycentric velocity $\mathbf{v}$, that is $\rho\mathbf{vv}$. The Newton's second law of motion states that the change in the momentum of a body is equal to the resultant of all forces, mass force $\mathbf{F}$ and surface force σ, acting on that body. If $\mathbf{F}_i$ is the force exerted per unit mass of component i, so that we have

$$\mathbf{F} = \frac{1}{\rho}\sum_{i=1}^{n} \rho_i \mathbf{F}_i = \sum_{i=1}^{n} w_i \mathbf{F}_i \tag{34}$$

The mass forces may be the gravitational force, the force due to rotational motion of a system, and the Lorentz force that is the proportional to the vector product of the molecular velocity of component i and the magnetic field strength. The normal stress tensor σ produces a surface force. No shear stresses occur (τ = 0) in a fluid, which is in mechanical equilibrium.

The time derivative of the momentum density is given by

$$\frac{\partial}{\partial t}(\rho \mathbf{v}) = -\nabla \cdot (\rho \mathbf{vv}) + \nabla \cdot \sigma + \rho \mathbf{F} \tag{35}$$

By taking account the following relations

$$\sigma = -P\boldsymbol{\delta} + \tau \tag{36}$$

$$\nabla \cdot (P\boldsymbol{\delta}) = \nabla P \tag{37}$$

Eq. (35) becomes

$$\frac{\partial}{\partial t}(\rho\mathbf{v}) = -\nabla \cdot (\rho \mathbf{vv}) - \nabla P + \nabla \cdot \tau + \rho \mathbf{F} \tag{38}$$

The term δ is the unit tensor. The terms on the right-hand side represent the change of momentum due to the convection momentum flow $\nabla \cdot (\rho\mathbf{vv})$, the pressure force ∇P, the viscous force $\nabla \cdot \tau$, and the mass force $\rho\mathbf{F}$, respectively. The momentum balance equation for a coordinate system moving along with the fluid is given by

$$\rho \frac{D\mathbf{v}}{Dt} = -\nabla P + \nabla \cdot \tau + \rho \mathbf{F} \tag{39}$$

The left-hand side of Eq. (39) contains the center-of-mass acceleration $d\mathbf{v}/dt$.

The state of mechanical equilibrium is characterized by the vanishing acceleration $d\mathbf{v}/dt = 0$. Usually, the mechanical equilibrium is established faster than those of thermodynamic processes, for example in the initial state when diffusion or thermal diffusion is considered. In the case of diffusion in a closed system, the acceleration may not be zero because of changing molecular weights of species, however this acceleration is very small, and the corresponding pressure gradient is negligible; the viscous part of the stress tensor also vanishes $\tau = 0$. The momentum balance, Eq. (39), is limited to the *momentum conservation equation*

$$\nabla P = \sum_{i=1}^{n} \rho_i \mathbf{F}_i = \rho \mathbf{F} \tag{40}$$

Eq. (40) shows that the pressure gradient is equal to the sum of the mass forces acting on the substance in a unit volume.

1.3. The energy balance equations

The time variation of the total energy e per unit volume is subject to a law of conservation, and given in terms of convection flow $\rho e\mathbf{v}$ and conduction flow $\mathbf{J}_e$

$$\frac{\partial(\rho e)}{\partial t} = -\nabla \cdot (\rho e\mathbf{v}) - \nabla \cdot \mathbf{J}_e \tag{41}$$

The total specific energy of a substance e is defined by

$$e = u + \frac{1}{2} v^2 + e_p \tag{42}$$

and comprises the specific internal energy u, the specific kinetic energy $(\frac{1}{2})\, v^2$, and the specific potential energy e_p.

The conduction flow of the total energy $\mathbf{J}_e$ consists of the conduction flow of the internal energy $\mathbf{J}_u$, the potential energy flow $\sum_{i=1}^{n} e_{pi}\mathbf{j}_i$ due to diffusion of components, and the work of surface forces $-\mathbf{v} \cdot \boldsymbol{\sigma}$ per unit surface area, and expressed as follows

$$\mathbf{J}_e = \mathbf{J}_u + \sum_{i=1}^{n} e_{pi}\mathbf{j}_i - \mathbf{v}\cdot\boldsymbol{\sigma} \tag{43}$$

The divergence of the total energy flow is expressed by

$$\nabla\cdot\mathbf{J}_e = \nabla\cdot(\mathbf{J}_u + \sum_{i=1}^{n} e_{pi}\cdot\mathbf{j}_i - \mathbf{v}\cdot\boldsymbol{\sigma}) \tag{44}$$

By using the diffusion flow, $\mathbf{j}_i = \rho_i(\mathbf{v}_i - \mathbf{v})$, the time variation of the potential energy of a unit volume of the fluid is given by

$$\frac{\partial(\rho\, e_p)}{\partial t} = -\nabla\cdot(\rho\, e_p\mathbf{v} + \sum_{i=1}^{n} e_{pi}\mathbf{j}_i) - \rho\,\mathbf{F}\cdot\mathbf{v} - \sum_{i=1}^{n}\mathbf{j}_i\cdot\mathbf{F}_i + \sum_{i=1}^{n} e_{pi}M_i\sum_{j=1}^{l}\nu_{ij}J_{r,j} \tag{45}$$

where the mass force $\mathbf{F}_i$ is associated with the specific potential energy e_{pi} of component i as

$$\mathbf{F}_i = -\nabla e_{pi} \tag{46}$$

with the properties of conservative mass forces

$$\frac{\partial e_{pi}}{\partial t} = 0, \quad \frac{\partial \mathbf{F}_i}{\partial t} = 0 \tag{47}$$

The last term on the right-hand side of Eq. (45) is zero if the potential energy is conserved for the chemical reaction j

$$\sum_{i=1}^{n} e_{pi}M_i\nu_{ij} = 0 \tag{48}$$

The balance equation for the kinetic energy is obtained by scalar multiplication of the momentum balance, Eq. (39), and the mass average velocity, and given by

$$\rho\frac{D\left(\frac{1}{2}\rho v^2\right)}{Dt} = -\nabla\cdot(Pv) + \nabla\cdot(\mathbf{v}\cdot\tau) + P(\nabla\cdot\mathbf{v}) - \tau : (\nabla\mathbf{v}) + \rho\,\mathbf{v}\cdot\mathbf{F} \tag{49}$$

The time variation of the kinetic energy per unit volume (for a motionless reference frame) is

$$\frac{\partial\left(\frac{1}{2}\rho v^2\right)}{\partial t} = -\nabla\cdot(\frac{1}{2}\rho v^2 \mathbf{v}) - \nabla\cdot(Pv) + \nabla\cdot(\mathbf{v}\cdot\tau) + P(\nabla\cdot\mathbf{v}) - \tau : (\nabla\mathbf{v}) + \rho\,\mathbf{v}\cdot\mathbf{F} \tag{50}$$

In Eqs. (49) and (50), the relation $\sigma = -P\boldsymbol{\delta} + \tau$ is used. In Eq. (50), the term $-\nabla\cdot\left(\frac{1}{2}\rho v^2 \mathbf{v}\right)$ is the convection transport of kinetic energy, $\nabla\cdot(Pv)$ is the work of the pressure, $\nabla\cdot(\mathbf{v}\cdot\tau)$ is the work of the viscous forces, and $\rho\mathbf{v}\cdot\mathbf{F}$ is the work of the mass forces. Part of the kinetic energy $P(\nabla\cdot\mathbf{v})$ is transformed reversibly into internal energy, and the part $-\tau : (\nabla\mathbf{v})$ is the transformed irreversibly, and is dissipated.

The total change in kinetic and potential energies per unit volume is obtained by adding Eqs. (50) and (45) under the conditions of conservation of energy given in Eq. (48), and we obtain

$$\frac{\partial\rho\left(\frac{1}{2}v^2 + e_p\right)}{\partial t} = -\nabla\cdot[\rho(\frac{1}{2}v^2 + e_p)\mathbf{v} - \mathbf{v}\cdot\boldsymbol{\sigma} + \sum_{i=1}^{n} e_{pi}\mathbf{j}_i] - \sigma : (\nabla\mathbf{v}) - \sum_{i=1}^{n}\mathbf{j}_i\cdot\mathbf{F}_i \tag{51}$$

Subtraction of Eq. (51) from the total energy-conservation equation, Eq. (41), yields the rate of change of the internal energy for an observer at rest

$$\frac{\partial(\rho u)}{\partial t} = -\nabla\cdot(\rho u\mathbf{v}) - \nabla\cdot\mathbf{J}_u - P(\nabla\cdot\mathbf{v}) + \tau : (\nabla\mathbf{v}) + \sum_{i=1}^{n}\mathbf{j}_i\cdot\mathbf{F}_i \tag{52}$$

The term $-\nabla\cdot(\rho u\mathbf{v})$ is the divergence of the convection internal energy flux, $-\nabla\cdot\mathbf{J}_u$ is the divergence of the conduction internal energy flux, $-P(\nabla\cdot\mathbf{v})$ is the reversible increment of internal energy due to volume work, $-\tau : (\nabla\mathbf{v})$ is the irreversible increment of internal energy due to viscous dissipation, and $-\sum_{i=1}^{n}\mathbf{j}_i\cdot\mathbf{F}_i$ is the transport of potential energy by diffusion flows. Eq. (52) can be rewritten in terms of the substantial derivative

$$\rho\frac{Du}{Dt} = -\nabla\cdot\mathbf{J}_u - P(\nabla\cdot\mathbf{v}) + \tau : (\nabla\mathbf{v}) + \sum_{i=1}^{n}\mathbf{j}_i\cdot\mathbf{F}_i \tag{53}$$

The internal energy balance equation for the fluid is based on the momentum balance equation. Introducing the assumption of local thermodynamic equilibrium will enable us to introduce the thermodynamic relationships linking intensive quantities in the state of equilibrium and to derive the internal energy balance equation on the basis of equilibrium partial quantities. By assuming that the diffusion is a slow phenomenon, $-\sum_{i=1}^{n}\mathbf{j}_i / \rho << \rho v^2$, the change of the total energy of all components per unit volume is expressed by

$$\frac{\partial}{\partial t}\left[\sum_{i=1}^{n}\rho_i(\overline{u}_i + \frac{1}{2}v_i^2 + e_{pi})\right] = \frac{\partial}{\partial t}\left[\rho(u + \frac{1}{2}v^2 + e_p)\right] \tag{54}$$

This form is based on the concept of local thermodynamic equilibrium. From Eq. (54) the convection flow of the total energy is expressed by

$$\begin{aligned}\sum_{i=1}^{n}\rho_i(\overline{u}_i + \frac{1}{2}v_i^2 + e_{pi})\mathbf{v}_i &= \sum_{i=1}^{n}\overline{u}_i\mathbf{J}_i + \rho(u + \frac{1}{2}v^2 + \rho e)\mathbf{v} + \sum_{i=1}^{n}e_{pi}\mathbf{J}_i \\ &= \rho\, e\mathbf{v} + \sum_{i=1}^{n}\overline{u}_i\mathbf{J}_i + \sum_{i=1}^{n}e_{pi}\mathbf{J}_i\end{aligned} \tag{55}$$

Eq. (55) contains the convection flux of total energy and energy changes due to the diffusion flows. If $\mathbf{J}_q^{'}$ is the conduction energy flow of pure heat conduction without a flow of internal energy due to diffusion of the substance, the total energy conservation given in Eq. (41) becomes

$$\frac{\partial(\rho e)}{\partial t} = -\nabla\cdot(\rho e\mathbf{v}) - \nabla\cdot(\mathbf{J}_q^{'} + \sum_{i=1}^{n}\overline{u}_i\mathbf{j}_i + \sum_{i=1}^{n}e_{pi}\mathbf{j}_i - \mathbf{v}\cdot\boldsymbol{\sigma}) \tag{56}$$

The quantities $\mathbf{J}_u, \mathbf{J}_q^{'}$, and $\mathbf{j}_i$ are related by

$$\mathbf{J}_u = \mathbf{J}_q^{'} + \sum_{i=1}^{n}\overline{\mathbf{u}}_i\mathbf{j}_i \tag{57}$$

The second term on the right-hand-side of this equation represents the net flow of internal energy transported along with the diffusion of substances.

1.4. The entropy balance equations

The entropy balance equation is given in the general form of Eq. (13)

$$\frac{\partial(\rho s)}{\partial t} = -\nabla \cdot (\rho s \mathbf{v}) - \nabla \cdot \mathbf{J}_s + \Phi \tag{58}$$

Eq. (58) shows that the rate of change of the entropy per unit volume of substance is due to the convection entropy flow $\rho s \mathbf{v}$, the conduction entropy flow $\mathbf{J}_s$, and the entropy generation Φ. The conduction entropy flow is defined as

$$\mathbf{J}_s = \frac{\mathbf{J}_q^{''}}{T} + \sum_{i=1}^{n} \bar{s}_i \mathbf{j}_i \tag{59}$$

The conduction entropy flow consists of two contributions, which are due to the heat flow in the entropy balance equation $\mathbf{J}_q^{''}$ (reduced heat flow that is the difference between the change in energy and the change in enthalpy due to the matter flow), and the diffusion flow $\mathbf{j}_i$. With the substantial derivative and using Eq. (59), we obtain the entropy balance equation

$$\rho \frac{Ds}{Dt} = -\nabla \cdot \left(\frac{\mathbf{J}_q^{''}}{T} + \sum_{i=1}^{n} \bar{s}_i \mathbf{j}_i \right) + \Phi \tag{60}$$

To determine the relationship between the heat flow $\mathbf{J}_q^{''}$ and the conduction internal energy flux $\mathbf{J}_q^{'}$, and to determine the entropy source strength Φ, the entropy balance equation must be derived based on the local thermodynamic equilibrium.

2. ENTROPY GENERATION EQUATION

Assuming that the local thermodynamic equilibrium holds, we can write the Gibbs relation in terms of specific quantities

$$Tds = du + Pdv - \sum_{i=1}^{n} \mu_i dw_i \tag{61}$$

Eq. (61) can be applied to a fluid element moving with the mass average velocity $\mathbf{v}$, and replace the differential operators by substantial time derivative operators

$$\rho \frac{Ds}{Dt} = \frac{\rho}{T}\frac{Du}{Dt} + \frac{\rho P}{T}\frac{Dv}{Dt} - \frac{\rho}{T}\sum_{i=1}^{n} \mu_i \frac{Dw_i}{Dt} \tag{62}$$

The individual terms on the right-hand-side of Eq. (62) are substituted by Eq. (53)

$$\rho \frac{Du}{Dt} = -\nabla \cdot \mathbf{J}_u - P(\nabla \cdot \mathbf{v}) + \tau : (\nabla \mathbf{v}) + \sum_{i=1}^{n} \mathbf{j}_i \cdot \mathbf{F}_i$$

by Eq. (30)

$$\rho P \frac{Dv}{Dt} = P(\nabla \cdot \mathbf{v})$$

and by Eq. (32)

$$\rho \sum_{i=1}^{n} \mu_i \frac{D(w_i)}{Dt} = -\mu_i(\nabla \cdot \mathbf{j}_i) + \sum_{j=1}^{l} A_j J_{r,j}$$

where the affinity A of a chemical reaction j is defined as

$$A_j = -\sum_{i=1}^{n} M_i \mu_i \nu_{ij} \tag{63}$$

After the substitutions of equation above, Eq. (62) becomes

$$\rho \frac{Ds}{Dt} = -\frac{\nabla \cdot \mathbf{J}_u}{T} + \frac{1}{T}\tau : (\nabla \mathbf{v}) + \frac{1}{T}\sum_{i=1}^{n} \mathbf{j}_i \cdot \mathbf{F}_i + \frac{1}{T}\sum_{i=1}^{n} \mu_i(\nabla \cdot \mathbf{j}_i) - \frac{1}{T}\sum_{j=1}^{l} A_j J_{r,j} \tag{64}$$

Using the following transformations

$$\frac{\nabla \cdot \mathbf{J}_u}{T} = \nabla \cdot \left(\frac{\mathbf{J}_u}{T}\right) + \frac{1}{T^2}\mathbf{J}_u \cdot \nabla T$$

$$\frac{\mu_i}{T}(\nabla \cdot \mathbf{j}_i) = \nabla \cdot \left(\frac{\mu_i}{T}\mathbf{j}_i\right) - \mathbf{j}_i \cdot \nabla\left(\frac{\mu_i}{T}\right)$$

the entropy balance given in Eq. (64) becomes

$$\rho \frac{Ds}{Dt} = -\nabla \cdot \left(\frac{\mathbf{J}_u - \sum_{i=1}^{n} \mu_i \mathbf{j}_i}{T}\right) - \frac{1}{T^2}\mathbf{J}_u \cdot \nabla T - \frac{1}{T}\sum_{i=1}^{n} \mathbf{j}_i \cdot \left[T\nabla\left(\frac{\mu_i}{T}\right) - \mathbf{F}_i\right] \tag{65}$$
$$+ \frac{1}{T}\tau : (\nabla \mathbf{v}) - \frac{l}{T}\sum_{j=1}^{l} A_j J_{r,j}$$

Comparison of Eqs. (60) and (65) yields an expression for the conduction entropy flow

$$\mathbf{J}_s = \frac{\mathbf{J}_q^{''}}{T} + \sum_{i=1}^{n} \bar{s}_i \mathbf{j}_i = \frac{1}{T}\left(\mathbf{J}_u - \sum_{i=1}^{n} \mu_i \mathbf{j}_i\right) \tag{66}$$

Using the relation between the chemical potential and enthalpy given by

$$\mu_i = \bar{h}_i - T\bar{s}_i = \bar{u}_i + P\bar{v}_i - T\bar{s}_i \tag{67}$$

we can relate the second law heat flow $\mathbf{J}_q^{''}$, the conduction energy flow $\mathbf{J}_u$, and the pure heat flow $\mathbf{J}_q^{'}$ as follows

$$\mathbf{J}_q^{''} = \mathbf{J}_u - \sum_{i=1}^{n} \bar{h}_i \mathbf{j}_i = \mathbf{J}_q^{'} - \sum_{i=1}^{n} P\bar{v}_i \mathbf{j}_i \tag{68}$$

The heat flow can be defined in various ways if diffusion occurs in multicomponent fluids. The concept of heat flow emerges from a macroscopic treatment of the energy balance or the entropy balance. The internal energy of a substance is related to the molecular kinetic energy and the potential energy of the intermolecular interactions. If a molecule travels without colliding with other molecules, the loss of kinetic energy is due to the diffusion. If the kinetic energy loss is the result of molecular collisions, it is classified as heat conduction.

However changes in the potential energy of intermolecular interactions are not uniquely separable into those two cases. There is an ambiguity in defining the heat flow for open systems. We may split u into a diffusive part and a conductive part in several ways and define various number of heat flows. In the molecular mechanism of energy transport, the energy of a system is associated with the kinetic energy of the molecules and with the potential energy of their interaction. The kinetic energy changes in volume V are easily separated. If a molecule leaves the volume, the kinetic energy loss may be due to the diffusion. If kinetic energy loss occurs in the system because a molecule at the surface of volume transfers energy by collision to a molecule outside V, then this loss may be called heat flow. However, the potential energy of molecular interactions is the sum of the potential energies of interaction for each molecular pair. When some molecules leave the element volume and other molecules collide at the surface with molecules outside volume, they produce a complicated change in the potential energy of volume. These changes cannot be uniquely separated into contributions of pure diffusion and of molecular collision.

From Eqs. (60), (65) and (66), the entropy source strength or the rate of local entropy generation per unit volume Φ is defined by

$$\Phi = \mathbf{J}_u \cdot \nabla\left(\frac{1}{T}\right) - \frac{1}{T}\sum_{i=1}^{n} \mathbf{j}_i \cdot \left[T\nabla\left(\frac{\mu_i}{T}\right) - \mathbf{F}_i\right] + \frac{1}{T}\tau : (\nabla v) - \frac{1}{T}\sum_{j=1}^{l} A_j J_{r,j} \tag{69}$$

From Eq. (69) the following forces X_i are defined:

- Heat transfer

$$X_q = \nabla\left(\frac{1}{T}\right) \tag{70}$$

- Mass transfer

$$X_i = \frac{\mathbf{F}_i}{T} - \nabla\left(\frac{\mu_i}{T}\right) \tag{71}$$

where $\nabla\left(\frac{\mu_i}{T}\right) = \nabla\left(\frac{\mu_i}{T}\right)_T + h_i \nabla\left(\frac{1}{T}\right)$

- Viscous dissipation

$$X_v = -\frac{1}{T}(\nabla \mathbf{v}) \tag{72}$$

- Chemical reaction

$$\frac{A_j}{T} = \sum_{i=1}^{n} \frac{M_i \mu_i}{T} v_{ij} \qquad (j=1,2,\ldots,l) \tag{73}$$

Eq. (69), first derived by Jaumann in 1911, expresses the rate of entropy generation as the sum of four distinctive contributions as a result of the products of flows and forces that are:

- Entropy generation associated with heat transfer

$$\Phi_q = \mathbf{J}_u X_q \tag{74}$$

- Entropy generation due to mass transfer

$$\Phi_d = \sum_{i=1}^{n} \mathbf{j}_i X_i \tag{75}$$

- Entropy generation as a result of viscous dissipation of fluid

$$\Phi_v = \tau : X_v \tag{76}$$

- Entropy generation arising from chemical reactions

$$\Phi_c = \sum_{j=1}^{l} J_{r,j} A_j \tag{77}$$

Eq. (69) consists of three sums of products of tensors that are scalars with rank zero Φ_0, vectors with rank one Φ_1, and tensor with rank two Φ_2

$$\Phi_0 = \frac{1}{T}\tau(\nabla \cdot \mathbf{v}) - \frac{1}{T}\sum_{j=1}^{l} J_{r,j} A_j \geq 0 \tag{78}$$

$$\Phi_1 = \mathbf{J}_u \cdot \nabla\left(\frac{1}{T}\right) + \frac{1}{T}\sum_{i=1}^{n} \mathbf{j}_i \cdot \left[\mathbf{F}_i - T\nabla\left(\frac{\mu_i}{T}\right)\right] \geq 0 \tag{79}$$

$$\Phi_2 = \frac{1}{T}\boldsymbol{\tau}' : (\nabla \mathbf{v})'^{S} \geq 0 \tag{80}$$

where $\mathbf{t} : (\nabla \mathbf{v}) = \boldsymbol{\tau}' : (\nabla \mathbf{v})'^{S} + \tau(\nabla \cdot \mathbf{v})$. The tensor $(\nabla \mathbf{v})'$ is the sum of a symmetric $(\nabla \mathbf{v})'^{S}$ and anti symmetric part $(\nabla \mathbf{v})'^{a}$, and the double dot product of these is zero.

Eq. (79) represents the heat and mass transfer since they are vectorial processes with the rank 1. In Eq. (79) the conduction energy flow can be replaced by the heat flow $\mathbf{J}_q''$ using Eq. (68) and total potential μ^* comprising the chemical potential and the potential energy per unit mass of component

$$\mu^* = \mu_i + e_{pi} \tag{81}$$

The isothermal gradient of total potential is defined as

$$\nabla_T \mu_i^* = \nabla \mu_i + \bar{s}_i \nabla T + \nabla e_{pi} \tag{82}$$

From the force expression for mass transfer, we have

$$T\nabla\left(\frac{\mu_i}{T}\right) - F_i = \nabla\mu_i - \mu_i \frac{\nabla T}{T} + \nabla e_{pi} = \nabla_T \mu_i^* - \bar{h}_i \frac{\nabla T}{T} \tag{83}$$

Using Eqs. (81) to (83), we can rearrange Eq. (79) as

$$\Phi_{s1} = \mathbf{J}_q^{''} \cdot \nabla\left(\frac{1}{T}\right) - \frac{1}{T}\sum_{i=1}^{n} \mathbf{j}_i \cdot \nabla_T \mu_i^* \geq 0 \tag{84}$$

Since the only n-1 diffusion fluxes are independent, we have

$$\sum_{i=1}^{n} \mathbf{j}_i \cdot \nabla_T \mu_i^{'} = \sum_{i=1}^{n-1} \mathbf{j}_i \cdot \nabla_T (\mu_i^* - \mu_n^*) \tag{85}$$

Introducing Eq. (85) into Eq. (84) we have

$$\Phi_{s1} = \mathbf{J}_q^{''} \cdot \nabla\left(\frac{1}{T}\right) - \frac{1}{T}\sum_{i=1}^{n-1} \mathbf{j}_i \cdot \nabla_T (\mu_i^* - \mu_n^*) \geq 0 \tag{86}$$

Therefore the force shown in Eq. (51) becomes

$$X_i = \frac{1}{T}\nabla_T (\mu_i^* - \mu_n^*) \tag{87}$$

The dissipation function can be obtained from Eqs. (78)-(80), and we have

$$\Psi = T\Phi = T(\Phi_0 + \Phi_1 + \Phi_2) = \Psi_0 + \Psi_1 + \Psi_2 \tag{88}$$

If the dissipation function is chosen to identify the independent forces and flows, the following equations can be used

$$\Psi_0 = \tau(\nabla \cdot \mathbf{v})\left(\frac{1}{T}\right) - \sum_{j=1}^{l} A_j J_{r,j} \geq 0 \tag{89}$$

$$\Psi_1 = -\mathbf{J}_q^{''} \cdot \nabla \ln T - \frac{1}{T}\sum_{i=1}^{n-1} \mathbf{j}_i \cdot \nabla_T (\mu_i^* - \mu_n^*) \geq 0 \tag{90}$$

$$\Psi_2 = \boldsymbol{\tau}' : (\nabla \mathbf{v})'^S \geq 0 \tag{91}$$

Eq. (90) can be modified using the following transformation on Eq. (71)

$$TX_i = T\nabla\left(\frac{\mu_i}{T}\right) - F_i = \nabla\mu_i - \frac{\mu_i}{T}\nabla T + \nabla e_{pi} = \nabla\mu_i^* - \frac{\mu_i}{T}\nabla T \tag{92}$$

and

$$\sum_{i=1}^{n} \mathbf{j}_i \cdot \nabla\mu_i^* = \sum_{i=1}^{n-1} \mathbf{j}_i \cdot \nabla(\mu_i^* - \mu_n^*) \tag{93}$$

Eq. (90) can be expressed using Eqs. (92) and (93)

$$\Psi_1 = -J_s \cdot \nabla T - \sum_{i=1}^{n-1} j_i \cdot \nabla(\mu_i^* - \mu_n^*) \geq 0 \tag{94}$$

As shown by Prigogine, for diffusion in mechanical equilibrium, any other velocity can replace the center-of-mass velocity, and the dissipation function does not change. When diffusion fluxes are considered relative to various velocities, the thermodynamic forces remain the same only the values of the phenomenological coefficients change.

The dissipation or the entropy generation equation is used to identify the forces and flows, which are used in the linear phenomenological equations relating

flows and forces. If the Onsager reciprocal relations are to be applied to the phenomenological coefficients, the forces and flows must be chosen appropriately.

REFERENCES

[1] S. Wisniewski, B. Staniszewski and R. Szymanik, Thermodynamics of Nonequilibrium Processes, D. Reidel Publishing Company, Dordrecht, 1976.

[2] R.B. Bird, W.E. Stewart and E.N. Lightfoot, Transport phenomena. 2nd ed. Wiley, New York, 2002.

[3] I. Prigogine, Introduction to Thermodynamics of Irreversible Processes, Wiley, New York, 1967.

[4] R.W. Fox and A.T. McDonald, Introduction to Fluid Mechanics, 5th ed., Wiley, New York, 2000.

[5] G.A.J. Jaumann, Closed system of physical and chemical differential laws. Wien. Akad. Sitzungsberichte (Math-Nature Klasse) 129 (1911) 385.

Chapter 5

Entropy and exergy

1. ENTROPY

Mathematical statement of the second law is associated with the definition of entropy *S*, $dS = \delta Q_{\text{rev}} / T$. Entropy is a thermodynamic potential, and is not conserved; it gives a quantitative measure of irreversibility. For reversible processes; *dS* is an exact differential of the state function entropy, and the final result of the integration does not depend on the path of the process or on how it is materialized, provided that both the initial and final states are stable equilibrium states.

Entropy of a closed adiabatic system remains the same in a reversible process, and increases during an irreversible process. A system and its surrounding create an isolated system where the sum of the entropies of all bodies involving a reversible change remains the same, and increases during irreversible processes. Entropy can be used to distinguish between reversible and irreversible processes. Both entropy increase and time are intimately associated with the behavior of natural phenomena, and the fundamental law for closed, adiabatic system, $dS/dt > 0$, may be regarded as the pointer of the arrow of time.

The directional properties and the increase of entropy in natural adiabatic processes have lead to various interpretations of entropy: Clausius believed that the laws of thermodynamics have a universal validity. Entropy is a measure of the work value of the energy contained in the system, and the maximal entropy (thermodynamic equilibrium) means that the energy has zero work value, while low entropy means that the energy has relatively high work value. The energy of the world remains constant but its usability diminishes with every increase in the worlds' entropy. Schrodinger associated the concept of entropy with biological systems and stated that the existence of all living beings is based on entropy export, but it is also closely connected with information processing.

Entropy at any stage of an irreversible process is determined by assuming that the entropy is a unique function of the external and internal variables, regardless the energy and work capacity of the system. For a set of specified external variables, we can reach the same internal variables in various ways, both reversible and irreversible. However entropy is determined only by the set of local variables characterizing the state of the system. This assumption enables one to devise an ideal process that would bring the system reversibly to any configuration through which a system passes during an irreversible change. Therefore, the value of entropy corresponds to entropies of the real system at a certain time.

The fundamental form of the rate of increase in entropy in an irreversible process is expressed as the sum of the products of the thermodynamic forces and the flows within the system.

1.1. Entropy balance

In every nonequilibrium system, an entropy effect either within the system or through the boundary of system exists. Entropy is an extensive property, and if a system consists of several parts, the total entropy is equal to the sum of the entropies of each part. Entropy balance for any system undergoing any irreversible process can be expressed as

Change in the total entropy of the system	=	Total entropy in	-	Total entropy out	+	Total entropy generated

Entropy balance in the rate form is given by

$$\Delta \dot{S}_{\text{system}} = (\dot{S}_{\text{in}} - \dot{S}_{\text{out}}) + \Phi \tag{1}$$

where Δ shows the net change within the system including Φ, which is the total rate of volumetric *entropy generation* due to various processes, and it is not a property of the system. The first term on the right-hand side in Eq. (1) shows the rate of net entropy transfer by heat and mass (Fig. 1).

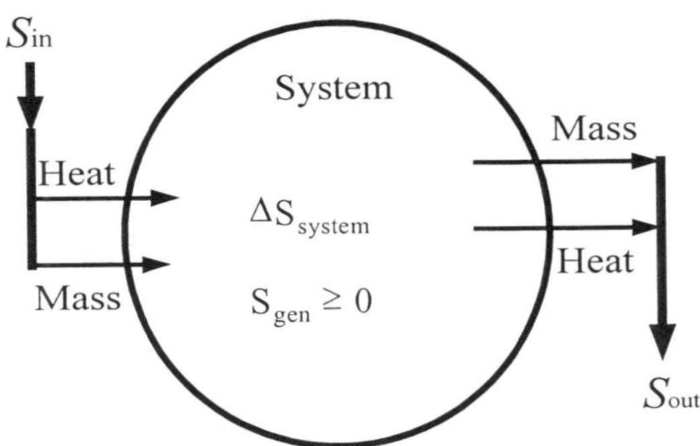

Fig. 1. Mechanism of entropy transfer for general system.

The value of entropy generation cannot be negative, however the changes in entropy of the system may be positive, negative or zero. The entropy of an isolated system during an irreversible process always increases, which is called the *increase of entropy principle*. Entropy change can be determined without detailed information of the process. For a reversible process the entropy generation is zero, and the entropy change of a system is equal to the net entropy transfer. The entropy balance is analogous to energy balance relation. Some concepts of entropy are:

- Processes can follow certain directions and paths; direction of a process must comply with the increase of entropy principle that is the positive entropy generation. This principle might force chemical reactions to undergo without reaching completion.

- Entropy generation is a measure of dissipated useful energy and degradation of the performance of engineering systems, such as transport and rate processes; and the dissipation depends on the extent of irreversibilities present during a process.

- Entropy is a nonconserved property; it is conserved during an ideal reversible process only.

- A reversible adiabatic process is *isentropic*, and a substance will have the same entropy values at inlet and outlet. Engineering systems such as pumps, turbines, nozzles, and diffusers are adiabatic operations, and performance of them will be high when the irreversibilities, such as friction, produced in the process, is reduced, and hence operated under isentropic conditions.

- *Isentropic efficiency* of a turbine η_T at steady state is defined as the ratio of the actual work output W_a of the turbine to the work output of isentropic operation W_s

$$\eta_T = \frac{W_a}{W_s} \tag{2}$$

- Isentropic efficiency of a compressor η_C is the ratio of isentropic work required to the actual work input

$$\eta_c = \frac{W_s}{W_a} \tag{3}$$

- Entropy does not exist in various forms.

Heat and mass flows can transfer entropy. Entropy exchange through the system boundary represents the entropy gained or lost by a system during a process. No entropy is transferred by work. According to the first law of thermodynamics, there is no difference between heat and work. According to the second law, however, an energy exchange accompanied by entropy transfer is the heat transfer, and energy exchange that is not accompanied by entropy transfer is the work. General entropy balance relations for a control volume is given in terms of the rate of entropy change due to the heat transfer, mass flow, and entropy generation Φ

$$\Delta \dot{S}_{cv} = \sum \frac{\dot{Q}}{T} + \sum \dot{m}_{\text{in}} s_{\text{in}} - \sum \dot{m}_{\text{out}} s_{\text{out}} + \Phi \tag{4}$$

For a general steady state flow process, the rate form becomes

$$\Phi = \sum \dot{m}_{\text{out}} s_{\text{out}} - \sum \dot{m}_{\text{in}} s_{\text{in}} - \sum \frac{\dot{Q}}{T} \tag{5}$$

Entropy analysis can quantify the level of energy quality. The definition of conversion efficiency based on the first law only may be misleading. The first law, and generally energy conservation cannot identify losses of work, possible improvements in energy converting processes, and the effective use of resources. One of the examples is the adiabatic throttling process.

Minimization of entropy generation requires thermodynamics, fluid mechanics, heat and mass transfer, kinetics, material properties, constraints, and geometry to establish the relationships between the physical configuration and the entropy generation. Generally, minimization of entropy generation is pursued through the changes in design and operating conditions.

2. EXERGY

Mass and energy are never lost in any physical transformation process; to determine what is lost in resource transformation processes we need to utilize the second law of thermodynamic, which ensures that a part of accessible work potential is always lost in any real process. Accessible work potential is called the *exergy* defining the maximum amount of work that may be performed theoretically by bringing a resource into equilibrium with its surrounding through a reversible process. Therefore exergy is a function of both the physical properties of a resource and its environment; it may be removed from a resource by loss or by transfer to other resources. In all real processes exergy transfer is always accompanied by exergy loss. Although energy transfer generally accompanies exergy transfer, the transfer of exergy may in theory occur with or without the net energy transfer.

Thermodynamic exergy is an extensive property, and corresponds to the amount of mechanical work, which may be reversibly extracted from a system and combines energy and entropy, and hence the first and second laws of thermodynamics. In contrast to energy, exergy is not conserved; it is destroyed in any irreversible processes, and internal energy is lost within the system

Exergy is a thermodynamic potential, and may be described as a general-purpose power (fuel), which can produce changes in physical systems. The concept of 'energy utilizable' (Gouy, 1889), then 'available energy' (Keenan, 1932), and finally 'exergy' are a measure of the power to drive an engine. Exergy is also a link between thermodynamics and information theory. In a broad base exergy is a useful concept not only in engineering but also for a proper resource management and reducing environmental destruction. Exergy expresses simultaneously the quantity and quality of energy; quality is the ability of the energy to be converted into mechanical work under the conditions determined by the natural environment. External exergy loss occurs if the waste product of the process is discharged to the environment due to the deviation of thermal parameters and chemical composition between the product and the components of the environment. The thermal state and chemical composition of the natural environment represent a reference level (dead state) for the calculation of exergy.

In the natural environment, however, there are the components of states differing in their composition or thermal parameters from thermodynamic equilibrium state, and these components of the environment can derive thermal and chemical processes; therefore they should be regarded as natural resources with positive exergy. Only for the commonly appearing components of the environment, a zero value of exergy may be accepted. The correct definition of the reference level is essential for the calculation of external exergy losses. The most probable chemical interaction between the components of the waste product and the environment occurs with the participation of common components of the environment. Therefore the determination of external exergy losses requires the assumption of a reference if common species appearing in the environment for every chemical element.

As the flow processes are common in industry, exergy of the mass flow crossing the system boundary is important. The main components of the exergy are the kinetic exergy, potential exergy, physical exergy and chemical exergy. Kinetic and potential exergies are expressed by the kinetic and potential energy calculated in relation to the environment. Physical exergy results from the deviation of temperature and pressure from the environmental values. Chemical exergy is defined at the environmental temperature and pressure, and results from the deviation of composition in comparison with the commonly appearing components of the environment.

Energy is conserved in every process and cannot be destroyed but can be converted to another form. A certain amount of energy does not directly related to its capacity to cause a change. For example, the same amount of energies may have different capacity to cause a change, because of the varying available energy of the system. The available energy, exergy X, is a measure of a process's maximum capacity to cause a change. The capacity exists because the process is in nonequilibrium state. To accomplish changes some exergy has to be consumed irreversibly. The maximum work output of any process is obtained, if the process is brought into equilibrium with the environment (dead state or reference state) reversibly. The actual work output is much smaller due to the process irreversibility. The work losses in a continuous process can be evaluated if the exergy before and after the process is determined. At the dead state, both the system and surrounding possess energy but no exergy, and hence there is no spontaneous change within the system or the surrounding. Exergy is a unifying concept of many kinds of energy, such as heat, mechanical work and chemical energy.

The exergy expression can be derived from the energy and entropy balances for the composite system shown in Fig. 2, and given by

$$X = (E - U_o) + P_o(V - V_o) - T_o(S - S_o) \tag{6}$$

where E is the total energy ($E = U + KE + PE$), U, V and S denote the internal energy, volume, and entropy of the system, respectively. The terms with indices o are the values of the same properties when the system were at the dead state. The terms KE and PE are the kinetic and potential energies, respectively. Exergy is the maximum available theoretical work that can be performed by a composite system, if the closed system is transformed to the dead state. Some properties of exergy are:

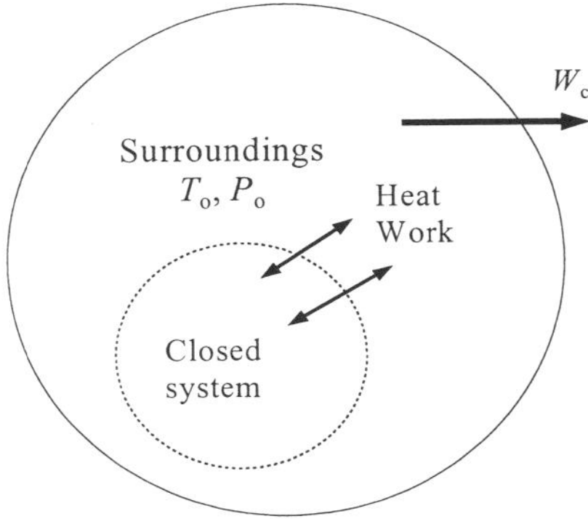

Fig. 2. Combined system.

- Exergy is measured with respect to the environment; therefore it is attributed to the composite system. If the environment is a reference state with $X = 0$, exergy becomes a property of the system.

- If the system is not at the dead state then the system would undergo a spontaneous change.

- The value of exergy loss cannot be negative.

- Exergy is destroyed by irreversibilities in the system. If a system is allowed to undergo a spontaneous change to the dead state without a device to perform work, then exergy is completely destroyed.

- Exergy can also be interpreted as the minimum theoretical work input necessary to change the system from the dead state to the specified sate.

Specific exergy x based on a unit mass is given by

$$x = (u - u_o) + P_o(v - v_o) - T_o(s - s_o) + \frac{v^2}{2} + gz \tag{7}$$

The kinetic ($v^2/2$) and potential energies (gz) are measured relative to the surrounding and contribute fully to the magnitude of exergy. Using Eq. (6), the change in exergy between two states of a closed system is expressed as

$$X_2 - X_1 = (E_2 - E_1) + P_o(V_2 - V_1) - T_o(S_2 - S_1) \tag{8}$$

where p_o and T_o show the pressure and temperature of the surroundings.

2. 1. Exergy balance

The decrease of exergy of a system during a process can be expressed as

Change in the total exergy of the system	=	Total exergy in	-	Total exergy out	-	Total exergy destroyed

Exergy balance consists of internal exergy loss, and represents the results of exergy analysis. Exergy losses may be distributed in the volume; if some irreversible processes occur simultaneously in the same region, then the partition of exergy losses should be determined.

The exergy balance of a closed system (Fig. 3) is obtained from energy and entropy balances between the states 1 and 2

$$E_2 - E_1 = \int_1^2 \delta Q - W \tag{9}$$

$$S_2 - S_1 = \int_1^2 \left(\frac{\delta Q}{T}\right)_b + S_{\text{gen}} \tag{10}$$

where W and Q denote work and heat transferred between the system and its surroundings, respectively, T_b is the temperature on the system boundary, and S_{gen} shows the entropy generation by internal irreversibilities.

For deriving the exergy balance, first we multiply the entropy balance by the temperature T_o and subtract from the energy balance, and we obtain

$$X_2 - X_1 = \int_1^2 \left(1 - \frac{T_o}{T_b}\right) \delta Q - [W - P_o(V_2 - V_1)] - T_o S_{\text{gen}} \tag{11}$$

Eq. (11) is analogous with the entropy balance of the second law. The first term in this expression shows the exergy transfer accompanying heat, the second is the exergy transfer accompanying work, and the third is the destruction of exergy due to irreversibilities within the system. In accordance with the second law, the exergy destruction is positive in an irreversible process and vanishes in a reversible process. The change in exergy of a system can be positive, negative, or zero. When the temperature of the process where heat transfer occurs is less than the temperature of the environment, the transfer of heat and exergy are oppositely directed. Also work and the accompanying exergy transfer can be in the same direction or oppositely directed.

For an isolated system, there is no transfer of exergy between the system and its surrounding, hence the change of exergy is equal to exergy destroyed

$$\Delta X = -T_o S_{\text{gen}} \tag{12}$$

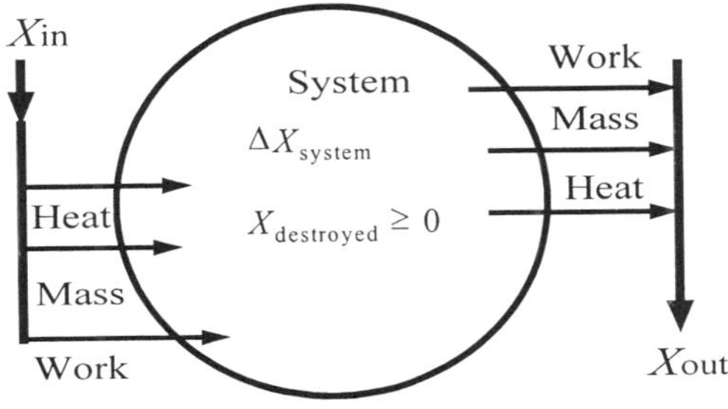

Fig. 3. Mechanism of exergy transfer for a general system.

This equation shows the *decrease of exergy principle*, which states that the exergy of an isolated system during an irreversible process always decreases, and remains constant for a reversible process only. This is in line with the increase of entropy principle, and can be regarded as an alternative statement of the second law.

Exergy balance can also be expressed in exergy rate form

$$\frac{dX}{dt}=\sum_j\left(1-\frac{T_o}{T_j}\right)\dot{Q}_j-\left(\dot{W}-P_o\frac{dV}{dt}\right)-\Psi \tag{13}$$

where Ψ represents the rate of exergy destruction $\Psi = T_o(dS_{gen}/dt)$.

If we consider the exergy of a change from a given reference state (where exergy is zero), the work attainable in a real process would be

$$W = X - T_o\Delta S_{total} \tag{14}$$

If the total entropy change vanishes, as in a reversible process all exergy will transfer to work, hence exergy defines an upper limit to the work that is extractable from any process.

If heat Q is transferred between two reservoirs with temperatures T and T_o, the exergy becomes

$$X = -T_o\Delta S_{\text{total}} = T_o\left|\frac{Q}{T_o}-\frac{Q}{T}\right| = Q\left|1-\frac{T_o}{T}\right| \tag{15}$$

Eq. (15) is a generalization of the Carnot relation. The ratio between the exergy and the heat X/Q is called the *exergy factor*. When $T < T_o$ there is a lack of energy in the system; the value of X/Q strongly increases for low temperatures; when T approaches the absolute zero temperature, 0 K, then X/Q approaches infinite; at higher temperatures X/Q becomes closer to unity. Therefore exergy reflects the quality of energy; heat or cold is more expensive and valuable when it is needed the most. Since the exergy depends on the state of environment, waste heat carries a higher exergy in winter than in summer.

The exergy of sensible heat with temperature T is expressed as

$$X = Q\left(1-\frac{T_o}{T-T_o}\ln\frac{T}{T_o}\right) \tag{16}$$

If a power generation unit discharges 2000 MW of waste heat at $T = 310$ K from a cooling water at a local ambient temperature $T_o = 300$ K, Eq. (16) shows that the waste exergy discharged with the slightly warmed water is approximately 33 MW. This waste energy causes an increase of the local environment temperature, which may gradually change the local ecology.

The exergy of light relates to the exergy power per unit area of black body radiation $\dot{x}$, and given by

$$\dot{x} = \dot{e}\left[1 + \frac{1}{3}\left(\frac{T_o}{T}\right)^4 - \frac{4}{3}\frac{T_o}{T}\right] \tag{17}$$

where $\dot{e}$ is the energy power emission per unit area and equals to σT^4, and $\sigma \approx 5.67\,10^8$ W K^{-4} m^{-2} is the Stefan Boltzmann constant. Since the earth receives sunlight with $T_{\mathrm{sun}} = 6000$ K, and for the environmental temperature $T_{\mathrm{earth}} = 300$ K the exergy factor becomes

$$\frac{\dot{x}}{\dot{e}} = \left[1 + \frac{1}{3}\left(\frac{300}{6000}\right)^4 - \frac{4}{3}\frac{300}{6000}\right] \approx 0.933 \tag{18}$$

The exergy of material substances can be calculated by assuming that the pressure P and the temperature T are constant and equal to ambient conditions of P_o and T_o, and the exergy is expressed by

$$X = \sum_i n_i(\mu_i - \mu_{io}) \tag{19}$$

where μ_i and μ_{io} are the chemical potentials of substance i in its present state and in its environmental state respectively, and n_i is the number of moles. The chemical potential μ_i may be expressed in terms of concentration c

$$\mu_i = \mu_i^o + RT\ln c_i \tag{20}$$

where μ_i^o is the standard state chemical potential. Substitution of Eq. (20) into Eq. (19) yields

$$X = \sum_i n_i(\mu_i^o - \mu_{io}^o) + RT_o\sum_i n_i \ln\frac{c_i}{c_{io}} \tag{21}$$

where μ_{io}^o and c_{io} are the chemical potential and concentrations of component i, respectively at the environmental conditions (dead state). For a single component system, Eq. (21) becomes

$$X = n\left(\mu^o - \mu_o^o + RT_o\ln\frac{c}{c_o}\right) \tag{22}$$

Combustion reactions often cause extensive exergy destruction; exergy calculations show that the entropy generation can cause loss of considerable potential work due to reaction; an electrochemical membrane reactor or a fuel cell could reduce exergy destruction considerably.

For pure components, the chemical exergy consists of the exergy that can be obtained by diffusing the components to their reference concentration c_{io} with a partial pressure of P_{io}, for example for an ideal gas we obtain

$$X = nRT_o \ln \frac{P_i}{P_{io}} \tag{23}$$

where P_i and P_{io} refer to the partial pressure of the gas in the emission and in the environment, respectively. Eq. (23) shows that the exergy use may cause ecological and environmental effects such as the global warming and acid rains due to the carbon dioxide emission. When a substance does not exist in the environment, it must also react with reference substances to reach the equilibrium with the environment. The reaction exergy at reference state equals to the standard Gibbs energy change.

Exergy also appears as the information capacity; the free energy of the information that a system posses is $kT \ln I$, where I is the information we have about the state of the system, and k is the Boltzmann constant. A relation between the exergy and information is expressed by

$$X = k \ln 2 T_o I \tag{24}$$

Transformation of information from one system to another is often almost entropy-free energy transfer, and the information capacity I in binary units are expressed as a function of the probability P

$$I = \frac{1}{\ln 2}\left(\sum_{j=1}^{\Omega} P_j \ln P_j - \sum_{j=1}^{\Omega} P_j^o \ln P_j^o \right) \tag{25}$$

where Ω is the number of possibilities, P_o is the probability at equilibrium, i.e., no knowledge, and P is the probability when we have some information of the system. Information here is used as a measure of order or structure, it is as fundamental as energy and matter. By small amount of exergy as information, processes converting large amounts of energy and matter can be controlled. The exergy carried as information is usually a very small part of the total exergy of any system, and it should be treated as a structural exergy. This becomes even more obvious in living systems.

The green plants convert exergy from sunlight into chemical exergy, via photosynthesis producing biomass, which then passes through different food chains in the ecosystems. At every tropic level exergy is consumed, and the decomposition in organisms is on the very lowest level in this food chain. The exergy that cannot be utilized by these organisms form peat or sediment that gradually becomes stocks of oil or coal deposits. The toxic matter accumulated over the years in these organisms and plants is also removed from the biosphere. Stocks in the living and dead organic matter on the earth represent different forms of stored exergy, which may build up an extensive exergy potential, such as a clean environment and air in the atmosphere, which is of crucial importance for the evolution of life into further diversity and complexity; this leads to a natural evolution, which is self creating and sustainable.

The exergy use in the society originates to about 60% from fossil fuels, the rest is composed of mainly wood, firewood, food, hydropower and nuclear radioactive deposits; the contributions to the exergy use from solar, wind and wave power is rather insignificant, so that the society's exergy use is not sustainable. As long as the level of limited amount of exergy stocks on the earth is kept stable, i.e., the output does not exceed the input, then we have a sustainable situation.

The superiority of biological systems relies on the difference in information transfer techniques between biological and physical systems. Information must be stored and transported safely; in biological systems the transfer is supplemented with a continuous debugging or control of the information, because information is operated by living systems. Biological systems operate on a microscopic level, i.e., with specific molecular structures and the unique positions of single atoms of the DNA (deoxyribonucleic acid) molecule that makes system far more exergy efficient than technological systems.

On the other hand, technological information systems work with macroscopic structures such as printing, orientation of magnetic particles on a media, or transistors in an electrical circuit. The information transfer efficiency of the most advanced microprocessor is still several times poorer than that of bacteria.

Living systems survive and evolve by transforming solar exergy into complex, highly ordered structures that are directed and controlled by the information of the genes in a suitable living conditions. The information to the genes is transferred from generation to generation by the DNA replication.

2.2. Flow exergy

In an open system, *flow exergy* accounts exergy transfer due to mass flow and flow work, and the *specific flow exergy* x_f is expressed as

$$x_f = (h - h_o) - T_o(s - s_o) + \frac{v^2}{2} + gz \qquad (26)$$

where h and s represent the specific enthalpy and entropy respectively, at the inlet or outlet; h_o and s_o represent the respective values at the dead state. The flow exergy and enthalpy have the similar interpretations; each term is associated with the mass flow and flow work. The flow work rate is expressed as $\dot{m}(Pv)$, where $\dot{m}$ is the mass flow rate, P is the pressure and v is the specific volume at the inlet or exit.

The exergy rate balance for a control volume is given by

$$\frac{dX_{cv}}{dt} = \sum_j \left(1 - \frac{T_o}{T_j}\right)\dot{Q}_j - \left(\dot{W}_{cv} - P_o \frac{dV_{cv}}{dt}\right) + \sum_i \dot{m}x_{fi} - \sum_e \dot{m}x_{fe} - \Psi \tag{27}$$

where the first three terms on the right hand side represent the rate of exergy transfer, and the last term is the rate of exergy destruction. The term $\dot{Q}_j$ shows the heat transfer rate through the boundary where the instantaneous temperature is T_j, while the term $\dot{W}_{cv}$ shows the energy transfer rate by work other than flow work. The terms $\dot{m}x_{fi}$ and $\dot{m}x_{fe}$ denote the exergy transfer rates accompanying mass flow and flow work at the inlet i and exit e, respectively.

For a control volume at steady state, exergy rate balance becomes

$$0 = \sum_j \left(1 - \frac{T_o}{T_j}\right)\dot{Q}_j - \left(\dot{W}_{cv}\right) + \sum_i \dot{m}x_{fi} - \sum_e \dot{m}x_{fe} - \Psi \tag{28}$$

This equation shows that the rate of exergy transferred into the control volume must exceed the rate of exergy transferred out, and the difference is the exergy destroyed due to irreversibilities. Exergy concepts for some steady state process are:

- Energy is conserved in the throttling valve, while exergy is destroyed because of the expansion of fluid.
- Thermodynamic inefficiency is a concern for heat exchangers: Exergy is destroyed by irreversibilities associated with pressure drop due to fluid friction and stream-to-stream heat transfer due to the temperature differences between the streams.
- In a steam turbine of power plant, exergy transfers are due to work and heat from the control volume and due to exergy destruction within the control volume.
- In a waste heat recovery system, we might reduce the heat transfer irreversibility by designing a heat-recovery steam generator with a smaller stream-to-stream temperature difference, and/or reduce friction by designing

a turbine with a higher isentropic efficiency. However, thermodynamic performance alone would not determine the best option, because other factors for example, cost must also be considered.

- A cost-effective design could be obtained from the trade-off between any resulting reduction of exergy destruction and potential increase in operating cost.

2.3. Exergetic (second law) efficiency

Energy supplied by the heat transfer $\dot{Q}_{\text{in}}$ is either utilized $\dot{Q}_u$ or lost to the surroundings $\dot{Q}_l$, and an efficiency η is defined as

$$\eta = \frac{\dot{Q}_u}{\dot{Q}_{\text{in}}} \tag{29}$$

Exergetic efficiency ε is (exergy recovered)/(exergy supplied), and may be expressed as

$$\varepsilon = \eta \left(\frac{1 - T_o / T_u}{1 - T_o / T_{\text{in}}} \right) \tag{30}$$

Generally, the value of exergetic efficiency is less than unity even when $\eta = 1$. Eq. (30) shows that the temperatures T_{in} and T_u are also important beside the value of η, because exergy utilization would increase as the temperature at utilization of energy approaches the temperature of inlet energy.

The rate of exergy loss accompanying the heat loss $\dot{Q}_l$ is given as $(1 - T_o / T_l)\dot{Q}_l$, which is a measure of thermodynamic value of the heat loss, and depends significantly on the operating temperature.

Exergetic efficiency expressions can take different forms, some of those for various engineering steady-state processes are:

- Turbine with adiabatic operation

$$\varepsilon = \frac{\dot{W}_t}{\dot{m}(x_{f,\text{in}} - x_{f,\text{out}})} \tag{31}$$

where $\dot{W}_t$ shows the work produced by the turbine.

- Compressor or pump with work input $\dot{W}$ and adiabatic operating conditions

$$\varepsilon = \frac{\dot{m}(x_{f,\text{out}} - x_{f,\text{in}})}{(-\dot{W})} \qquad (32)$$

- Heat exchanger without mixing at adiabatic conditions with both streams at temperatures above T_o

$$\varepsilon = \frac{\dot{m}_c(x_{f,\text{out}} - x_{f,\text{in}})_c}{\dot{m}_h(x_{f,\text{in}} - x_{f,\text{out}})_h} \qquad (33)$$

where $\dot{m}_c$ and $\dot{m}_h$ represent the mass flow rates of cold and hot streams, respectively.

- Adiabatic mixer with a hot stream 1 and cold stream 2 entering, and stream 3 leaving the system

$$\varepsilon = \frac{\dot{m}_2(x_{f,3} - x_{f,2})}{\dot{m}_1(x_{f,1} - x_{f,3})} \qquad (34)$$

- For an adiabatic chemical reaction at constant pressure, the enthalpy remains constant. The loss in exergy is given by the exergy of reactants x_1 and exergy of the reaction products x_2

$$W_l = x_1 - x_2 = T_o(s_1 - s_2) \qquad (35)$$

and

$$\varepsilon = 1 - \frac{W_l}{x_1} = 1 - \frac{T_o(s_1 - s_2)}{x_1} \qquad (36)$$

For a combustion reaction undergoing in a well-insulated chamber with no work produced, exergetic efficiency can be written as

$$\varepsilon = 1 - \frac{\Psi}{X_F} \qquad (37)$$

where X_F is the rate of exergy entering with the fuel, and Ψ is the exergy destruction.

The process improvement for a typical hydrogen production unit from methane reforming by combining the pinch analysis and the thermodynamic analysis using the exergy and load distribution method has been reported [12]. Also a simple relationship between the exergy efficiencies of the individual unit operations in a process and its overall efficiency of ammonia plant has been reported [13]. This relationship is based on the concepts of primary and transformed exergies; both of these are consumed within the system but primary exergy enters the system through its boundaries, transformed exergy is produced within its boundaries. The primary exergy load $X_{p,i}$ is the fraction of the total primary exergy. The transformed exergy load $X_{t,i}$ is the ratio of the transformed exergy to the total primary exergy. The following relationship between the individual efficiencies η_i, and the overall efficiency η is given by

$$\eta = \sum_i [X_{p,i}\eta_i - X_{t,i}(1-\eta_i)] \tag{38}$$

The primary exergy loads have the constraint

$$\sum_i X_{p,i} = 1 \tag{39}$$

Eq. (38) shows that by increasing a local efficiency η_i or decreasing a transformed exergy load $X_{t,i}$, the overall efficiency η of a process may increase as long as this does not cause any opposite and larger effect through a change in other parameters in the relationship. An effective way of improving the overall exergy efficiency of a process is to increase the primary load of the units with the largest efficiencies at the expense of those with the lowest efficiencies. The exergy efficiency is defined according to the second law

$$\eta = \frac{X_{\text{out}}}{X_{\text{in}}} \tag{40}$$

However the formalism of Eq. (38) is independent of the way in which the efficiency is expressed. The second law does not always represent adequately the thermodynamic performance of an operation; the intrinsic efficiency, for example takes into account the transiting exergy X_{tr}, and is defined by

$$\eta_{in} = \frac{X_{\text{out}} - X_{tr}}{X_{\text{in}} - X_{tr}} = \frac{X_p}{X_c} \tag{41}$$

The transiting exergy is the part of the exergy entering a unit operation; it traverses without undergoing any transformation; therefore, it is not consumed by the operation. The terms X_c and X_p are the exergies actually consumed and produced respectively within the operation. In Eq. (38) the intrinsic efficiency is used.

A heat exchanger network can be treated as a single heat transfer operation, which can be characterized by its minimum overall heat supply and withdrawal requirements. Those minimum requirements can be determined by pinch analysis. In a heat exchanger network system, as seen in Fig. 4, to apply the exergy load distribution, the intrinsic exergy efficiency η_{in}, primary exergy X_p, and transformed exergy load X_{tr} are computed and used in Eq. (38). The exergy flow corresponding to the heat supplied to the process X_{sq} and the heat withdrawn X_{wq} are expressed as

$$X_{sq} = Q_{H,\min}(1 - T_o / T_H) \tag{42}$$

$$X_{wq} = Q_{C,\min}(1 - T_o / T_C) \tag{43}$$

The terms $Q_{H,\min}$ and $Q_{C,\min}$ are the energy load targets, T_H and T_C are the temperature of hot and cold utilities, and T_o is the temperature of the environment. Assuming that $T_c = T_o$, the term X_{wq} becomes zero and the intrinsic efficiency is expressed as

$$\eta_{in} = \frac{\sum_i X_{w,ci} - X_{s,ci}}{\sum_j (X_{s,hj} - X_{w,hj}) + X_{sq}} \tag{44}$$

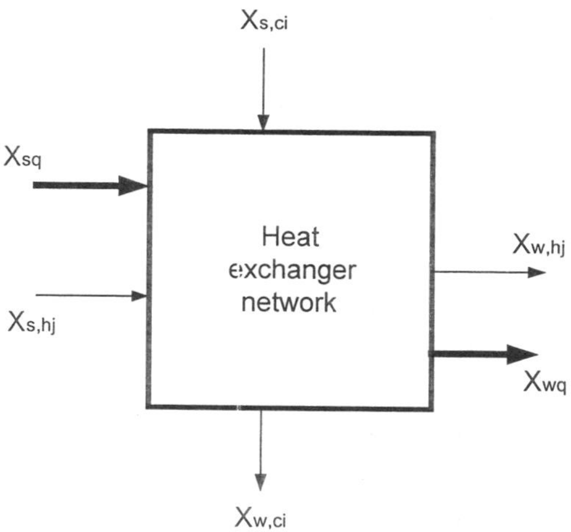

Fig. 4. Heat exchanger network representation as a single operation.

The primary X_p and transformed exergy X_t loads are

$$X_p = \frac{X_{sq}}{X_c} \tag{45}$$

$$X_t = \frac{\sum_j X_{s,hj} - X_{w,hj}}{X_c} \tag{46}$$

The term X_c is the exergy consumed by the overall processes. The pinch analysis can be combined with exergy analysis to reduce the number of unit operations for which the thermodynamic parameters must be computed, and hence the computing work can be reduced to a minimum. Using the combined method, an actual process close to optimization could be further improved with or without design modification or equipment changes.

2.4. Chemical exergy

The chemical exergy is the maximum theoretical work that can be obtained from a combined system of a combustion cell and its surrounding (Fig. 2). Fuel enters the combustion cell operating at steady state at temperature T_o and pressure P_o; oxygen enters from the environment at T_o and partial pressure $y_{O_2} P_o$, where y_{O_2} is the mole fraction of oxygen; the fuel and oxygen undergoes combustion reaction completely and produce water vapor and carbon dioxide. The reaction for a hydrocarbon is given by

$$C_aH_b + (a + b/4)O_2 = aCO_2 + (b/2)H_2O \tag{47}$$

For a steady-state operation, the energy rate balance over the control volume of the cell is expressed, ignoring kinetic and potential energy effects, as

$$\dot{W}_c = \dot{N}_F[\dot{Q}_c + h_F + (a + b/4)h_{O_2} - ah_{CO_2} - (b/2)h_{H_2O}] \tag{48}$$

where $\dot{N}_F$ is the molar flow rate of the fuel, and h_F is the molar enthalpy of the fuel. Eq. (48) represents the work produced by the combined system of combustion cell plus environment, and heat is transferred to environment. Eq. (48) can also be expressed in terms of the enthalpies of products and reactants h_P and h_R, respectively

$$\dot{Q}_c - \dot{W}_c = \dot{N}_F(h_P - h_R) \tag{49}$$

When enthalpy of formation data are lacking, the enthalpy of combustion, which is known experimentally, can be used, since some fuels are normally composed of several chemical compounds. The heating value of a fuel is equal to the enthalpy of combustion; lower heating value is obtained when all the water formed is a vapor. The entropy rate balance for the combustion cell can also expressed as

$$0 = \sum_j \frac{\dot{W}_j}{T_j} + \dot{N}_F[s_F + (a+b/4)s_{O_2} - as_{CO_2} - (b/2)s_{H_2O} + \Phi] \tag{50}$$

The entropies of the mixture components can be calculated using the appropriate partial pressures

$$s_i(T,P_i) = s_i^o(T) - R\ln\frac{P_i}{P_{\text{ref}}} \tag{51}$$

where $s_i^o(T)$ is the absolute entropy of component i at temperature T and P_{ref}. Eliminating the heat transfer rate between Eqs. (48) and (50), we have

$$\begin{aligned}\dot{W}_c = {} & \dot{N}_F[h_F + (a+b/4)h_{O_2} - ah_{CO_2} - (b/2)h_{H_2O}] \\ & -T_o\dot{N}_F[s_F + (a+b/4)s_{O_2} - as_{CO_2} - (b/2)s_{H_2O} + \Phi]\end{aligned} \tag{52}$$

The specific enthalpies can be determined from the temperature T_o, and the specific entropies can be calculated by the temperature, pressure and composition of the environment. Once the environment is specified, all enthalpy and entropy terms are fully defined regardless of the process within the control volume. The term $T_o\Phi$ depends on the nature of the process and the irreversibility. The maximum theoretical work is obtained when there is no irreversibility, and hence the following chemical exergy $\dot{X}_{ch}$ is obtained

$$\begin{aligned}\dot{X}_{ch} = {} & \dot{N}_F[h_F + (a+b/4)h_{O_2} - ah_{CO_2} - (b/2)h_{H_2O}] \\ & -T_o[s_F + (a+b/4)s_{O_2} - as_{CO_2} - (b/2)s_{H_2O}]\end{aligned} \tag{53}$$

This equation can also be expressed in terms of Gibbs function $g = h - Ts$ of respective substances

$$\begin{aligned}\dot{X}_{ch} = {} & \dot{N}_F[g_F + (a+b/4)g_{O_2} - gh_{CO_2} - (b/2)g_{H_2O(g)}] \\ & + RT\ln\left[\frac{(y_{O_2})^{a+b/4}}{(y_{CO_2})^a (y_{H_2O})^{b/2}}\right]\end{aligned} \tag{54}$$

The specific Gibbs functions are evaluated at the temperature T_o and pressure P_o of the environment, and are given by

$$g(T_o, P_o) = g_f^o + [h(T_o, P_o) - h(T_{\text{ref,}} - P_{\text{ref}})] \\ -[Ts(T_o, P_o) - T_{\text{ref}} s(T_{\text{ref,}} - P_{\text{ref}})] \qquad (55)$$

where g_f^o is the Gibbs function of formation at specified conditions.

Standard chemical exergy values, in units of kJ/kmol, are based on a standard exergy reference environment with T_o and P_o such as 298.15 K (536.67 R) and 1 atm, respectively, and consisting of a set of reference substances with standard concentrations of gaseous, liquid and solid components. The standard chemical exergy tables often simplify the applications of exergy principles.

2.5. Depletion number

Exergetic resource depletion may cause the environmental change. By reducing resource depletion, on-going environmental transformation due to industrial systems will be reduced. Resource depletion may be slowed down by reducing the consumption and quantified by the *Depletion number Dp*, that is a nondimensional indicator of resource depletion X_{Dp} per unit consumption X_C.

$$Dp = \frac{\dot{X}_{Dp}}{\dot{X}_C} \qquad (56)$$

Biological systems have evolved to allow a sustainable consumption to occur with little or no depletion. The depletion number therefore provides a measure of system progress or maturity, and is a useful basis for studying the evolution of industrial resource use patterns and the implementation of resource conservation strategies.

The depletion number may be expressed as a function of three indicators showing the level of implementation of resource conservation strategies:

(i) Exergy cycling fraction ψ, which is a measure of recycling that accounts for both the throughput and quality change aspects of resource consumption and upgrading.

(ii) Exergy efficiency η, which is a universal measure of process efficiency that accounts for the first and second laws principles.

(iii) Renewable exergy fraction Ω, which is a measure of the extent to which resources supplied to an industrial system are derived from renewable sources. The industrial systems consume resources by means of supporting processes associated with the supply and the removal of resources. Therefore the temporal and spatial boundary conditions are important in defining the universal relations among renewable and nonrenewable sources conservation strategies. The boundary conditions will determine which resources and processes constitute an industrial system. The spatial boundary conditions are mainly geographical and resource-specific, while the temporal boundary conditions define the scope of time for the exergy transfer and loss in processes.

Definition of depletion number (Eq. 44) may be expressed by

$$Dp = \frac{\dot{X}_{Dp}}{\dot{X}_C} = 1 + \frac{\dot{X}_{Dsl}}{\dot{X}_C} + \psi\left[\frac{(1-\Omega_{RU})}{\eta_{RU}} - 1\right] + \frac{\dot{X}_{TV}}{\dot{X}_C}\left[\frac{(1-\Omega_{VU})}{\eta_{VU}} - 1\right] \tag{57}$$

where $\dot{X}_{Dsl}$ is the exergy dissipation rate, Ω_{RU} and η_{RU} are the renewable exergy fraction and transfer efficiency for recovered resource upgrade process, respectively, $\dot{X}_{TV}$ is the exergy transfer rate to nonrenewable source, and Ω_{VU} and η_{VU} are the renewed exergy fraction and transfer efficiency, respectively for nonrenewable resource upgrade process. Two structural constants $\alpha_{Ds\text{-}C}$ and $\alpha_{V\text{-}C}$ are defined as

$$\alpha_{Ds-C} = \frac{\dot{X}_{Dsl}}{\dot{X}_C(1-\psi)} \tag{58}$$

$$\alpha_{V-C} = \frac{\dot{X}_{TV}}{\dot{X}_C(1-\psi)} \tag{59}$$

With these definitions Eq. (57) becomes

$$Dp = 1 + \psi\left[\frac{(1-\Omega_{RU})}{\eta_{RU}} - 1\right] + (1-\psi)\left\{\alpha_{Ds-C} + \alpha_{V-C}\left[\frac{(1-\Omega_{VU})}{\eta_{VU}} - 1\right]\right\} \tag{60}$$

Eq. (60) expresses the depletion number as a function of a system's structural constants, the exergy efficiency and renewed exergy fraction of the individual resource upgrade processes, and the extent of resource cycling.

The approach is based on the thermodynamic principles; generalized depletion number may be developed by including numerous consumption processes, such as incomplete cycling or partial upgrading, and the direct reuse of resources without upgrade.

Recycling may reduce the need for resource and the exergy requirements of manufacturing processes. The depletion number may be used to determine the conditions to reduce resource depletion. Generally increasing resource cycling reduces depletion due to less exergy transfer from other sources. For example producing aluminum from bauxite requires 27400 MJ/ton of exergy transfer, while converting the recycled aluminum to feedstock aluminum requires far less exergy transfer [12].

REFERENCES

[1] A. Bejan, Entropy Generation Through Heat and Fluid Flow, Wiley, New York, 1982.

[2] R.B. Bird, W.E. Stewart and E.N. Lightfoot, Transport phenomena, 2nd ed., Wiley, New York, 2002.

[3] A. Bejan, Advanced Engineering Thermodynamics, Wiley, New York, 1988.

[4] A. Bejan, Entropy Generation Minimization, CRC Press, Boca Raton, 1996a.

[5] M.J. Moran and H.N. Shapiro, Fundamentals of Engineering Thermodynamics, 4th ed., Wiley, New York, 2000.

[6] Y.A. Cengel and M.A. Boles, Thermodynamics; An Engineering Approach, 4th ed., McGraw-Hill, New York, 2002.

[7] M.A. Rosen, I. Dincer, Exergy Int. J., 1 (2001) 3.

[8] G. Wall, M. Gong, Exergy Int. J., 1 (2001) 128.

[9] M. Gong, G. Wall, Exergy Int. J., 1 (2001) 217..

[10] A. Bejan, Exergy Int. J., 1 (2001) 269.

[11] J. Szargut, In Finite-Time Thermodynamics and Thermoeconomics, Eds. S. Sieniutcyz, P. Salamon, Taylor & Francis, New York, 1990.

[12] L. Connely and C.P. Koshland, Exergy Int. J., 1 (2001) 234.

[13] M. Sorin and J. Paris, Computers Chem. Engng, 21, Suppl. (1997) 23.

[14] M. Sorin and V.M. Brodyansky, Energy, 11 (1992) 1019.

Chapter 6

Using the second law of thermodynamics

INTRODUCTION

The aspects of first law of thermodynamics are accepted and widely used, however, the second law of thermodynamics is comparatively less appreciated and thought to be of a theoretical concept rather than a practical engineering tool. This causes the limited use of thermodynamics in design of engineering processes. The first law of thermodynamics mainly deals with energy, a conserved property, regardless the quality of it, while the second law is mainly related to the quality of energy, and entropy and exergy, which are nonconserved properties. The first-law efficiency or thermal efficiency is defined as the work output to total heat input, and it cannot describe the best performance of the process. On the other hand, the second-law efficiency relates the actual performance to the best possible performance under the same conditions.

In thermodynamics, reversible work for a process is defined as the maximum useful work output. The difference between the reversible work and actual work is due to irreversibility, which causes the wasted work (energy). For example, the head loss in a pipe flow represents the conversion of mechanical energy to unwanted increase in internal energy and the loss of energy via heat transfer. For a frictionless flow only the Bernoulli equation would predict no energy loss. If the adapted operating conditions cause excessive entropy generation, they are not compatible for an existing design, and the system will not be capable of delivering the maximum useful output. Using the second law in engineering applications, one can identify the major sources of irreversibility and minimize or rearrange them in order to maximize the performance of the process.

From the combination of the first and second laws of thermodynamics and the general balance equations, the linear nonequilibrium thermodynamics approach has been developed; it is mainly based on the following four postulates: (i) The systems are not far away from equilibrium; the gradients or the thermodynamic forces are not too large; this is called quasi-equilibrium postulate. Within the system local thermodynamic equilibrium holds. (ii) All flows in the system are expressed as a linear function of all the forces involved; the proportionality constants in these equations are called the phenomenological coefficients. (iii) The matrix of the phenomenological coefficients is symmetric provided that the conjugate flows and forces identified by a suitable dissipation function or an entropy generation equation. (iv) In an anisotropic system no coupling of flows and forces occurs; this is called the Curie-Prigogine principle, which states that if the tensorial order of the flows and forces differs by an odd number the coupling is not allowed.

To analyze the transport and rate processes with the linear nonequilibrium thermodynamics approach, conjugate flows and forces are identified by the dissipation function or the volumetric rate of entropy generation obtained from the general balance equations, including the entropy balance, and the Gibbs relation; the conjugate forces and flows are related linearly through the phenomenological coefficients, which obey the Onsager reciprocal relations; these coefficients are closely related to the conventional transport and rate coefficients. As a direct consequence of the Onsager relations, the number of unknown coefficients is reduced. The phenomenological coefficients are also highly instrumental in defining the coupled phenomena. For example the coupled processes of heat and mass transport give rise to the Soret effect that is the mass diffusion due to heat transfer, and the Dufour effect that is the heat transport due to mass diffusion. The field of linear nonequilibrium thermodynamics yields a new insight into the transport and rate processes, and the coupled processes in physical, chemical, electrochemical, and biological systems. For example we can identify the cross coefficients of the coupling between the mass diffusion (vectorial process) and chemical reaction (scalar process) in an anisotropic membrane wall; therefore the linear nonequilibrium thermodynamics theory has a unifying approach to various processes usually studied under separate disciplines.

There exist a large number of 'phenomenological laws', such as Fick's law between flow of a substance and its concentration gradient, and the mass action law between reaction rate and chemical concentrations or affinities. When two or more of these phenomena occur simultaneously in a system, they may couple and cause new effects, such as facilitated and active transport in biological systems. In active transport a substrate can be transported against the direction imposed by its thermodynamic force. If this coupling does not take place, such "uphill" transport would be in violation of the second law of thermodynamics; hence, dissipation due to either diffusion or chemical reaction can be negative, only if these two processes couple and produce a positive total dissipation.

1. SECOND LAW ANALYSIS

The Gouy-Stodola theorem states that the lost available energy (work) is directly proportional to the entropy generation in a nonequilibrium phenomenon. Transport phenomena and chemical reactions are nonequilibrium phenomena and irreversible processes. The second law of thermodynamics is applicable to all physical, chemical, biological processes, as well as to heat and work conversions. In general energy crosses system or process boundaries in the forms of work and heat. Thermal energy is transferred due to a temperature driving force, which can never be totally converted to work. The second law can be used to quantify the thermodynamic equivalence of heat to work through exergy and availability; it is often directional; it can give specific insights to the design of a process, the operating conditions, or the improvement of an existing technology.

With the help of the linear nonequilibrium thermodynamics theory it may be possible to design thermodynamically optimum processes. This trend is called the second law analysis, in which dissipation loss, or exergy destruction is calculated from the rate of entropy generation or from the exergy balance in the system. The entropy generation approach is especially important in terms of the process optimality as the each process contributing to the entropy generation can be identified and evaluated separately. The map of the volumetric entropy generation rate identifies the regions within the system where excessive entropy generation occurs due to the irreversible processes. Through the minimization of the excessive irreversibilities by the modified operating conditions or design parameters, a thermodynamic optimum can be achieved for a required task. The information on the trade-offs between the various contributions to the rate of entropy generation may be helpful for thermodynamically optimum design criteria of the system.

Availability is $A = H - T_oS$, is a measure of departure from the ambient or dead state. As shown in Fig. 1, in a heat exchanger, the temperature of the hot stream decreases, and its availability goes down, while the temperature of a cold stream increases, and availability increases. So that the availability is transferred from the hot stream to the cold stream, and some of it expended to allow the heat transfer processes to occur in a finite time and cost in a heat exchanger.

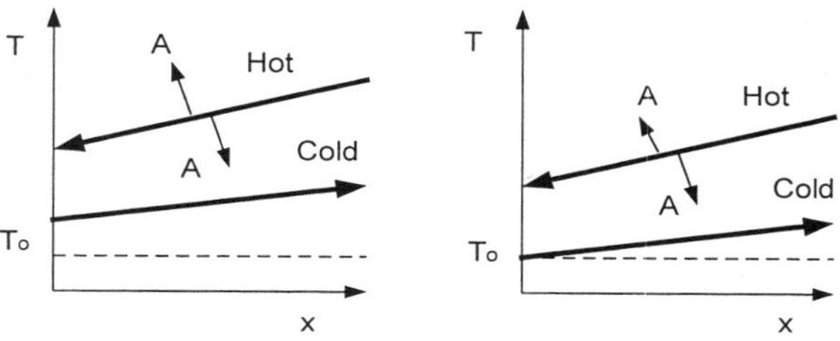

Fig. 1. Heat transfer above and at the ambient temperature T_o.

1.1 Optimization problem

Mainly the method of modeling, analysis and optimization in engineering begins with deciding on the system geometry, architecture, components, and manner in which the components are connected. Engineering analysis involves the mathematical description of the assumed system and its performance. Finally, the optimization is the simulation of system under various conditions, which are most favorable for maximum performance, e.g., minimum entropy generation, or minimum cost.

The search for an optimal design may be considerably more challenging, since one may contemplate a very large number of geometrical aspects, and boundary and initial conditions in a fluid flow network. In practice, one may assume a number of alternative configurations, optimize their performance, and compare the optimized alternatives. Selecting the best among them needs a strategy to lead to more suitable configurations with least irreversibility or cost.

In complex designs, the designer has to deal with the increasing number of degrees of freedom. The thermodynamic optima of individual processes may be robust and useful shortcuts in the optimization of larger and more complex systems. This approach is important in engineering systems, and in physical and biological systems with well-adapted extensions and modifications in a continuous manner leading to the better and more complex designs.

Three optimization procedures can be discussed in a general design problem: (i) combination and permutation; (ii) mathematical modeling; and (iii) second law approach. These should be considered closely within the frame of definition of optimization. Engineering design is usually associated with the exact economic optimum leading to a global minimum. On the other hand, only simple problems, such as determining the minimum value of a parabolic equation, have single optimum solutions. For complex engineering problems either there is no a single optimum solution or it is difficult to find a global optimum. The complex problems may involve various processes coupled to each other with various sources of irreversibilities. The level of irreversibilities introduces thermodynamics imperfections eventually decreasing the overall performance of the system. Some of the optimization procedures are:

(i) In a heat exchanger network system, one can calculate the total annual cost for a possible combination of heat exchangers that can accomplish the required heating and cooling loads. The approach of minimum total annual cost approach does not take into account the significant differences between solutions that have similar total annual costs. For example, we may find significant differences on control, operability, safety, and environmental impact in a complex network. The particular combination with the minimum cost can then be identified as the global minimum. Even for a simple network problem with n_h of hot streams and n_c of cold streams, the number of evaluations required for the global economic optimum would be $(n_h \times n_c)!$, and calculation of every possible design combination and permutation may not be the best way to optimize a design.

(ii) The mathematical modeling can improve our understanding of a thermal process, and is the key to a good process design. However mathematical modeling usually deals with the optimization of the design parameters for a specified process. For example an ethanol-water mixture can be separated by distillation, or by a semi-permeable membranes, and an appropriate mathematical model to optimize the distillation may be the wrong choice because one of the other separation methods would be more suitable. Improvements in a design are often a result of changes in the process, and mathematical modeling does not usually address such changes. Engineer should be able to make the final design decision, after careful consideration of the results of mathematical modeling.

(iii) One important tool to evaluate the processes is to use the second law approach, which guides to improve a process design. For example, suppose that a high-pressure steam has been unnecessarily throttled before being used in a heat exchanger. This can be identified as an error, because availability in the steam is destroyed without any gain. So that elimination of the throttle in the operation will improve the design, and the operating cost of the heat exchanger will be reduced without any increase in the use of fuel. For example, nitric acid is produced from the oxidation of ammonia, which is an exothermic reaction; hence it is possible to utilize or export power emitted from the process; this yields a reduction in the capital cost. Some second law guidelines are [36]:

- Excessively large or small thermodynamic driving forces in process operations should be avoided; minimize the mixing of streams with large differences in temperature, pressure or chemical composition.
- Do not discharge heat at high temperature to the environment.
- Do not heat refrigerated streams with hot streams.
- When choosing streams for heat exchange, try to match streams where the final temperature of one is close to the initial temperature of the other.
- Extremely large or small amount of flows may not be easy to manage economically.
- When exchanging heat between two streams, the exchange is more efficient if the flow heat capacities of the streams are similar, otherwise consider splitting the stream with the larger flow heat capacity.
- Hot or cold sources with temperatures far from the ambient are valuable.
- Minimize the throttling of fluid flow, steam, or other gases.
- Use the exergy balance or exergy destruction calculations to evaluate the utilization of energy and as a guide for process modifications.
- These guidelines can be used as the strategies to design and optimize the processes, such as power plants, heat exchangers, and other thermal

systems. This may require suitable trade-offs between the use of energy and capital by identifying and eliminating the design parameters and the operating conditions that cause excessive entropy generation.

In next section, the second law analysis is briefly presented in the following areas: heat and fluid flow, heat and mass transfer, chemical reactions and reacting flows, and separation. Related references are listed at the end of the chapter.

2. HEAT AND FLUID FLOW

The second law analysis has been extensively utilized in the field of heat and fluid flow. In an early study, Bejan [1-3] presented the basic approach, methodology and applications of the second law analysis in the thermal engineering field.

Local rate of entropy generation per unit volume Φ in a convective heat transfer is given in a two dimensional Cartesian coordinates as

$$\begin{aligned}\Phi dxdy = &\frac{q_x + \frac{\partial q_x}{\partial x}dx}{T + \frac{\partial T}{\partial x}dx}dy + \frac{q_y + \frac{\partial q_y}{\partial y}dy}{T + \frac{\partial T}{\partial y}dy}dx - \frac{q_x}{T}dy - \frac{q_y}{T}dx \\ &+ \left(s + \frac{\partial s}{\partial x}dx\right)\left(v_x + \frac{\partial v_x}{\partial x}dx\right)\left(\rho + \frac{\partial \rho}{\partial x}dx\right)dy \\ &+ \left(s + \frac{\partial s}{\partial y}dy\right)\left(v_y + \frac{\partial v_y}{\partial y}dy\right)\left(\rho + \frac{\partial \rho}{\partial y}dy\right)dx \\ &- sv_x\rho\, dy - sv_y\rho\, dx + \frac{\partial(\rho s)}{\partial t}dxdy\end{aligned} \tag{1}$$

The first four terms on the right-hand side of Eq. (1) account for the entropy transfer due to heat transfer, the next four terms represent the entropy convected into and out of the system, and the last term represents the rate of entropy accumulation in the control volume. Dividing Eq. (1) by *dx dy*, the local rate of entropy generation becomes

$$\begin{aligned}\Phi = &\frac{1}{T}\left(\frac{\partial q_x}{\partial x} + \frac{\partial q_y}{\partial y}\right) - \frac{1}{T^2}\left(q_x\frac{\partial T}{\partial x} + q_y\frac{\partial T}{\partial y}\right) + \rho\left(\frac{\partial s}{\partial t} + v_x\frac{\partial s}{\partial x} + v_y\frac{\partial s}{\partial y}\right) \\ &+ \mathrm{s}\left[\frac{\partial \rho}{\partial \mathrm{t}} + v_x\frac{\partial \rho}{\partial x} + v_y\frac{\partial \rho}{\partial y} + \rho\left(\frac{\partial v_x}{\partial x} + \frac{\partial v_y}{\partial y}\right)\right]\end{aligned} \tag{2}$$

The last term on the right-hand side of this equation vanishes based on the mass conservation principle

$$\frac{D\rho}{Dt} + \rho \nabla \cdot \mathbf{v} = 0 \tag{3}$$

where *D/Dt* is the substantial derivative. Therefore in vectorial notation the volumetric rate of entropy generation can be expressed as

$$\Phi = \frac{1}{T}\nabla \cdot \mathbf{q} - \frac{1}{T^2}\mathbf{q} \cdot \nabla T + \rho \frac{ds}{dt} \tag{4}$$

From the canonical relation of $du = Tds - Pd(1/\rho)$, we obtain

$$\rho \frac{ds}{dt} = \frac{\rho}{T}\frac{du}{dt} - \frac{P}{\rho T}\frac{d\rho}{dt} \tag{5}$$

The first law of thermodynamics expressed locally in the convection of a Newtonian fluid is given by

$$\rho \frac{du}{dt} = -\nabla \cdot q - P(\nabla \cdot v) + \boldsymbol{\tau} : \nabla \mathbf{v} \tag{6}$$

Introducing Eq. (6) into Eq. (5), and combining the resulting equation with Eq. (4) we have the following equation for an incompressible flow

$$\Phi = -\frac{1}{T^2}(\mathbf{q} \cdot \nabla T) + \frac{1}{T}(\boldsymbol{\tau} : \nabla \mathbf{v}) \tag{7}$$

The term $\tau : (\nabla \mathbf{v})$ represents the conversion of mechanical energy into thermal energy and called the viscous dissipation heating that occurs in all flow systems. This heat source can be considerably high in flows with large viscosity and large velocity gradients, as in high-speed flight, in rapid extrusion and in lubrication. The viscous dissipation is always positive for Newtonian fluids as it may be written in terms of a sum of squared terms of the velocity gradients and the viscosity in the following form

$$\tau : \nabla \mathbf{v} = \frac{1}{2}\mu \sum_i \sum_j \left[\left(\frac{\partial v_i}{\partial x_j} + \frac{\partial v_j}{\partial x_i}\right)\right]^2 = \mu \Theta \tag{8}$$

where μ is the viscosity, and Θ is the viscous dissipation function (in s^{-2}). When the index i takes on the values 1, 2, 3, the velocity components v_x, v_y, v_z and the rectangular coordinates x_i become x, y, z. Using the Fourier law $\mathbf{q} = -k\nabla T$ and Eq. (8), Eq. (7) becomes

$$\Phi = \frac{k}{T^2}(\nabla T)^2 + \frac{\mu}{T}\Theta \tag{9}$$

For a two dimensional Cartesian coordinate system, Eq. (9) can be expressed as

$$\Phi = \frac{k}{T^2}\left[\left(\frac{\partial T}{\partial x}\right)^2 + \left(\frac{\partial T}{\partial y}\right)^2\right] + \frac{\mu}{T}\left\{2\left(\frac{\partial v_x}{\partial x}\right)^2 + 2\left(\frac{\partial v_y}{\partial y}\right)^2 + \left(\frac{\partial v_x}{\partial y} + \frac{\partial v_y}{\partial x}\right)^2\right\} \tag{10}$$

Eq. (10) shows that the local irreversibilities are due to heat and viscous effects. The value of entropy generation Φ is positive and finite when temperature and velocity gradients are present in the medium.

1.1.1. Case studies

The rate of entropy generation for convection heat transfer in ducts, and the trade offs between the irreversibilities due to heat transfer and friction are widely investigated. The inlet temperature difference between the fluid and the wall is an important design criterion and should be optimized. The map of volumetric entropy generation rate and the distribution of irreversibility ratio at various thermal boundary conditions can be used to evaluate the system.

A heat exchanger changes the mutual thermal energy (exergy) levels between two or more fluids in thermal contact without external heat and work interactions. The second law analysis is widely applied to heat exchangers and heat-exchanger networks, and a set of heuristic for optimal design has been prepared. The analysis can estimate the quality of the heat exchanged in a heat exchanger with arbitrary flow arrangements, inlet temperature ratio, and flow heat capacity rate ratio. The thermodynamic method is becoming a common tool for evaluating the efficiency based on the second law. Reducing the irreversibilities from a design yields a thermodynamically improved system.

Thermomechanical Coupling

Couette flow provides the simplest model for the analysis of heat transfer for flow between two coaxial cylinders or parallel plates. The Couette flow is important in lubrication, polymer and food processing. The tangential annular flow is a model for a journal and its bearing in which one surface is stationary while the other is rotating, and the clearance between the surfaces is filled with a lubricant oil of high viscosity. For such a system the viscous-energy-dissipation appears as a heat source term in the energy equation, which must be solved to predict the temperature distribution in the narrow gap of a Couette device. Heat transfer and friction in a Couette flow are accompanied by the entropy generation that shows the amount of useful energy dissipated in the process.

The Gouy-Stodola theorem links the lost available work to the entropy generation; hence it relates the economic implications of the different irreversibilities in a process. The entropy generation rate in a steady, Newtonian fluid is calculated for two different geometries, and given in the next section [9].

For a plane geometry and Cartesian coordinates system, the rate of entropy generation Φ of an incompressible, Newtonian fluid is given by

$$\Phi = \frac{k}{T^2}\left(\frac{dT}{dy}\right)^2 + \frac{\mu}{T}\left(\frac{du}{dy}\right)^2 \tag{11}$$

Fig. 2 shows a Couette flow of a fluid of constant density ρ, viscosity μ, and thermal conductivity k between parallel plates. The bottom plate is at rest, while the top plate is moving at a constant velocity u_1. The upper and lower plates are kept at uniform temperatures T_1 and T_2, respectively.

Equation of motion for fully developed flow in the x direction is given by

$$-\frac{d}{dy}\left[\mu\left(\frac{du}{dy}\right)\right] = \left(-\frac{dP}{dx}\right) \tag{12}$$

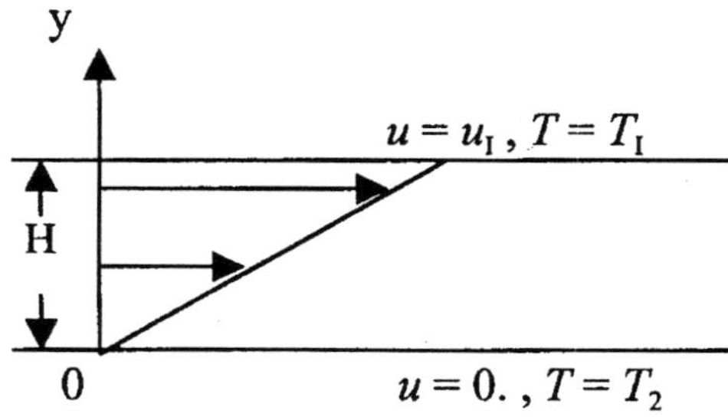

Fig. 2. The plane Couette flow. Reprinted with permission from Elsevier, Int. J. Heat Mass Transfer 43 (2000) 4205.

The boundary conditions are $u = u_1$ at $y = H$, and $u = 0$ at $y = 0$. Eq. (12) can be integrated twice and the velocity profile (Fig. 3) is obtained as

$$U = \frac{u}{u_1} Y\left[1 + A(1 - Y)\right] \tag{13}$$

where

$$A = \frac{(-dP/dx)H^2}{2\mu u_1}; \text{ and } Y = \frac{y}{H}$$

For the case of $(-dP/dx) = 0$, known as simple Couette flow, the velocity is linear across the fluid. For a negative pressure drop the velocity is positive, and for a pressure increase the velocity can become negative that leads to backflow. At the point of reversal $du/dy = 0$ at $y = 0$. This occurs when $-dP/dx = -2\mu u_1/(H^2)$. The velocity gradient du/dy is obtained from Eq. (13), and given by

$$\frac{du}{dy} = \frac{(-dP/dx)}{2\mu}(H - 2y) + \frac{u_1}{H} \tag{14}$$

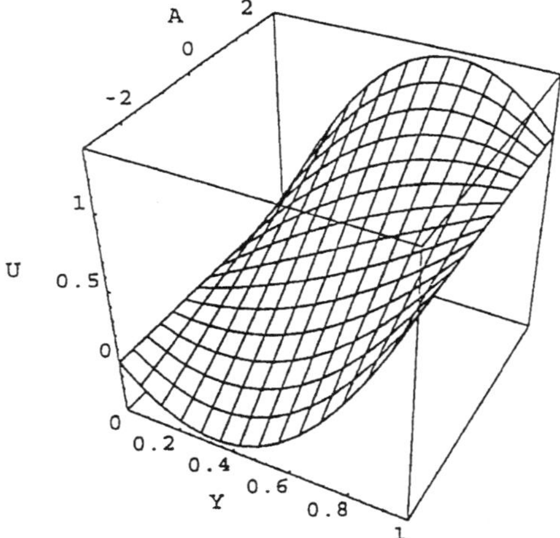

Fig. 3. Dimensionless velocity field U for the plane Couette flow for $H = 0.005$ m. Reprinted with permission from Elsevier, Int. J. Heat Mass Transfer 43 (2000) 4205.

The energy equation for laminar and hydrodynamically developed flow is

$$\frac{d^2T}{dy^2} = \frac{\mu}{k}\left(\frac{du}{dy}\right)^2 \tag{15}$$

with the boundary conditions of $T = T_2$ at $y = 0$, and $T = T_1$ at $y = H$. Substitution of Eq. (14) into Eq. (15) yields the temperature distribution given by

$$\begin{aligned}\theta = \frac{T - T_2}{T_1 - T_2} = \frac{(-dP/dy)Y}{12k(T_1 - T_2)}[aH^4(1 - Y^3) - 2bH^3(1 - Y^2) + 6cH^2(1 - Y)] \\ + Y\left[1 + \frac{1}{2}Br(1 - Y)\right]\end{aligned} \tag{16}$$

where

$$a = \frac{(-dP/dy)}{\mu};\quad b = \frac{(-dP/dy)H}{\mu} + \frac{2u_1}{H}$$

$$c = \frac{(-dP/dy)H^2}{4\mu} + u_1;\quad Br = \frac{\mu u_1^2}{k(T_1 - T_2)}$$

Here Br is the Brinkman number, which is a measure of the viscous heating as compared to the heat conducted through the gap of Couette device. The temperature gradient can be obtained from Eq. (16), and given by

$$\begin{aligned}\frac{dT}{dy} = \frac{(-dP/dy)}{12k}[aH^3(1 - 4Y^3) - 2bH^2(1 - 3Y^2) + 6cH(1 - 2Y)] \\ + \frac{T_1 - T_2}{H}\left[1 + \frac{1}{2}Br(1 - 2Y)\right]\end{aligned} \tag{17}$$

Fig. 4 shows the temperature profile for $T_2 = 300$ K and $-2.0 < Br < 8.0$. The rise of temperature, in the middle part of Couette device, is considerably large for high values of Br. Inserting Eqs. (14) and (17) into Eq. (11) yields an expression for the volumetric entropy generation rate for a Couette flow.

For a circular Couette flow (Fig. 5) the entropy generation rate for an incompressible Newtonian fluid held between two coaxial cylinders is given by

$$\Phi = \frac{k}{T^2}\left(\frac{dT}{dr}\right)^2 + \frac{\mu}{T}\left[r\frac{d}{dr}\left(\frac{u_\theta}{r}\right)\right]^2 \tag{18}$$

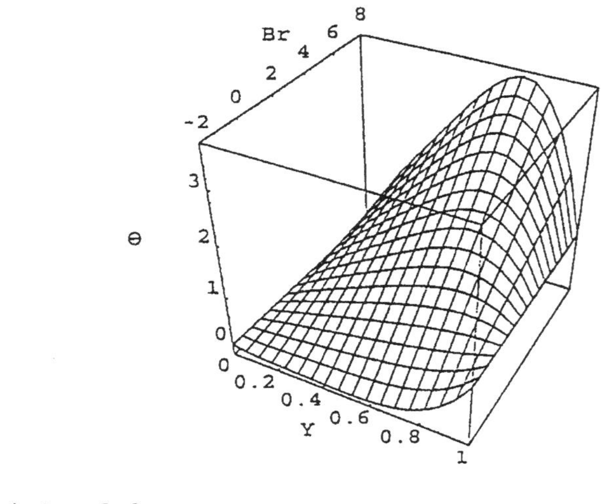

Fig. 4. Dimensionless temperature profile θ for the plane Couette flow for u_1 = 5.0 m/s, H = 0.005 m. Reprinted with permission from Elsevier, Int. J. Heat Mass Transfer 43 (2000) 4205.

The circular Couette flow between concentric cylinders is in the θ direction only, and it satisfies that $u_r = u_z = 0;\ u_\theta = u_\theta(r);\ T = T(r)$. The inner cylinder is stationary while the outer cylinder is rotating with an angular velocity w. It is assumed that the annular flow is laminar and there are no end effects, and the steady state is reached with the controlled angular velocity.

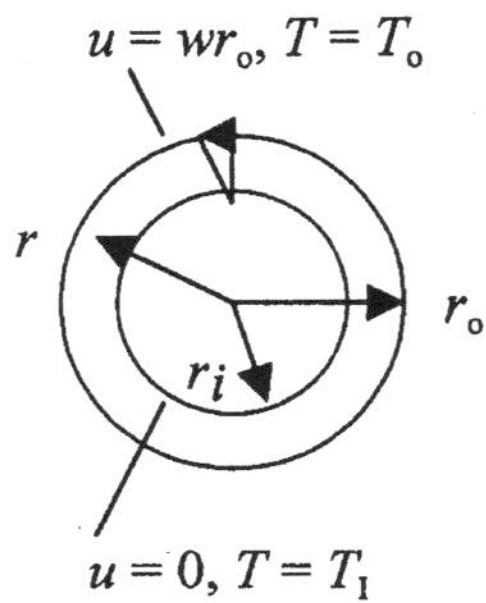

Fig. 5. The circular Couette flow. Reprinted with permission from Elsevier, Int. J. Heat Mass Transfer 43 (2000) 4205.

The velocity distributions is

$$\frac{u_\theta}{wr_o} = \frac{1}{R}\frac{(r^2 - r_i^2)}{(r_o^2 - r_i^2)} \tag{19}$$

where $R = \dfrac{r}{r_o}$

The second term on the right hand side of Eq. (18), related to the velocity gradient, which can be obtained from Eq. (19), and given by

$$\frac{\mu}{T}\left[r\frac{d}{dr}\left(\frac{u_\theta}{r}\right)\right]^2 = \frac{\mu}{T}\frac{4u_1^2 r_o^2 r_i^4}{r^4 (r_o^2 - r_i^2)} \tag{20}$$

With the surface temperatures of T_o and T_i of the outer and inner cylinders respectively, the temperature profile is given by

$$\theta = \frac{T - T_i}{T_o - T_i} = (B+1)\left(1 - \frac{\ln R}{\ln n}\right) + B\left(\frac{\ln R}{n^2 \ln n} - \frac{1}{R^2}\right) \tag{21}$$

with the dimensionless quantities of

$$n = \frac{r_i}{r_o}\ ;\ B = Br\frac{n^4}{(1-n^2)^2}\ ;\ \text{and}\ Br = \frac{\mu w^2 r_o^2}{k(T_o - T_i)}$$

where Br is the Brinkman number for the annulus. Eq. (21) satisfies the boundary conditions of $\theta = 0$ at $R = n$, and $\theta = 1$ at $R = 1$. The temperature gradient can be obtained from Eq. (21), and given by

$$\frac{dT}{dr} = (T_o - T_i)\left(\frac{2Br_o^2}{r^3} + \frac{B}{rn^2 \ln n} - \frac{B+1}{r \ln n}\right) \tag{22}$$

By substituting Eqs. (20) and (22) into Eq. (18), an expression for the entropy generation rate for the tangential annular flow can be obtained.

The first terms on the right hand sides of Eq. (11) and (18) show the entropy generation due to the heat transfer $\Phi_{\Delta T}$, while the entropy generation due to the fluid friction $\Phi_{\Delta P}$ is shown by the second terms, hence the rate of entropy generation expression has the following basic form

$$\Phi = \Phi_{\Delta T} + \Phi_{\Delta P} \tag{23}$$

The irreversibility distribution ratio is defined as

$$Be = \Phi_{\Delta T} / \Phi \tag{24}$$

and was named as the Bejan number *Be*. *Be* = 1 is the limit at which all the irreversibility is due to the heat transfer. The irreversibility due to the heat transfer dominates when *Be* >> ½, while *Be* << ½ is the case where the irreversibility due to the friction dominates.

The effects of *A* and *Br* on the irreversibility distributions are shown in Fig. 6 for unused engine oil with $k = 0.14$ W m^{-1} K^{-1}, $\nu = 0.839\ 10^{-4}$ m^2 s^{-1}, $\rho =$ 864.04 kg m^{-3}. The temperature dependence of viscosity, density and thermal conductivity is neglected. Except for the simple Couette flow where $A = 0$, the peaks of *Be* appear in the middle part of Couette device due to development of maximum temperature in that region.

The increase of *Be* indicates competition of the irreversibilities caused by the heat transfer and friction. At high *Re*, the distribution of *Be* is relatively more uniform than that of lower *Re*. For circular Couette device, the Reynolds number ($Re = wr_o^2 / \nu$) at the transition from laminar to turbulent flow is strongly dependent on the ratio of the gap to the radius of the outer cylinder, 1 - *n*. The critical Reynolds number reaches a value about 50000 at 1 - *n* = 0.05. With various operational conditions such as the gap of Couette device, the Brinkman number, and the boundary conditions, the distribution of the irreversibility ratio can be controlled.

Packed Systems

Introduction of a suitable packing into the fluid flow passage enhances the wall-to-fluid heat transfer considerably, hence reducing the entropy generation due to heat transfer but increasing the entropy generation due to fluid-flow friction. The net entropy generation provides a new criterion in analyzing such systems seen in some thermal systems operations and catalytic chemical reactors.

Local rate of entropy generation per unit volume Φ of an incompressible Newtonian fluid for a two-dimensional annular flow is expressed in terms of the velocity and temperature profiles

$$\Phi = \frac{k}{T^2}\left[\left(\frac{\partial T}{\partial r}\right)^2 + \left(\frac{\partial T}{\partial z}\right)^2\right] + \frac{2\mu}{T}\left(\frac{\partial u}{\partial z}\right)^2 . \tag{25}$$

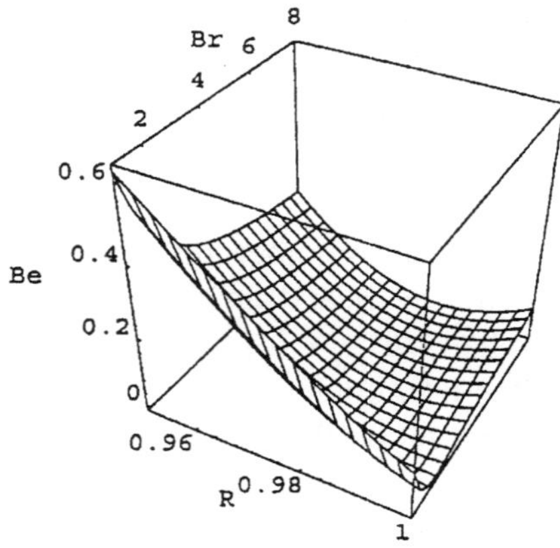

(a) w= 1048 s^{-1} ; Re= 5000

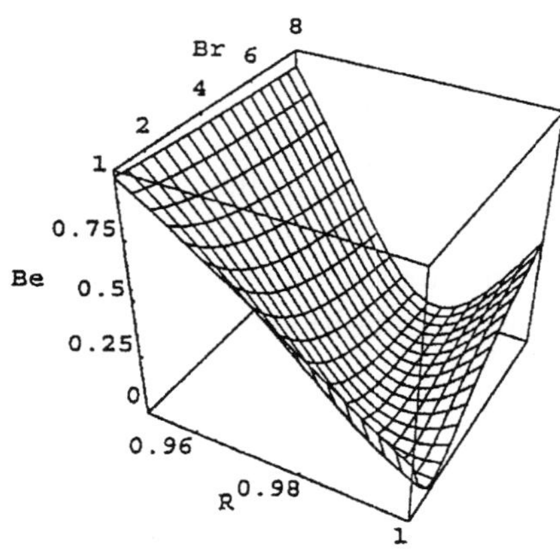

(b) w= 8390 s^{-1} ; Re= 40000

Fig. 6. The Bejan number *Be* for the circular Couette flow for $T_i = 300$ K, $r_o = 0.02$ m, $r_i = 0.019$ m. Reprinted with permission from Elsevier, Int. J. Heat Mass Transfer 43 (2000) 4205.

Here k and μ are the thermal conductivity and dynamic viscosity of the fluid, respectively. The terms u and T denote the velocity and temperature of the fluid. The first term on the right-hand side of Eq. (25) shows the entropy generation due to finite temperature differences in axial z and in radial r directions, while the second term is due to the fluid friction. Entropy generation profiles may be constructed using Eq. (25) if the temperature and the velocity fields are known in the heat transfer medium.

Assuming the fully developed velocity and temperature profiles for the control volume of an annular packed bed, the energy equation is expressed by

$$\frac{1}{r}\frac{\partial}{\partial r}\left(r\frac{\partial T}{\partial r}\right)=\frac{u}{\alpha_e}\left(\frac{dT_b}{dz}\right) \tag{26}$$

Here α_e is the effective thermal diffusivity of the bed, and T_b is the bulk fluid temperature. The axial thermal conduction in the bed has been neglected in Eq. (26). It is assumed that the plug flow conditions ($u = u_{av}$) and essentially radially flat superficial velocity profiles prevail through the cross section of the packed flow passage. The uniform heat fluxes at each of the two surfaces provide the necessary boundary conditions with positive heat fluxes when the heat flows into the fluid

at $r = r_o$ $\qquad -k_e\,(\partial T/\partial r) = q_o$ = constant (27)

at $r = r_i$ $\qquad -k_e\,(\partial T/\partial r) = q_i$ = constant (28)

Eq. (26) can be directly integrated as the term dT/dz = constant. The linearity of the energy equation suggests that superposition methods may be employed to build solutions for asymmetric heating by adding the two fundamental solutions: (1) the outer wall heated with the inner wall insulated and (2) the inner wall heated with the outer wall insulated. The fundamental solutions are [12]

$$f_1 = T - T_o = \frac{q_o}{h}\left[(StZR'+1)+\frac{hR'r_o^2}{2k_eD_H}\left(R^2-1-2r^{*2}\ln R\right)\right] \tag{29}$$

$$f_2 = T - T_o = \frac{q_i}{h}\left[(StZr^*R'+1)+\frac{hr^*R'r_o^2}{2k_eD_H}\left(R^2-r^{*2}+2\ln r^*-2\ln R\right)\right] \tag{30}$$

where

$$R=\frac{r}{r_o};\; r^*=\frac{r_i}{r_o};\; R'=\frac{2r_o}{r_o+r_i}\;;\; Z=\frac{z}{r_o-r_i}$$

St and h are the Stanton number and the heat transfer coefficient, respectively. The temperature profile for the annular packed bed with asymmetric heating can be obtained by adding the fundamental solutions of f_1 and f_2, and we have

$$T = T_o(1 + \tau A), \tag{31}$$

where

$$A = StZR'(r^* + n) + n + 1$$
$$+ \frac{Nuk_f r_o^2 R'}{2k_e D_p D_H}[R^2(r^* + n) - 2r^*(1 + nr^*)\ln R + r^*(2\ln r^* - r^{*2}) - n]$$

$$\tau = \frac{Q/h}{T_o} = \frac{T_w - T_b}{T_o} \; ; \quad n = \frac{q_o}{q_i}$$

The hydraulic diameter of the annular bed is $D_H = 2\,(r_o - r_i)$, and D_p shows the packing diameter.

The Nusselt number and the effective thermal conductivity k_e for the annular packed bed are expressed by

$$Nu = \frac{hD_p}{k_f} = 5.9Re^{0.44} \tag{32}$$

$$k_e = k_f(0.6 + 0.157PrRe). \tag{33}$$

The heat transfer parameters have been derived for annular packed in the range $200 < Re < 800$ and $D = D_H / D_P = 6$.

The average fluid temperature is obtained from

$$T_{av} = \frac{2\int_{r_i}^{r_o} rTdr}{\int_{r_i}^{r_o} rdr}. \tag{34}$$

The term (dT/dz) may be calculated from the simple energy balance

$$q_i(1 + n)[2\pi(r_o + r_i)]dz = GC_p\pi(r_o^2 - r_i^2)dT \tag{35}$$

or directly from the differentiation of Eq. (31) with respect to axial distance z as

$$\frac{dT}{dz} = \frac{2q_i R'}{Pek_f}\frac{(n + r^*)}{D}. \tag{36}$$

where Pe is the Peclet number, and G is the mass flux. The temperature gradient in the radial direction may be obtained from Eq. (31), and given by:

$$\frac{\partial T}{\partial r} = \frac{q_i R' r_o}{k_e D_H} \left[R(n+r^*) - \frac{r^*}{R}(1+nr^*) \right]. \tag{37}$$

The velocity may be related to the pressure by inviscid-flow behavior $(dP/\rho) = -d(u_b^2/2)$, and using the Bernoulli equation and Ergun equation the velocity gradient in the flow direction is expressed by

$$\frac{du}{dz} = \frac{[C_1(1-\varepsilon)^2 + C_2(1-\varepsilon)Re]\mu}{\varepsilon^3 D_p^2 \rho}. \tag{38}$$

The constants C_1 and C_2 by taking into account the effect of confining walls are reported as

$$C_1 = 130 \text{ and } C_2 = \frac{D}{0.335D + 2.28}. \tag{39}$$

The Reynolds number is based on the packing diameter $Re = GD_p / \mu$.

By substituting Eqs. (30)-(38) into Eq. (25), the following expression for the volumetric entropy generation for the packed annulus can be derived

$$\Phi = \frac{k_f}{T^2} \left[\left(\frac{q_i R' r_o}{k_e D_H} \right)^2 \left[R(n+r^*) - \frac{r^*}{R}(1+nr^*) \right]^2 + \left(\frac{2D_p q_i R'}{Pe k_f D_H} \right)^2 (n+r^*)^2 \right] + \frac{2\mu}{T} \left(\frac{130(1-\varepsilon)^2 \mu + C_2(1-\varepsilon)\mathrm{Re}\,\mu}{\varepsilon^3 D_p^2 \rho} \right)^2 \tag{40}$$

Here the terms on the right-hand side of Eq. (40) shows the entropy generation due to heat transfer $\Phi_{\Delta T}$ and due to fluid friction $\Phi_{\Delta P}$, respectively; hence the entropy generation expression has the following basic form $\Phi = \Phi_{\Delta T} + \Phi_{\Delta P}$. The volumetric entropy generation rate is positive and finite as long as temperature or velocity gradients are present in the medium.

The dimensionless entropy generation profile can be obtained as

$$J = \Phi \frac{k_f T_o^2}{Q^2} \tag{41}$$

where $Q = q_i + q_o$.

In the second law analysis of convective heat transfer, there are two new dimensionless parameters. One of them is the dimensionless temperature difference

$$\tau = \frac{Q/h}{T_o} = \frac{T_w - T_b}{T_o} \tag{42}$$

The other dimensionless parameters are the irreversibility distribution ratios, which are given by

$$\phi = \Phi_{\Delta P} / \Phi_{\Delta T} \tag{43}$$

$$Be = \Phi_{\Delta T} / \Phi = (1+\phi)^{-1} \tag{44}$$

where Be is the Bejan number (Be). $Be = 1$ is the limit at which the irreversibility due to heat transfer dominates, $Be = 0$ is the opposite limit at which the irreversibility due to fluid friction is the dominating effect. In Eq. (40) local entropy generation has been expressed in terms of τ, R, Z and D including the properties of the fluid ρ and C_p.

The rate of entropy generation over the cross section S'' may be calculated by the following integration

$$S'' = \int_{r_i}^{r_o} \Phi r dr \tag{45}$$

Heat transfer to a fluid flowing in an annulus has a technical importance because either or both of the surfaces can be heated or cooled independently. A specified task will determine the entropy generation, and the Bejan number profiles will help to determine the dominant contribution to the rate of entropy generation.

Fluid flow and the wall-to-fluid heat transfer in a packed duct are of interest in chemical fixed bed reactors, packed separation columns, heat exchangers, and some heat storage systems. In this analysis wall effect on the velocity profile is taken into account in the calculation of entropy generation in a packed duct with the top wall heated and the bottom wall cooled (Fig 7). It is assumed that the difference of wall-to-fluid bulk temperature is small enough, and that there is no considerable change in physical properties of the fluid, no axial conduction, and no natural convection.

The following relations show an approximate expression for the velocity profile in terms of the average velocity in a packed bed with $H/d_p > 5$ and T = 293 K [11]

$$U = \frac{u}{u_{av}} = \beta\{1-(1-a_3 Y)e^{a_1 Y} - [1-a_3(1-Y)]e^{a_1(1-Y)}\} \tag{46}$$

where the term β denotes the deformation factor that depends on the ratio H/d_p and Re_p, and can be expressed as

$$\beta = \left(1 + \frac{2(1-a_2+a_2 a_3)}{a_1} + \frac{2a_3(1-a_2)}{a_1^2}\right)^{-1} \tag{47}$$

where

$$a_1 = aH/d_p,\ a_2 = e^{a_1},\ a_3 = nH/d_p,\ Y = y/H$$

$$n = -1083 + 201.6a_4 - 3737a_4^{0.5} + 5399a_4^{1/3} \qquad 1 < Re_p < 1000$$

$$a = 4n/(4-n),\ a_4 = \ln Re_p + 4,\ Re_p = u_{av} d_p / \nu$$

The dimensionless velocity profile is shown in Fig. 8.

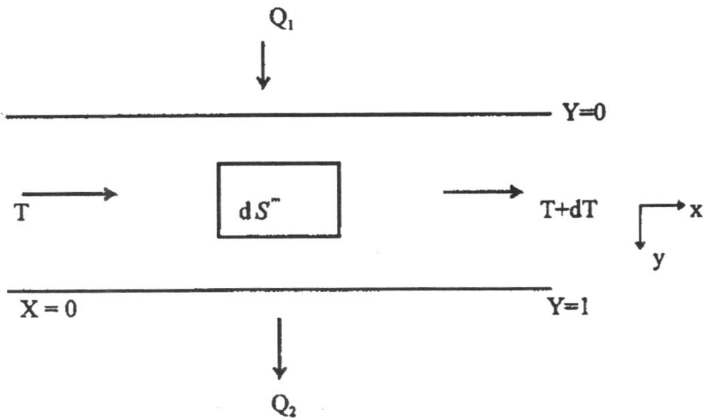

Fig. 7 Control volume of the rectangular packed bed with asymmetric heat effects. Reprinted with permission from Elsevier, Int. J. Heat Mass Transfer 42 (1999) 2337.

The energy equation for a fully established laminar flow is

$$\frac{\partial}{\partial y}\left(\frac{\partial T}{\partial y}\right)=\frac{u}{\alpha_e}\left(\frac{dT_b}{dx}\right) \tag{48}$$

where α_e is the effective thermal diffusivity, and T_b is the bulk air temperature

$$T_b(x)=\frac{\int_A uTdA}{\int_A udA} \tag{49}$$

The heat flux at the upper and lower surface specifies the temperature gradient at the wall, and the necessary boundary conditions

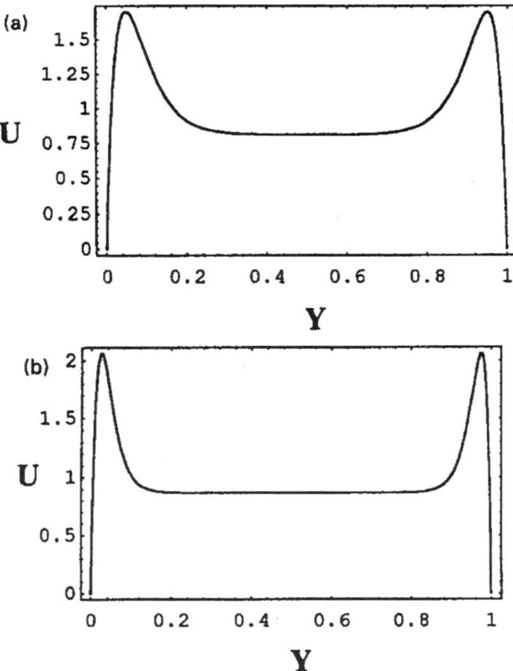

Fig. 8. Velocity profiles in the packed bed for (a) $H/d_p = 5$, $Re_p = 309$, $\beta = 0.816$; (b) $H/d_p = 10$, $Re_p = 154$, $\beta = 0.871$. Reprinted with permission from Elsevier, Int. J. Heat Mass transfer 42 (1999) 2337.

$$y = 0; \quad -k_e\left(\frac{\partial T}{\partial y}\right) = Q_1 = \text{constant}$$

$$y = H; \quad -k_e\left(\frac{\partial T}{\partial y}\right) = Q_2 = \text{constant} \tag{50}$$

The linearity of the energy equation suggests that the superposition method may be applied top build solutions for asymmetric heating by adding two fundamental solutions belonging the top and bottom walls. For a constant heat flux, a simple energy balance is

$$Qdx = \rho u_{av} H C_p dT_b \tag{51}$$

yields the temperature gradient in the flow direction.

Using Eqs. (46) to (51) the velocity and temperature profiles can be calculated, and used to determine the entropy generation in the packed bed considered

$$\Phi = \frac{k_e}{T^2}\left[\left(\frac{\partial T}{\partial y}\right)^2 + \left(\frac{\partial T}{\partial x}\right)^2\right] + \frac{\mu}{T}\left(\frac{du}{dy}\right)^2 \tag{52}$$

Here the first term on the right-hand side shows the entropy generated due to heat transfer, and the second terms shows the entropy generated due to fluid friction. Eq. (52) expresses the rate of entropy generation in terms of H/d_p, heat duty Q, Reynolds number, and Stanton number $St = h / \rho u_{av} C_p$.

Nondimensionless entropy generation N can be obtained as

$$N = \Phi \frac{k_e T_o^2}{(Q_1 - Q_2)^2} \tag{53}$$

Distribution of entropy generation is calculated for airflow with constant properties of $k = 0.026$ W/(m K), $\mu = 1.84\ 10^{-5}$ kg/(m s), $\rho = 1.17$ kg/m^3, $Pr = 0.7$, and shown in Fig. 9 for an inlet air temperature of $T_o = 297$ K.

When there is no packing in the flow passage, the temperature profiles can be calculated similarly using the parabolic velocity profiles of $u = 6u_{av}(Y - Y^2)$ with the superposition approach, and the following temperature profile is obtained for a rectangular bed with asymmetric heating

$$T^* = T_o(1 - \tau\psi^*) \tag{54}$$

where, $X = x/H$, and $\quad ^* = \dfrac{hH}{2k_f}[(r-1)(Y^4 - 2Y^3 - 2Y + r)] + (1-r)(StX + 1)$

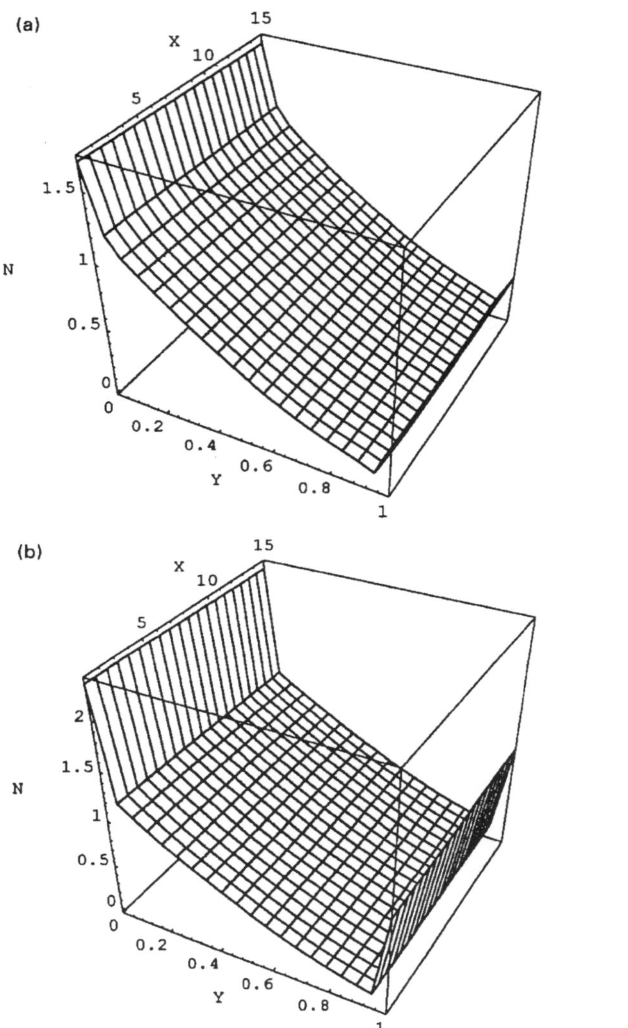

Fig 9. Distribution of entropy generation N, in the packed bed (a) $H/d_p = 10$, $Q_1 = 66$ W/m^2, $r = 0.2$, $Re_p = 154$, $\beta = 0.871$; (b) $H/d_p = 20$, $Q_1 = 66$ W/m^2, $r = 0.2$, $Re_p = 774$, $\beta = 0.905$. Reprinted with permission from Elsevier, Int. J. Heat Mass transfer 42 (1999) 2337.

Using the velocity and temperature gradients the dimensionless entropy generation for the empty bed is obtained, and shown in Fig. 10. Comparison of Fig. (9) and Fig. (10) indicates that outside the wall region, the packed bed entropy generation rate is uniformly distributed which is the thermodynamic optimality criterion. The empty bed has the typical *S* shape of the entropy generation rate profile.

Usually minimizing the entropy generation leads to increasing the equipment size, for example adding transfer area to a heat exchanger may not produce an economic optimum, then we may consider other possible optimal configurations. For a single process where the distribution of the flows and forces is determined by the boundary conditions such as the temperatures and flow rates at the entrance of a heat exchanger (Fig. 11), the optimization is the problem of deciding on the best operating conditions such as the best flow configurations. Once the operating conditions are set, the distribution of irreversibilities will be fixed inside the exchanger.

One optimum is based on the uniformly distributed entropy generation rate of a heat exchanger, mixer, or separator with best configuration. Consider the example of countercurrent and cocurrent heat exchangers shown in Fig. 11. Temperature profiles show that the driving force ΔT, or $1/\Delta T$ is more uniformly distributed in countercurrent then that of cocurrent flow operation. Since the

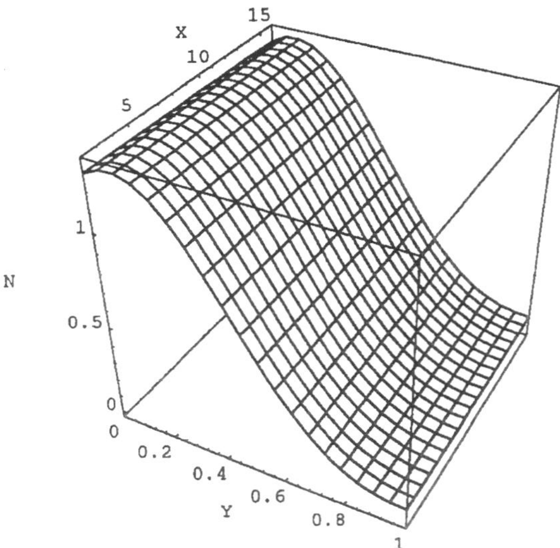

Fig. 10. Distribution of entropy generation N profile in the empty bed for $Re_H = 1641$, $St = 0.005$, $Q_1 = 34$ W/m^2, $r = 0.2$. Reprinted with permission from Elsevier, Int. J. Heat Mass transfer 42 (1999) 2337.

driving force is directly related to the entropy generation, it is also uniformly distributed; this is the basic thermodynamic reason for a countercurrent to be better than that of cocurrent operation. In order to explain this, a duty of the exchanger will be specified in terms of the flow rate Q and inlet and outlet temperatures T_1 and T_2 of cold streams. The duty is the amount of heat transferred from the hot fluid to cold fluid. The heat exchangers are identical except the flow arrangements. The cocurrent exchanger will require a higher flow rate and/or higher temperature of hot fluid, and hence the operating cost will be higher than that of the countercurrent exchanger. Alternatively, cocurrent exchanger will require a larger heat transfer area, and hence higher investment for a specified flow rate and inlet temperature of the hot fluid. Therefore, countercurrent exchanger is capable of minimizing either operating cost or investment cost compared with the cocurrent exchanger.

Some of the options for a thermodynamic optimum are to improve an existing design towards to be less irreversible, and to distribute the irreversibilities uniformly over the space and time. This approach needs to relate the distribution of irreversibilities to the minimization of entropy generation based on the theorem stating that in the range of validity of linear nonequilibrium thermodynamics, dissipative process will have all driving forces uniformly distributed in space and time. For a transport of a single substance, the local rate of entropy generation is expressed as

$$\Psi = T\Phi = JX \tag{55}$$

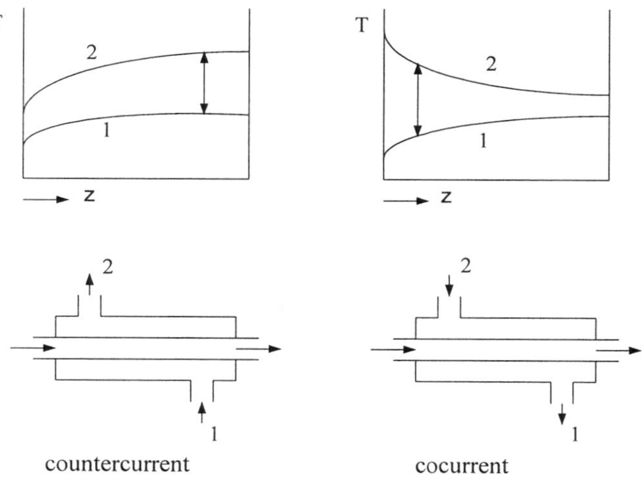

Fig. 11. Heat exchangers with countercurrent and cocurrent operations.

where J is the local flow of substance, and X is the conjugate driving force. Assuming that the linear phenomenological relations hold between the flow and force, we have

$$J = LX \tag{56}$$

where L is a phenomenological coefficient assumed as constant and positive. The total entropy generation is the integral of Φ over the time and space variables

$$P = \int\int \Phi dVdt = L\int\int X^2 dVdt \tag{57}$$

The total flow is the integral over time and space of the local flow

$$J = L\int\int XdVdt = LVtX_{av} \tag{58}$$

where X_{av} is the average driving force.

The following relations are from Tondeur and Kvaalen [54]. The total entropy generation is obtained from Eq. (57) as

$$P_{av} = JX_{av} \tag{59}$$

The difference between the general case and the average value is given by

$$P - P_{av} = L[\int\int X^2 dVdt - (X_{av})^2 Vt] = LVt[(X^2)_{av} - (X_{av})^2] \tag{60}$$

The square bracket on the right-hand side of Eq. (60) is the difference between the mean square and the square mean of the force distribution, and is the variance of X. We therefore have

$$\frac{P - P_{av}}{Vt} > 0 \tag{61}$$

or

$$P_{av}(\text{equipartioned}) < P(\text{arbitrary}) \tag{62}$$

The entropy production P_{av} of a process with uniform driving force is smaller than that of a nonuniform situation with the same size, and duration of the same average driving force and the same overall load J. Eqs. (55) and (56) show that the local flow and the local rate of entropy production Φ will be constant when X

is uniform. Therefore, equipartition of forces is analogous to equipartition of flows or of the entropy generation. The flow and the entropy generation in matrix forms are expressed by

$$\mathbf{J} = [\mathbf{L}] \cdot \mathbf{X} \tag{63}$$

$$\Phi = \mathbf{X}^{\mathrm{T}} \cdot \mathbf{J} = \mathbf{X}^{\mathrm{T}} \cdot [\mathbf{L}] \cdot \mathbf{X} \tag{64}$$

where $[\mathbf{L}]$ is the symmetric matrix of phenomenological coefficients due to the Onsager reciprocal relations. The total entropy generation P and the total flow $\mathbf{J}$ (specified duty) are expressed by

$$P = \int\int \mathbf{X}^{\mathrm{T}} \cdot [\mathbf{L}] \cdot \mathbf{X} \cdot dVdt \tag{65}$$

$$\mathbf{J} = \int\int [\mathbf{L}] \cdot \mathbf{X} dVdt = Vt[\mathbf{L}] \cdot \mathbf{X}_{av} \tag{66}$$

where $\mathbf{X}_{av}$ is the average driving force vector, and elements of it are the averages of the individual driving forces. Using the average force the total entropy generation becomes

$$P_{av} = \mathbf{J} \cdot \mathbf{X}_{av}^{T} \tag{67}$$

The excess entropy generation $P\text{-}P_{av}$ is obtained as

$$P - P_{av} = \int\int (\mathbf{X}^{\mathrm{T}} \cdot [\mathbf{L}] \cdot \mathbf{X} - \mathbf{X}_{av}^{\mathrm{T}} \cdot [\mathbf{L}] \cdot \mathbf{X}_{av}) dVdt \tag{68}$$

Eq. (68) can also expressed as

$$P - P_{av} = \int\int (\mathbf{e}^{\mathrm{T}} \cdot \mathbf{e} - \mathbf{e}_{av}^{\mathrm{T}} \cdot \mathbf{e}_{av}) dVdt = [(\mathbf{e}^{\mathrm{T}} \cdot \mathbf{e})_{av} - \mathbf{e}_{av}^{\mathrm{T}} \cdot \mathbf{e}_{av}]Vt \tag{69}$$

where $\mathbf{e} = [\mathbf{R}] \cdot \mathbf{X}$, and $[\mathbf{L}]$ is the positive matrix and may be decomposed into a product of matrices

$$[\mathbf{L}] = [\mathbf{R}]^{\mathrm{T}} \cdot [\mathbf{R}] \tag{70}$$

The quantity of excess entropy generation is positive by the Cauchy-Schwartz inequality, (similar to inequality in Eq. 62), indicating that $P > P_{av}$.

Since the minimization of entropy generation is not always an economic criterion, it is necessary to relate the overall entropy generation and its distribution to the economy of the process. To do that, we may consider various processes with different structures and operating configurations:

3. HEAT AND MASS TRANSFER

As a unified approach, the direct variational method for finding the nonstationary fields of temperature, concentration, pressure, and velocities is important. The thermodynamic approach can be used to investigate the qualitative properties of the trajectories of stationary-coupled heat and mass transfer processes in flows without chemical reactions. The Liapunov functions can be constructed on the basis of the entropy balance as some thermodynamic potential of the flow processes, similar to the appropriate counterparts of the usual thermodynamic potentials of closed systems.

The second law analysis has been widely used for combined forced convective heat and mass transfer. Mainly two-dimensional duct with various boundary conditions for laminar and turbulent flows are investigated in the search for the thermodynamic optimum. An optimum Reynolds number and the best configuration of a certain geometry are determined after minimizing the entropy generation, which results from a trade-off between the irreversibilities due to heat and mass transfer in a fluid flow.

The following relations for the entropy generation in external flows over a plate or a cylindrical geometry are from Poulikakos and Johnson [30]. A general expression for the rate of entropy generation is

$$\Phi = \frac{1}{T_\infty^2}\int_A q(T - T_\infty)dA + \int_A J_i \frac{R}{M_i}\left(\frac{T_s - T_\infty}{T_\infty} + \frac{c_{i,s} - c_{i,\infty}}{c_{i\infty}}\right)dA + \frac{Fu_\infty}{T_\infty} \tag{71}$$

where T_∞ is the free stream temperature, q is the heat flow, A is the surface area of the body, J_i is the mass flow of species i per unit area, R is the gas constant, M_i is the molecular weight of species i, $c_{i,s}$ is the concentration of species i at the body surface, $c_{i\infty}$ is the concentration of species i at free stream conditions, F is the total drag force exerted on body, and u_∞ is the free stream velocity. In Eq. (71), it has been assumed that the temperature difference $|T_s - T_\infty|$ and the concentration difference $|c_{i,s} - c_{i,\infty}|$ are small. The irreversibilities are due to heat transfer across a finite temperature difference, mass transfer across a finite difference in the chemical potential of species, and flow friction. The drag force on the plate is

$$\frac{F}{W} = \int_0^L (1/2)\rho_\infty u_\infty^2 C_f dx \tag{72}$$

where C_f is the local friction coefficient. The temperature and concentration differences are related to the respective flows as

$$T - T_\infty = \frac{q}{h}; \quad c_{i,s} - c_{i,\infty} = \frac{J_i}{k_m} \tag{73}$$

where h and k_m are the local heat and mass transfer coefficients respectively.

For laminar and turbulent flows, we need appropriate correlation equations for the friction coefficient, heat transfer coefficient, and mass transfer coefficient. For laminar flow in the ranges of $5\ 10^6 > Re > 10^3$, and Pr and $Sc > 0.5$, we have the following relations for the coefficients

$$C_f = 0.664\, Re^{-0.5} \tag{74}$$

$$h = 0.458\, Re^{0.5} Pr^{0.33} \frac{k}{x} \tag{75}$$

$$k_m = 0.458\, Re^{0.5} Sc^{0.33} \frac{D_i}{x} \tag{76}$$

For turbulent flow in the ranges of $5\ 10^5 < Re < 10^7$ and $60 > Pr > 0.6$, $3000 > Sc > 0.6$, we have

$$C_f = 0.0592\ Re^{-0.2} \tag{77}$$

$$h = 0.296\ Re^{0.8} Pr^{0.33} \frac{k}{x} \tag{78}$$

$$k_m = 0.296\, Re^{0.8} Sc^{0.33} \frac{D_i}{x} \tag{79}$$

Introducing Eqs. (72)-(79) into Eq. (71) and after performing the integration, the entropy generation rate can be obtained for a laminar flow on a flat plate.

$$\begin{aligned} \Phi = 1.456 \frac{W}{L^2} Re_L^{-0.5} & \left[\left(\frac{q^2}{T_\infty^2} + \frac{J_i q R}{M_i T_\infty} \right) \frac{Pr^{-0.33}}{k} + \frac{J_i^2 R Sc^{-0.33}}{M_i^2 C_{i,\infty} D_i} \right] \\ & + 0.664\, Re_L^{0.5} \left(\frac{\rho_\infty v_\infty U_\infty^2}{T_\infty} \right) \end{aligned} \tag{80}$$

A Similar expression for the entropy generation for a turbulent flow is obtained from Eq. (71) and Eqs. (77)-(79), and given by

$$\Phi = 28.15\frac{W}{L^2}Re_L^{-0.8}\left[\left(\frac{q^2}{T_\infty^2}+\frac{J_i qR}{M_i T_\infty}\right)\frac{Pr^{-0.33}}{k}+\frac{J_i^2 RSc^{-0.33}}{M_i^2 C_{i,\infty}D_i}\right] + 0.037\,Re_{\mathrm{L}}^{0.8}\left(\frac{\rho_\infty \nu_\infty U_\infty^2}{T_\infty}\right) \tag{81}$$

where W and L are the width and length of the plate, D is the mass diffusivity coefficient, Sc and Pr are the Schmidt number and Prandtl number respectively.

By setting $\partial\Phi/\partial Re = 0$, the optimum Reynolds number, which yields minimum entropy generation can be determined at known values of q and J_i, and for a laminar flow over a plate it is given by

$$Re_{\mathrm{L,opt}} = \frac{2.19\left[\frac{1}{L^2}\left(\frac{q^2}{T_\infty^2}+\frac{J_i qR}{M_i T_\infty}\right)\frac{Pr^{-0.33}}{k}+\frac{J_i RSc^{-0.33}}{LM_i^2 C_{i,\infty}D_i}\right]}{\left(\frac{\rho_\infty \nu_\infty U_\infty^2}{T_\infty}\right)} \tag{82}$$

For a turbulent flows over a plate, optimum Reynolds number becomes

$$Re_{\mathrm{L,opt}} = 62.69\left\{\frac{\frac{1}{L^2}\left[\left(\frac{q^2}{T_\infty^2}+\frac{J_i qR}{M_i T_\infty}\right)\frac{Pr^{-0.33}}{k}+\frac{J_i^2 RSc^{-0.33}}{M_i^2 C_{i,\infty}D_i}\right]}{\left(\frac{\rho_\infty \nu_\infty U_\infty^2}{T_\infty}\right)}\right\}^{0.625} \tag{83}$$

For a cylindrical geometry, the rate of entropy generation is obtained as

$$\Phi = 0.462\frac{1}{L}Re_D^{-0.466}\left[\left(\frac{q^2}{T_\infty^2}+\frac{J_i qR}{M_i T_\infty}\right)\frac{Pr^{-0.33}}{k}+\frac{J_i^2 RSc^{-0.33}}{M_i^2 C_{i,\infty}D_i}\right] + 2.743\,Re_D^{0.754}\left(\frac{\rho_\infty \nu_\infty U_\infty^2}{T_\infty}\right) \tag{84}$$

For cylindrical geometry, the following empirical relations, which are valid for $40 < Re < 1000$, are used

$$C_f = 5.484\,Re_D^{-0.246} \tag{85}$$

$$h = 0.689 \ Re_D^{0.466} Pr^{0.33} \frac{k}{D} \tag{86}$$

$$k_m = 0.689 \, Re_D^{0.466} Sc^{0.33} \frac{D_i}{D} \tag{87}$$

The expression for the optimum Reynolds number is obtained as

$$Re_{\mathrm{D,opt}} = 1.57 \left\{ \frac{\frac{1}{L^2}\left[\left(\frac{q^2}{T_\infty^2} + \frac{J_i qR}{M_i T_\infty}\right)\frac{Pr^{-0.33}}{k} + \frac{J_i RLSc^{-0.33}}{M_i^2 C_{i,\infty} D_i}\right]}{\left(\frac{\rho_\infty v_\infty U_\infty^2}{T_\infty}\right)} \right\}^{0.82} \tag{88}$$

The optimal Reynolds number defines the operating conditions at which the cylindrical system performs a required heat and mass transport, and generates the minimum entropy. These expressions offer a thermodynamically optimum design when an external flow, and heat and mass transfer occur in a system.

Some expressions for the entropy generation due to heat and mass transfer processes in a multicomponent fluid take into account the coupling effect between heat and mass transfer. The resulting diffusion fluxes obey generalized Stefan-Maxwell relations including the effects of ordinary, forced, pressure and thermal diffusion.

3.1. Case Studies

For a fixed design, minimization of the process entropy generation may yield optimal solutions in some economic sense. Such a minimization is usually carried out under certain set of constraints. The theory of linear nonequilibrium thermodynamics and the Onsager relations are used for formulation of the transport problem with conjugate flows and forces.

The local rate of volumetric entropy generation is expressed as

$$\Phi = \int_V LX^2 dV \tag{89}$$

where L is the phenomenological coefficient and is not a function of the driving force X. The minimization problem is the optimization of the system with a finite size V, and the solution is the homogeneous distribution of the force over the system. The following relations are from Tondeur and Kvaalen [54]. Assuming a steady-state heat transfer operation with no momentum and mass transfer, the expression of total entropy generation is

$$\Phi = -\int_V \frac{1}{T^2}(J\nabla T)dV = -k\int_V \left(\frac{\nabla T}{T}\right)^2 dV \tag{90}$$

where the heat flux is obtained from the Fourier law

$$J = -k\nabla T \tag{91}$$

and k is the thermal conductivity assumed as a constant.

The entropy generation is a function of the temperature field. Then the minimization problem is to obtain the temperature distribution T(x) corresponding to a minimum of entropy generation Φ using the following Euler-Lagrange equations

$$\Phi - \Sigma \frac{d}{dx}\left(\frac{\partial \Phi}{\partial T}\right) = 0 \tag{92}$$

The minimization of entropy generation function is based on a constraint expressed in Eq. (90), so that Eq. (92) becomes

$$\Sigma\left[\frac{1}{T}\left(\frac{\partial T}{\partial x}\right)^2 - \frac{\partial T}{\partial x^2}\right] = 0 \tag{93}$$

For a heat exchanger, a characteristic direction related to the temperature field is the direction $Z(x)$ normal to the heat transfer area, and Eq. (93) yields

$$\frac{\partial}{\partial x}\left[\frac{1}{T}\frac{\partial T}{\partial x}\right]_{Z(x)} = 0 \tag{94}$$

and we obtain

$$\left(\frac{\nabla T}{T}\right)_{Z(x)} = \text{constant} \tag{95}$$

Eq. (95) shows that by keeping the driving force $\nabla T / T$ uniformly distributed along the space variables, the entropy generation will be minimum.

For an optimum design, we consider

$$\frac{\nabla T}{T} = \frac{\Delta T}{T} = \text{constant} \tag{96}$$

with $\Delta T = T_h - T_c$, or $\Delta T = (p-1)T_c$, and the temperature gradients is a function of the temperatures T_h and T_c of hot and cold streams, respectively

$$T_h = pT_c, \tag{97}$$

where p is a constant with value of $p > 1$. The following expression also produces a constant C

$$\frac{\delta\Phi}{\delta A} = \frac{U(T_h - T_c)^2}{T_h T_c} = C = \text{constant} \tag{98}$$

where δ denotes a small change, U is a total heat-transfer coefficient, and A is the heat-exchange area

$$\delta\Phi = \delta Q\left(\frac{1}{T_c} - \frac{1}{T_h}\right), \text{ and } \delta A = \frac{\delta Q}{U(T_h - T_c)}$$

From Eqs. (97) and (98), we have

$$p = \frac{(2 + C/U) + [(2 + C/U)^2 - 4]^{1/2}}{2} \tag{99}$$

The energy balance is expressed as

$$\dot{w}_h dT_h = \dot{w}_c dT_c \tag{100}$$

where the terms $\dot{w}_h, \dot{w}_c$ are the heat capacity flow rates of hot and cold streams, respectively. From Eqs. (97) and (100) we have

$$\frac{dT_h}{dT_c} = p = \frac{\dot{w}_c}{\dot{w}_h} = \frac{T_h}{T_c} \tag{101}$$

which are the matching conditions to minimize the entropy generation in any heat exchanger. For example, for a specified heat exchanger area, and hot stream inlet and outlet temperatures T_i and T_o, respectively, the minimum entropy generation is obtained when

$$\frac{\dot{w}_c}{\dot{w}_h} = \frac{T_i}{T_o} = \text{constant} \tag{102}$$

This approach can also be extended for a network of heat exchangers.

4. CHEMICAL REACTIONS AND REACTING FLOWS

The classical thermodynamics is successfully extended to near equilibrium regions where the linear relations relate the thermodynamic flows and forces. There is an attempt to extend the theory of linear nonequilibrium thermodynamic to nonlinear systems occurring far from equilibrium, such as open chemical reactions, which may include multiple stationary states, periodic and nonperiodic oscillations, chemical waves, and spatial patterns. A part of this attempt involves describing the nonlinear phenomena in terms of conventional thermodynamic functions such as entropy and rate of entropy production. Some examples are the determination of entropy of stationary states in a continuously stirred tank reactor, which may provide insight into the thermodynamics of open nonlinear systems; also the optimum operating conditions of multiphase combustion, obtained from the minimization of entropy generation and the lost available work, may yield the maximum net energy output per unit mass of the flow at the combustor exit. The methods of minimization of entropy generation, exergy analysis, and pinch analysis are currently among the major theoretical contributions to the field.

The rates of change of the reaction coordinates in terms of the affinities can be expressed in linear phenomenological equations, and the phenomenological coefficients obey the Onsager reciprocal relations. The reciprocal relations are valid both for states that are in chemical equilibrium and for states that are not.

The principle of equipartition of forces obtained by combining the nonequilibrium thermodynamics approach with Cauchy-Lagrange optimization procedure shows that the best trade-off between the entropy production and transfer area in transport processes is obtained when the thermodynamic driving forces are uniformly distributed over the transfer area [39,40]. This principle has been applied to conduction of mass, heat, and charge, and it can also be applied for improving chemical reactors, such as batch and plug-flow reactors as well as the coupled processes. In a rate controlled reaction, $\Delta G/T$ should be distributed uniformly through the space and time in the reactor system (ΔG is the change in Gibbs energy for a reaction).

One of the options to reduce the energy cost in chemical process industry is to increase the process reversibility by increasing equipment size. Engineer has to make a trade-off between the energy and area costs. Other trade-offs are possible between the system output and transfer area, and between the system output and energy consumption. Equipartition of forces principles suggests that, the best of these trade-offs for such processes are the ones with uniformly distributed thermodynamic driving forces over the transfer area.

The equipartition principle has been derived for transport processes with constant phenomenological coefficients, and requires the prediction or measurement of a large number of parameters before the entropy generation could be redistributed. In the special case of constant phenomenological coefficients, equipartitioned forces imply equipartitioned entropy generation. Numerical and empirical evidence in the literature support the theoretically derived equipartition of forces principle. For example mathematical models show that a cascade drying process with uniform driving force across every stage in the cascade yields a substantial decrease in the energy consumption [46].

The following relations are from Sauar et al. [39,40]. The local entropy generation of a reacting mixture in a system with gradients in temperature T, and the chemical potentials μ_i, is given by

$$\Phi = -J_q \nabla\left(\frac{1}{T}\right) - \frac{1}{T}\sum_i^n J_i \nabla \mu_{i,T} - \sum_j^r J_{r,j}\frac{\Delta G}{T} \tag{103}$$

here, J_q is the total heat flow, J_i is the mass flow of component i and $J_{r,j}$ is the scalar reaction rate (flow) of reaction j. For a reaction process $\Delta G/T$ means the change in the Gibbs energy or the driving force for the reaction, which can be used in the following linear phenomenological equations

$$J_{r,j} = -\sum_j^l L_j \frac{\Delta G_j}{T} \tag{104}$$

With a homogeneous reaction and mechanical equilibrium $\nabla P = 0$, consider a reactor consisting of a large number of n subsystems with equal volumes, and one and the same reaction is taking place in all subsystems. The subsystems can be considered to have a uniform composition and temperature. The reaction in the subsystem k is J_k and the driving force is $\Delta G_k / T$. The total system can be considered as a nonhomogeneous reactor with variations in temperature and compositions.

$$\Phi_{\min} = \sum_k^n \Phi_k V_k = \sum_k^n L_k \left(\frac{\Delta G_k}{T}\right)^2 V_k \tag{105}$$

For a specified total reaction rate.

$$\sum J_k V_k = -\sum L_k \frac{\Delta G_k}{T} V_k = \text{Constant} \tag{106}$$

the Cauchy-Lagrange method of constant multipliers yields

$$\frac{\partial \sum\limits_m \Phi_m}{\partial\left(\dfrac{\Delta G_k}{T}\right)} + \lambda \frac{\partial \sum\limits_m J_m}{\partial\left(\dfrac{\Delta G_k}{T}\right)} = 2L_k\left(\frac{\Delta G_k}{T}\right) + \lambda L_k = 0 \tag{107}$$

Thus we obtain

$$\frac{\Delta G_k}{T} = -\frac{\lambda}{2} \tag{108}$$

Eq. (108) implies that for a given total reaction rate and a given total volume, minimum entropy generation is obtained when the driving force $\Delta G/T$ for the reaction is equal in all n subsystems. According to the linear duality theory, the results of the optimization will be the same if we maximize the total reaction speed for a given entropy generation. Therefore, thermodynamically efficient reactor design that facilitates a uniform $\Delta G/T$ in all parts of the reactor volume should be recommended. This result is independent of local variations in the reaction rate.

Another consequence of Eq. (108) is that if we arrange the n subsystems in time instead of in space, than the collection of subsystems constitutes the reaction path of a batch reactor. For a specified conversion within a specified time and with the minimum total entropy generation, the sum of $J_k V_k$ for the n time intervals must again be constant, and the sum of $J_k\left(\dfrac{\Delta G_k}{T}\right)V_k$ should be minimized. The result is the same with Eq. (108), and implies the equipartition of forces. Hence for a given total conversion and a reaction time, minimum entropy generation is obtained in a batch reactor when the driving force $\Delta G/T$ is equal in all n time intervals. Similarly maximum total conversion is obtained for a given entropy generation and reactor time when the driving force $\Delta G/T$ is uniform.

Gibbs-Helmholtz relation is

$$\left[\frac{\partial(\Delta G/T)}{\partial(1/T)}\right]_P = \Delta H \tag{109}$$

If ΔH is constant, integrating of Eq. (109) from an equilibrium temperature T_{eq} to the optimal temperature T_{opt} yields

$$\frac{\Delta G_{\text{opt}}}{T_{\text{opt}}} = \Delta H\left(\frac{1}{T_{\text{opt}}} - \frac{1}{T_{\text{eq}}}\right) \tag{110}$$

There is no constant of integration due to boundary condition that both $\Delta G/T$ and $\Delta(1/T)$ are zero at equilibrium. However, ΔH most of the time will be temperature dependent. For example, in ammonia production from hydrogen and nitrogen, the goal is to maximize the output of ammonia at the exit. It is found that an approximately constant ΔT between the optimal path and the equilibrium temperature provides the optimal temperature profile, which reduces the exergy loss by approximately 60 % in the reactor.

Generalization of equipartition of forces principle to multiple, independent-rate controlled reactions, and application to multiphase and coupled phenomena, such as reactive distillations, may lead to improvement in both energy use and the investment costs.

4.1. Case studies

An example is the study of oxidation of sulfur dioxide to trioxide over a vanadium pentoxide catalyst

$$SO_2 + 1/2O_2 \rightarrow SO_3 \tag{111}$$

Sulfur trioxide is used to produce sulfuric acid, one of the most common chemicals used in industry. The reaction is strongly exothermic. Here a tubular reactor is considered. The following relations are from Kjelstrup et al. [19]. The entropy generation rate per unit volume of a chemical reactor is given by

$$\Phi = J_r\left(-\frac{A}{T}\right) = J_r X \tag{112}$$

where J_r is the reaction rate, and $(-A/T)$ is the thermodynamic driving force X, and A is the affinity that is the Gibbs energy of reaction

$$A = \Delta G_r \tag{113}$$

The minimization problem is formulated as a Lagrange optimization

$$\frac{\delta}{\delta X}\int(-J_r X + \lambda J_r)dz = 0 \tag{114}$$

where λ is the Lagrange multiplier for the constraint of constant output J

$$J = A_c \int J_r dz \tag{115}$$

where A_c is the cross-sectional area of the reactor. Since X can be related to concentration c we can differentiate with respect to c instead of X, and obtain from Eq. (114)

$$\frac{\Delta}{\delta c} \int (-J_r X + \lambda J_r) dz = 0 \tag{116}$$

A solution for Eq. (116) is possible for a particular expression for J_r.

Considering a nonlinear chemical reaction, a practical solution may be obtained by the integration of the Gibbs-Helmholtz equation from equilibrium to optimal state.

$$X_{\text{opt}} = -\int_{\text{eq}}^{\text{opt}} \Delta H_r d(1/T) \tag{117}$$

At equilibrium $A_{\text{eq}} = 0$. The equilibrium temperature for a given conversion is found by using the condition $J_r = 0$. When we know the enthalpy of reaction as a function of temperature, we can carry out the integral in Eq. (117). Since X_{opt} is known from Eq. (116), we can then find T_{opt}.

Optimum criterion with a constant output is the operation of reactor with highest possible temperature of cooling medium in the heat removal system. For a worked example of the nonadiabatical reactor, there are $N = 4631$ cylindrical tubes packed with catalyst and surrounded by a constantly boiling liquid at 703 K. Sulfur dioxide and air are fed into the reactor at a total pressure P_T, in volume fractions of $y_{SO_2} = 0.11$ and $y_{O_2} = 0.10$, and their inner diameter is 7 mm. The empirical expression of J_r takes into account the diffusion and reaction kinetics, and we have

$$J_r = -k \left(\frac{P_{SO_2}}{P_{SO_3}} \right)^{1/2} \left[P_{O_2} - \left(\frac{P_{SO_3}}{P_{SO_2} K_p} \right)^2 \right] \tag{118}$$

The rate and rate constant k have the dimension of mol/g of catalyst, while the partial pressures are given in bar. The equilibrium constant is based on partial pressures, K_p (bar$^{-0.5}$). The rate of reaction is negligible below 670 K and the temperature of the materials in the reactor should not exceed 880 K, therefore the temperatures of 670 and 880 K are practical boundaries of real operation. An expression for k is given by

$$k = \exp\left[-\frac{97800}{T} - 110\ln(1.8T) + 913\right] \tag{119}$$

where T is in K. Similarly K_p in terms of temperature is given by

$$K_p = \exp\left[\frac{11800}{T} - 11.2\right] \tag{120}$$

The volume flow into the reactor is 3590 kmol/h. By introducing the degree of conversion x, and the pressure P as variables we obtain

$$J_{SO_3} = k\rho\frac{1-x}{x}\left[P_{SO_2}\left(\frac{0.91-0.5x}{1-0.0055x}\right)\left(\frac{P}{P_T}\right) - \left(\frac{x}{(1-x)K_p}\right)^2\right] \tag{121}$$

where ρ is the mass density of the gas mixture.

The affinity of the reaction is

$$A = \mu_{SO_3} - \mu_{SO_2} - \frac{1}{2}\mu_{O_2} \tag{122}$$

and the driving force is given as

$$\frac{-A}{T} = -R\left[\ln\left[\left(\frac{P_T}{P_{SO_2}P}\right)\left(\frac{x}{1-x}\right)\left(\frac{1-0.055x}{0.91-0.5x}\right)^{0.5}\right] - \ln K_p\right] \tag{123}$$

Eqs. (121) and (123) show that J_r is a complicated function of P, T, and composition, and it cannot be expressed in the form of linear phenomenological equation with constant coefficients.

Estimations indicate that 80% of the conversion is accomplished in the first 20% of the reactor length. By solving Eq. (116) with Eqs. (122), and (123) we obtain the optimum force X_{opt} as a function of λ

$$X_{opt} = -\lambda - R\frac{R\Gamma\left[\dfrac{1}{x(1-x)} + \dfrac{0.225}{(1-0.055x)(0.91-0.5x)}\right]}{\left[\dfrac{\Gamma}{2x(1-x)} + \dfrac{0.45P_{SO_3}}{(100.055x)^2}\left(\dfrac{P}{P_T}\right) + \dfrac{2x}{(1-x)^3 K_p^3}\right]} \tag{124}$$

where

$$\Gamma = P_{SO_2}\left(\frac{0.91-0.5x}{1-0.055x}\right)\left(\frac{P}{P_T}\right)-\left(\frac{x}{(1-x)K_p}\right)^2$$

The results for X_{opt} can be combined with Eq. (114) for a simultaneous determination of the conversion x and P.

The optimal driving force is small, around 10 J/mol K, and it is almost constant through the reactor. However, the actual force is very high at the inlet (about 30 J/mol K), and it passes a minimum (about 10 J/mol K) in the central region before it becomes flat (about 16 J/mol K). This minimum is due to high temperature produced from the exothermic reaction.

The large deviation between the actual and the optimal force indicates that the lost work can be saved without changing the amount of product. modifying only the operating conditions that bring the temperature profile or the actual driving force close to the theoretical optimum can reduce entropy generation. One realistic solution would be maintaining a stepwise control of temperature by using heat exchangers at the right position. When the temperature is controlled, the entropy generation and hence the lost work in the reactor will also be controlled. When the optimal driving force is found for a given product rate, one may consider further optimization; for example the production rate can be further traded with reactor size.

Generally a chemical reaction has a nonlinear relation between the rate J_r and the driving force $-A/T$. Chemical reactors are often designed to operate at the maximum rate of reaction. Alternative path to that is a reactor operated with the minimum loss of useful work. The lost work per unit time in a chemical reactor is given by the Gouy-Stodola theorem, and is obtained by integrating the entropy generation rate over the reactor volume

$$W_{\text{lost}} = T_o A_c \int \Phi dx \tag{125}$$

where Φ is the entropy generation per unit volume, and A_c is the cross sectional area that is assumed to be constant, and x is the length. Eq. (125) implies a plug-flow reactor since the integration is over the volume; in a batch reactor integration is over the time. The reaction takes place at temperature T, while T_o shows the temperature of the surroundings. Any reduction in lost work can give economic gains, which is not uniquely dependent on the reductions in lost work, since it varies with energy and labor cost. So that it may be useful to undertake energy optimization studies separately from economical analyses.

Minimization of W_{lost} means minimization of integral (125) with the constraints of constant production rate J (mole/h)

$$J = A_c \int J_r dx \tag{126}$$

where J_r is the reaction rate (flow). The minimization without the constraint yield the meaningless result of $\Phi = 0$. It is a common practice to optimize under constraints and to use the Euler-Lagrange method.

When a chemical reaction is the only process in a system, the entropy generation is the products of flows J_r and conjugate forces X

$$\Phi = J_r\left(-\frac{A}{T}\right) = J_r X \tag{127}$$

Here, A is the affinity (the Gibbs energy of reaction) and T is the temperature. The flow for reaction is given by

$$J_r = L(X)X \tag{128}$$

where L is a phenomenological coefficient, which may be dependent on the force.

An Euler-Lagrange minimization problem is formulated as

$$-\frac{\delta}{\delta X}\int (J_r X + \lambda J_r)dx = 0 \tag{129}$$

where λ is the Lagrange multiplier to be obtained in terms of J_r.

For a required production rate, minimum lost work can be obtained by compatible operating conditions with the design parameters. This is achieved when $d\Phi / dJ_r$ is constant. Using Eq. (127), the minimum lost work is obtained when

$$X + J_r \frac{dX}{dJ_r} = \text{constant} \tag{130}$$

The solution of Eq. (129) is obtained with Eq. (130) when λ = constant. Eq. (130) means that the operating temperature should be parallel to the equilibrium temperature, when the enthalpy of reaction is constant. Finding the optimum driving force that gives the maximum reaction rate for a total entropy generation rate is mathematically equivalent to the problem of finding the driving force that gives minimum entropy generation for a constant total production; this means a possible trade off between the production rate and the energy cost.

Consider a chemical reaction far from equilibrium, given in Eq. (131), with the unknown phenomenological coefficients.

$$\mathrm{A} \Leftrightarrow \mathrm{B} \tag{131}$$

and the reaction rate is given by

$$J_r = k_1 c_A = k_2 c_B \tag{132}$$

where k_1 and k_2 are the constants, and c_A and c_B show the concentrations of A and B respectively. The following relations are from Kjelstrup et al. [19]. The normalized concentration *c,* due to mass conservation, is $c = c_A = 1 - c_B$. The output of B per unit time J is given as

$$J = A_c t \int [k_1 c - k_2 (1-c)] dx \tag{133}$$

The driving force of the reaction is

$$X = -\frac{1}{T}(\mu_A - \mu_B) = -R\left[\ln\left(\frac{1-c}{c}\right) - \ln K_{eq}\right] \tag{134}$$

where $K_{eq} = k_1/k_2$ is the equilibrium constant, and R the gas constant, and the mixture is assumed to be ideal. The rate J_r, given in Eq. (132) is related to the force, Eq. (134) in a nonlinear way, as it is the case for chemical reactions far from equilibrium. The total entropy generation is determined by using the local value of the reaction rate and conjugate driving force in Eq. (127). Since the force X can be expressed in terms of concentration c, Eq. (129) becomes

$$\frac{\partial}{\partial c}(J_r X + \lambda J_r) = 0 \tag{135}$$

The result is given by

$$R\left[\ln\left(\frac{1-c}{c}\right) - \ln K_{eq}\right] - \frac{RJ_r}{k_1 + k_2}\left(\frac{1}{1-c} + \frac{1}{c}\right) = \lambda \tag{136}$$

The first term on the right-hand side of Eq. (136) is the thermodynamic force, and it is not necessarily constant when we have minimum lost work. The optimum force that gives the minimum total entropy generation rate becomes

$$X_{opt} = \frac{RJ_r}{k_1 + k_2}\left(\frac{1}{1-c} + \frac{1}{c}\right) - \lambda \tag{137}$$

To find the value of λ as a function of J_r and the rate constants of the reaction, we solve $c(x)$ from Eq. (136) and substitute in Eq. (133). This value can be used with Eq. (137) to determine the value of the optimal force along the reactor.

The optimum value of Lagrange multiplier λ is obtained as follows. Eq. (129) gives

$$-X-\left(\frac{\partial X}{\partial \ln J_r}\right)=\lambda \tag{138}$$

By substituting Eq. (128) into Eq. (138), we have

$$X_{\text{opt}}=-\frac{\lambda}{2}+\frac{1}{2}(X+\lambda)\frac{\partial \ln L}{\partial \ln X} \tag{139}$$

The solution for λ is obtained as

$$\lambda=\frac{-2J+J_r A_c \int L^*(-X)^2 dx}{tA_c \int (L+(-X)L^*)dx} \tag{140}$$

where $L^*=-\dfrac{\partial L}{\partial X}$

L is independent of $-X$ means that L can only depend on the state variables. The coefficient can depend on the temperature, pressure, and some of the concentrations, and we must be able to vary ($-X$) independent of L.

The optimal force is constant when the phenomenological coefficients are independent of the force. For the reactions close to equilibrium, L is assumed to be independent of the forces, although it may vary along the path because the composition, temperature, and pressure vary along the reactor.

By assuming $L^* = 0$, Eq. (128) yields

$$-\left(\frac{\partial \ln J_r}{\partial X}\right)^{-1}=-\frac{A}{T} \tag{141}$$

From Eq. (139) we obtain

$$[X(x)]_{\text{EoF}}=-\frac{\lambda}{2} \tag{142}$$

Here the subscript EoF means the equipartition of the force, and the force is constant. From Eq. (140) the Lagrange multiplier λ is obtained as

$$\lambda=\frac{2J}{tA_c}[\int L(x)dx]^{-1} \tag{143}$$

The specified production rate in terms of λ becomes

$$J = -\frac{\lambda}{2} tA_c \int L(x)dx \tag{144}$$

and the constant force is obtained as

$$X_{\text{EoF}} = \left(\frac{J}{tA_c}\right)[\int L dx]^{-1} \tag{145}$$

The force distribution that gives minimum lost work is uniform over the length of the steady-state reactor or over the time in a batch reactor. Since the affinity $A < 0$ for a spontaneous reaction, λ is a negative constant. According to Eq. (138), the force is close to constant when we have

$$L^* << -\frac{L}{X} \tag{146}$$

Near equilibrium, the local entropy generation rate per volume is given by

$$\Phi_{\min}(x) = LX_{\text{EoF}}^2 = L\left(\frac{J}{tA_c}\right)^2 [\int L(x)dx]^{-2} \tag{147}$$

If the phenomenological coefficients are constant than Φ is constant through the reactor, and $\Phi_{\min}$ becomes the summation of local entropy generations.

$$\Phi = A_c \int \Phi_{i,\min} dx = \left(\frac{J}{tA_c}\right)^2 [\int L(x)dx]^{-1} \tag{148}$$

Eq. (148) shows that Φ becomes smaller for larger values of L(x) for a given reactor. Eq. (148) and can be used to determine the minimum lost work for a process, and Eq. (137) reduces to a constant force when the reaction is close to equilibrium. Consider the perturbation δc on concentration at equilibrium c_{eq}

$$c = c_{\text{eq}} + \delta c \tag{149}$$

The concentration at equilibrium is obtained for $J_r = 0$

$$c_{\text{eq}} = \frac{k_2}{k_1 + k_2} \tag{150}$$

Close to equilibrium, we have

$$J_r = (k_1 + k_2)\delta c \tag{151}$$

and

$$X = R\delta c \frac{(k_1 + k_2)^2}{k_1 k_2} \tag{152}$$

$$X = \frac{RJ_r}{k_1 + k_2}\left(\frac{1}{(1 - c_{\mathrm{eq}})} + \frac{1}{c_{\mathrm{eq}}}\right) > 0 \tag{153}$$

Eq. (152) shows a constant force, while Eq. (153) shows that even if $-X$ is constant J_r need not be so since c_{eq} varies along the path. The path of minimum lost work is therefore not a path of constant entropy generation rate per volume.

The problem of finding the driving force that gives the maximum reaction rate for a specified total entropy generation rate is mathematically equivalent to the finding the driving force that gives minimum entropy generation for a total production J.

We can expect minimum lost work by equipartition of forces when the reaction enthalpy varies little with composition, and if the reaction rate is a linear function of the conjugate force. Construction of particular operating path for a reactor with specific output requires knowledge of reaction kinetics. The path that gives the minimum lost work is needed to trade-off between energy efficiency and production rate for a given reactor. It may be profitable to increase reactor investments and lower the energy cost for the same product, or to lower the production rate for a given energy input. For example, a considerable (30%) reduction in the lost work can be achieved by lowering the production by only slightly (5%).

5. SEPARATION

Since the minimization of entropy generation is not always an economic criterion, it is necessary to relate the overall entropy generation and its distribution to the economy of the process. To do that, we may consider various processes with different operating configurations. Sometimes, an existing design may be modified to attain an even distribution of forces hence an even distribution of entropy generation. In the next section extraction and distillation are considered for thermodynamic analysis.

5.1. Extraction

Consider a simple mixer for extraction (Fig 12). In the minimal entropy generation, the size V, the time t and the duty J are specified, and average driving force is also fixed. Beside that we can define the flow rate Q and the inlet concentration of the solute, and at steady state outlet concentration is determined. The only unknown variables are the solvent flow rate and composition, and one of them is decision variable; specifying the flow rate will determine the solvent composition. Cocurrent and countercurrent flow configurations of the extractor can now be compared with the same initial specifications (V, t, J, Q, c). Cocurrent operation will yield a large entropy generation P_2 than the countercurrent operation P_1, and investigation of the implication of this on the decision variable is important. For a steady state and adiabatic operation, we have the following relations from Tondeur and Kvaalen [54]. For the processes 1 and 2 with the solvent flow rates of F_1 and F_2 we have

$$F_1 \Delta s_1 = -\Delta S + P_1 \tag{154}$$

$$F_2 \Delta s_2 = -\Delta S + P_2 \tag{155}$$

where ΔS is the total entropy change, and Δs_1 and Δs_2 are the changes in specific entropies of the solvent. Subtracting Eq. (155) from Eq. (154), we have

$$F_1 \Delta s_1 - F_2 \Delta s_2 = P_1 - P_2 < 0 \tag{156}$$

The load is defined as

$$J = F_1 \Delta c_1 = F_2 \Delta c_2 \tag{157}$$

where Δc is the concentration change of the solute in the solvent, respectively throughout the process. Combining Eqs. (156) and (157), we obtain

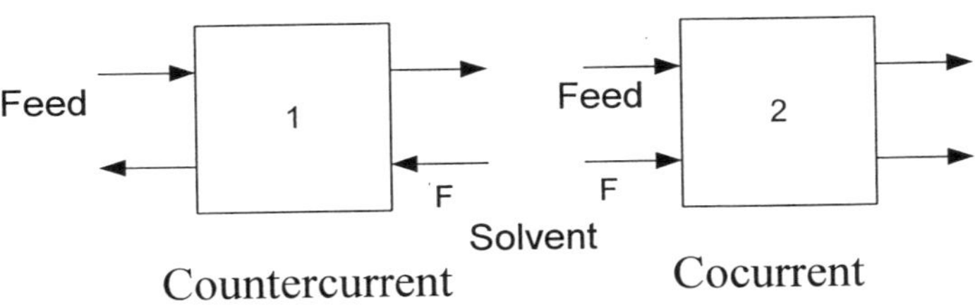

Fig. 12. Countercurrent and cocurrent extractor operations.

$$\frac{\Delta s_1}{\Delta c_1} < \frac{\Delta s_2}{\Delta c_2} \tag{158}$$

The specific entropy of a solvent increases with the solute concentration, and if the inlet solvent is the same, inequality in Eq. (158) yields $\Delta c_1 > \Delta c_2$, and hence Eq. (157) shows that $F_1 < F_2$. This means that the solvent-flow rate is smaller in the less dissipative operation, and the solvent at the outlet is more concentrated. That is the operating conditions of solvent determine the less dissipative operation. Whether this optimum is an overall economic optimum will mainly depend on the cost of the technology.

We can also compare the two processes with the same total entropy generations, the same size and duration, and the same phenomenological coefficients. Process 1 only has the equipartitioned forces, therefore the duties of these processes will be different. The total entropy generations for the processes are expressed as

$$P_1 = P_{av1} = L(X_{av1})^2(Vt) \tag{159}$$

$$P_2 = \int\int L \cdot X_2 dVdt > P_{av2} = L(X_{av2})^2(Vt) \tag{160}$$

Since $P_1 = P_2$, combination of Eqs. (159) and (160) yields $X_{av1}^2 > X_{av2}^2$, and hence $|J_1| > |J_2|$ that is the flow rate for the equipartitioned process 1 is larger than that of process 2 at a given size, duration, and entropy generation.

In another operating configuration, we can compare the respective size and durations for specified duty and entropy production, and Eqs. (159) and (160) are still valid, and we have $P_1 > P_{av2}$ and $J_1 = J_2$, which yield

$$\frac{P_1}{J_1} > \frac{P_{av2}}{J_2} \tag{161}$$

and thus

$$X_{av1} > X_{av2} \tag{162}$$

$$(Vt)_1 < (Vt)_2 \tag{163}$$

This result indicates that for given flow and entropy generation, the equipartitioned configuration is smaller in size for a specified operational time. Alternatively, it requires less contact time for a given size, and thus a higher throughput.

To determine an economic optimum, we assume the operating costs are a linear function of the solvent entropy change and entropy generation, and the investment costs are linear function of the space and time of the process. The total cost is expressed as

$$C_{\mathrm{T}} = aP + b + c\tau Vt = \int\int (aLX^2 + c\tau) dVdt + b \tag{164}$$

where τ is the amortization rate and *a, b*, and *c* are the constants related to the costs. Integral in Eq. (164) is subject to the constraint of a specified flow

$$J = \int\int LXdVdt \tag{165}$$

The variational technique can be used to minimize the total cost, and the Euler equation for the variable X is given by

$$\frac{\partial}{\partial X}(aLX^2 + c\tau + \lambda LX) = 0 \tag{166}$$

where λ is a Lagrange multiplier. Eq. (166) yields

$$2aLX + \lambda L = 0 \tag{167}$$

$$X = -\frac{\lambda}{2a} = \text{constant} \tag{168}$$

The obtained value of X that minimizes the total cost subject to J is a uniform distribution.

5.2. Distillation

Since it requires a large amount of heat in reboiler and discharges a similar amount of heat at the condenser, distillation column resembles a heat engine that requires an optimum operating condition. Therefore a distillation column is an energy intensive process, and heat causes the work for separating the components of a feed stream into products

The thermodynamic analysis may be an effective tool for identifying the possible improvements in distillation column design by understanding the thermodynamic inefficiencies in a column. Considering the distillation column as a heat engine, the lost work profiles can be determined to quantify the inefficiency in terms of the pressure drop, heat and mass transfer, and coupling between heat and mass transfer. The column efficiency may be related to the optimal feed conditions including the feed plate location, leading to the minimum irreversibility based on the utility requirements.

The thermodynamic optimization of a distillation column should lead to producing more uniform irreversibility distributions. This may be achieved through the column modifications, such as feed condition, feed stage location and use of intermediate exchangers in order to reduce irreversibility in sections with large driving forces and to increase irreversibility in sections with small driving forces. The pinch point calculations may also be used for determining the minimum energy demand for distillation with distributed components and side-product withdrawals. The minimum energy calculations for ideal and non-ideal distillations shows that the minimum energy requirement of a distillation column sets a thermodynamically defined operating limit.

Distillation columns generally operate far from their thermodynamically optimum conditions. In absorption, desorption, membrane separation, and rectification, the major irreversibility is due to the mass transfer. If the mass transfer is optimum, the conditions on the concentration profiles provide the minimal irreversibility leading to minimum the energy consumption. Analysis in a sieve tray distillation column reveals that the irreversibility on the tray is mostly due to the bubble-liquid interaction on the tray, and the mass transfer is the largest contributor to the irreversibility [34,35]. With additional investments in the column for the distillation of toluene and benzene, the entropy production rate (exergy loss) could be reduced by 7%, for the same operating conditions [33].

Nonequilibrium molecular dynamics simulations show that the assumption of local equilibrium in a column with heat and mass transfer is acceptable. The entropy production rate, described by the irreversible thermodynamics, gives a detailed mathematical formulation of the dissipated energy in a system at local thermodynamic equilibrium.

The following relations are from Ratkje et al. [33]. Dissipation function in a binary distillation, where the transport of heat and mass of two components takes place, is given by

$$\Psi = -J_q \nabla \ln T - \sum_1^2 J_i \nabla \mu_{i,T} \tag{169}$$

or the rate of volumetric entropy generation Φ is

$$\Phi = -J_q \frac{\nabla T}{T^2} - \frac{1}{T} \sum_1^2 J_i \nabla \mu_{i,T} \tag{170}$$

When the pressure is constant $\nabla \mu_{i,T} = \nabla \mu_i^c$, which is the concentration dependent part of the chemical potential gradient. Through the Gibbs-Duhem equation, we can relate the chemical potentials of heavy h and light l components in the gas phase as follows

$$\nabla \mu_h^c = -\frac{y_l}{y_h} \nabla \mu_l^c \tag{171}$$

where y_l and y_h are the mole fractions in the gas phase of the light and heavy components. From Eqs. (169) and (171) we obtain

$$\Phi = -J_q \frac{\nabla T}{T^2} - J_d y_l \frac{\nabla \mu_l}{T} \tag{172}$$

where J_d (in m^3 m^{-2} h^{-1}) is the relative mass flux across the interface

$$J_d = \frac{J_l}{y_l} - \frac{J_h}{y_h} \tag{173}$$

The phenomenological equations that follow from Eq. (172) are

$$J_q = -L_{qq} \frac{\nabla T}{T^2} - L_{ql} y_1 \frac{\nabla \mu_l}{T} \tag{174}$$

$$J_d = -L_{lq} \frac{\nabla T}{T^2} - L_{ll} y_l \frac{\nabla \mu_l}{T} \tag{175}$$

where L_{ji} is the local phenomenological coefficients, which can be determined from experiments. For isothermal conditions, the phenomenological coefficients for mass transfer are defined as

$$L_{ll} = -\left(\frac{J_d T}{y_l \nabla \mu_l} \right)_{\Delta T = 0} \tag{176}$$

Using the chemical force for mass transfer

$$y_l \frac{\nabla \mu_l}{T} = \frac{J_d}{L_{ll}} - \frac{L_{lq}}{L_{ll}} \frac{\nabla T}{T^2} \tag{177}$$

we obtain an expression for the heat flow

$$J_q = -\left(L_{qq} - L_{ql} \frac{L_{lq}}{L_{ll}} \right) \frac{\nabla T}{T^2} + \frac{L_{lq}}{L_{ll}} J_d \tag{178}$$

On the other hand Fourier's law of heat conduction without mass transfer is expressed as

$$\left(J_q\right)_{J_d = 0} = -k \nabla T \tag{179}$$

The thermal conductivity k is defined in terms of the phenomenological coefficients as follows

$$k = \left(L_{qq} - L_{ql} \frac{L_{lq}}{L_{ll}} \right) \frac{1}{T^2} \tag{180}$$

The total rate of entropy generation for a stage is obtained by integrating over the transport path between liquid and vapor.

$$\Phi = \int_V \Phi_v dV \tag{181}$$

The entropy generation rate is determined with quasi-steady state calculations. For this purpose the following constant gradient in the gas phase on each stage are used

$$\nabla T = \frac{\Delta T}{\Delta x} \tag{182}$$

$$\nabla \mu_1 = \frac{\Delta \mu_1}{\Delta x} \tag{183}$$

By assuming that y_l and T are approximately constant, and using Eq. (181), the entropy generation for a stage is obtained as

$$\Phi = -\frac{1}{T^2} \frac{\Delta T}{\Delta x} \int J_q dV - \frac{y_1}{T} \frac{\Delta \mu_1}{\Delta x} \int J_d dV \tag{184}$$

where $dV = dAdx$, A is the contact area, and x is the transfer distance on the stage. The expressions inside the integrals are the heat and mass transfer per unit time in the mixture volume on a stage, respectively, and can be calculated using Eqs. (174) and (175). Using Eq. (181), the entropy generation rate can be estimated.

The flow on a stage J_d can be calculated from the diffusion coefficients in the gas phase and from the energy balance. In an adiabatic distillation, the heat flow across the interface contains the latent heat and heat conducted away from the interface

$$(J_h + J_l)H_n = (J_h + J_l)H_{n-1} + J_h \Delta H_h + J_l \Delta H_l - k\Delta T / \Delta x \tag{185}$$

If the enthalpy of the mixtures on stage n and n-1, H_n, H_{n-1}, respectively, are similar, and the temperatures of these stages are close to each other, we have

$$J_l \Delta H_l \approx -J_h \Delta H_h \tag{186}$$

This means that the heat of vaporization is approximately equal to the heat of condensation. From Eqs. (177) and (186), the flow on a stage is expressed by

$$J_d = J_l \left(\frac{1}{y_l} + \frac{1}{y_h} \frac{\Delta H_l}{\Delta H_h} \right) \tag{187}$$

Diffusion of the light component is defined by Fick's law for the gas phase

$$J_l = -D \frac{\Delta c_1}{\Delta x} \tag{188}$$

where D is the diffusion coefficient of the light component and Δc_1 is the concentration difference of light component across Δx. The concentration difference is

$$\Delta c_l = c_{l,g} - c_{l,g}^* \tag{189}$$

where the concentration in the gas phase at the total pressure P_T is given by

$$c_{l,g} = y_l \frac{P_T}{RT} \tag{190}$$

At the liquid-vapor interface we have

$$c_{l,g}^* = y_l^* \frac{P_T}{RT} \tag{191}$$

where the mole fraction y_l^* is the inlet composition in the liquid. Introducing Eqs. (187) and (188) into Eq. (176), and with the assumptions of constant driving forces, we express the phenomenological coefficient of mass transfer as follows

$$\overline{L}_{ll} = \frac{T}{y_l \Delta\mu_l}\left[D\Delta c_l\left(\frac{1}{y_l} + \frac{1}{y_h}\frac{\Delta H_l}{\Delta H_h}\right)\right] \tag{192}$$

The average coefficient $\overline{L}_{ll}$ is defined as

$$\overline{L}_{ll} = \frac{\int_0^{\Delta x} L_{ll} dx}{(\Delta x)^2} \tag{193}$$

The phenomenological coefficients obtained from Eq. (192) may vary considerably from enriching section to stripping section, and this should be taken into account in the optimization criterion.

For an energy intensive process, the energy efficiency based on the second law of thermodynamics may be defined as

$$\eta = \frac{W_{\min}}{W_{\min} + \int_0^t \Phi dt} \tag{194}$$

where $W_{\min}$ is the minimum work required for a required separation over a reversible path. In a steady state distillation heat and mass transfer is coupled, and if the temperature field or chemical force is specified in the column, the other force would be defined. A maximum in the second law efficiency may be obtained by minimizing the entropy generation rate with respect to one of the forces. For example, assuming that the contribution due to the difference in chemical potential is dominant, the change of the entropy generation with respect to the chemical force can be studied.

The chemical driving force on a stage has in and out concentrations as boundary conditions. In the enriching section below the feed plate, a flow J_d from liquid to vapor occurs; while in the stripping section the direction of flow is from vapor to liquid. For a two stage in the column with specified inlet and outlet compositions, the entropy generation rates are given by

$$\Phi_1 = \int_V L_{ll,1} X_1^2 dA dx, \quad \Phi_2 = \int_V L_{ll,2} X_2^2 dA dx \tag{195}$$

where the net separation flow $J_{d,1}$, for example, for stage 1 is obtained as

$$\int_V J_{d,1} dA dx = \int_V L_{ll,1} X_1 dA dx \tag{196}$$

The chemical force is replaced by X in Eqs. (195) and (196). As the level of separation is fixed, the boundary conditions for the forces are specified, and an increase in the force in one stage must lead to a reduction in another stage. The sum of the entropy generation rates is

$$\Phi_1 + \Phi_2 = \int_V (L_{ll,1} X_1^2 + L_{ll,2} X_2^2)\, dAdx \tag{197}$$

The total flow J_d is given by

$$\int_V J_d dAdx = \int_V (J_{d,1} + J_{d,2}) dAdx = \int_V (L_{ll,1} X_1 + L_{ll,2} X_2)\, dAdx \tag{198}$$

It is desired to have an increase in the flow for a given entropy generation rate, and a reduction in the entropy generation rate for a specified separation; the yield y is defined as the benefit-cost ratio in an economic sense, and given by

$$y = \frac{J_{d,i}}{\Phi_i} = \frac{L_{ll} X_i}{L_{ll} X_i^2} = \frac{1}{X_i} \tag{199}$$

When the derivative of y with respect to X_i is higher in one stage than in another, the separation per entropy generation rate can be increased by increasing the driving force, and hence the entropy generation rate is adjusted by increasing or reducing the force. We can maximize the separation output per entropy generated, by redistributing of forces between the stages, and such a distribution is obtained with the differentiation

$$\frac{d(1/X_1)}{dX_1} = \frac{d(1/X_2)}{dX_2} \tag{200}$$

Eq. (200) yields

$$X_1 = X_2 \tag{201}$$

The equality of forces is independent of the individual values of the phenomenological coefficients, and means that the variation in the entropy generation rate along the column follows the variation in the phenomenological coefficient $\bar{L}_{ll}$. The reversible operation is a limit case achieved when X_1 and X_2 approach zero and y increases toward infinity. Therefore, the practical improvement in the second-law efficiency is to apply the relationship between dX_1 and dX_2. For a constant J_d we obtain

$$dX_1 = -\frac{\overline{L}_{ll,1}}{\overline{L}_{ll,2}} dX_2 \tag{202}$$

Eq. (202) relates the driving forces of the two stages. By knowing $\overline{L}_{ll}$ across the column, we can determine the possible locations for modifications. A uniform entropy generation rate corresponds to either minimum energy costs for a required separation and area investments, or minimum investments for specified energy cost, and leads to the thermodynamically optimum design and operation.

5.3 Case Studies

In a diathermal distillation with heat exchanger all along the column, it is possible to adjust the flow ratio of the phases and thus the slopes of operating lines, and the driving forces along the column. This modification may produce a mass transfer unit more reversible but requiring more transfer units and hence more column height and heat transfer area; it also affects the driving force distribution and entropy generation. In order to show that an optimal distribution exists, consider a steady state operation and that the forces are uniformly distributed. The following relations are from Tondeur and Kvaalen [54]. The investment cost C_i of a transfer unit is assumed to be linearly related to the size V, and the operating costs C_o are linearly related to the exergy consumption

$$C_v = C_i - C_{if} = AV \tag{203}$$

$$C_o = C_{of} + B\Delta X \tag{204}$$

where C_{if} is a fixed investment cost and C_{of} is a fixed operating cost, A and B are the cost parameters, and ΔX_c is expressed as

$$\Delta X_c = \Delta X_m + T_o P_{\text{av}} \tag{205}$$

Here T_o is a reference temperature (dead state, usually 298 K), and ΔX_m is a thermodynamic minimum value. The total flow $J = LVX_{av}$ can be written by combining with Eq. (203)

$$J = \frac{LX_{\text{av}} C_v}{A} \tag{206}$$

where C_v is the variable part of the investment cost. Eliminating the constant force between Eq. (205) and the total entropy generation P_{av}, we have

$$P_{\text{av}} = \frac{AJ^2}{LC_v} \tag{207}$$

Substituting Eq. (207) into Eq. (203) and the latter into Eq. (204), we obtain a relationship between the operating and investment costs

$$C_o = \frac{K_1}{C_v} + K_2 \tag{208}$$

where

$$K_1 = \frac{ABT_oJ^2}{L}, \text{ and } \quad K_2 = C_{of} + B\Delta X_m$$

K_1 and K_2 are constants in terms of the minimum thermodynamic work ΔX_m and the transfer coefficients L. The optimal size is obtained by minimizing the total cost of operating and investments costs, which is linearly amortized with amortization rate τ

$$C_T(C_i) = \tau C_i + C_o = \tau C_i + \frac{K_1}{C_i - C_{if}} + K_2 \tag{209}$$

The minimum of C_T is obtained as

$$\frac{dC_T}{dC_i} = 0 = \tau - \frac{K_1}{C_v^2} \tag{210}$$

Combining Eq. (207) and (210) and the definition of K_1 yields

$$\frac{BP_{\text{av}}}{(C_v)_{\text{opt}}} = \frac{BP_{\text{av}}}{AV_{\text{opt}}} = \frac{\tau}{T_o} \tag{211}$$

According to Eq. (211), the quantities BT_oP_{av}, which is related to irreversible dissipation and τV_{opt}, should be equal in any transfer unit. Generally operating costs are linearly related to dissipation, while investment costs are linearly related to the size of equipment. The optimum size distribution of the transfer units are obtained when amortization cost is equal to the cost of lost energy due to irreversibility. The cost parameters A and B may be different from one transfer unit to another. When $A = B$ all along the column, then $P_{\text{av}}/V_{\text{opt}}$ is a constant, and the optimal size distribution reduces to equipartition of the local rate of entropy generation. The optimal size of a transfer unit can be obtained from Eq. (210)

$$(C_v)_{\text{opt}} = C_{i,\text{opt}} - C_{if} = J\left(\frac{ABT_o}{L\tau}\right)^{1/2} \tag{212}$$

and from Eq. (203), we have

$$V_{\text{opt}} = J\left(\frac{BT_o}{AL\tau}\right)^{1/2} \tag{213}$$

By distributing the entropy production as evenly as possible along the space and time economical separation process would be designed and operated.

In a distillation column heat is supplied at a higher temperature source in reboiler, and then rejected at a lower temperature in condenser (Fig 13). Assuming the column as a reversible heat engine the net work available from thermal energy is

$$W_{\text{heat}} = Q_R\left(1-\frac{T_o}{T_R}\right)-Q_C\left(1-\frac{T_o}{T_C}\right) \tag{214}$$

where T_o is the ambient temperature. The temperature corrections describe the maximum fraction of theoretical work extracted from thermal energy at a particular ambient temperature.

The minimum separation work W_s required for separation is the net change in availability A ($A = H - T_oS$)

$$-W_s = \Delta A_s = A_{\text{prod}} - A_{\text{feed}} \tag{215}$$

The change of availability of separation is the difference between the work supplied by the heat and the work required for separation, which contains the lost work due to irreversibilities

$$\Delta A_s = W_{\text{heat}} - W_{\text{ts}} \tag{216}$$

where W_{ts} is the total work necessary for the separation. Minimizing the lost work due to irreversibility will minimize the total energy needed for separation. Mainly there are three sources contributing to the lost work; these are pressure drop due to fluid flows, heat transfer between the fluids with different temperatures, and mass transfer between the streams that are not in equilibrium.

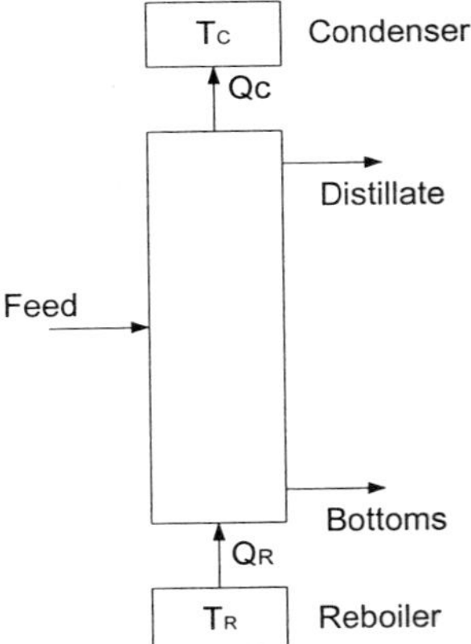

Fig. 13. Distillation column as a heat engine between reboiler and condenser.

- The lost work due to high-pressure drop (as high as 10 psi) is considerable at the condenser and reboiler systems, and is relatively less through the trays (0.1 psi or less). Change of pressure affects the distance from equilibrium, and causes the large temperature difference and hence utility costs between the condenser and reboiler of a distillation column.
- Large lost work due to mass transfer mainly occurs around the feed trays, where the streams with widely different compositions are mixed. The heavier component compositions are being reduced above the feed tray and lighter components are diminishing below. From the thermodynamic perspective, the feed tray location may be adjusted to balance the lost work. Commonly, the feed location is determined at the minimum utility loads and tray count or simply by taking into account light-key and heavy-key component compositions. The relative cost of the heating and cooling media will also influence the location of the feed stage.
- The lost work due to heat transfer occurs due to differences in temperature between the inlet streams of liquid and vapor on each tray, and is usually a large contributor to the total lost work. The heat transfer on each tray is determined from the vapor and liquid enthalpy profiles. Since the heat and mass transfer are coupled, any changes in heat transfer through the heating and cooling modifications will change the internal mass balances. A modifications regarding mass transfer will have the similar effects on heat transfer properties in the column. If a cheap utility is available, intermediate exchangers may be feasible, although the

number of trays will increase due to the operating line closer to the equilibrium curve. Lowering the duties of reboiler and condenser will reduce the overall flow in the column, and results a smaller diameter column design. The basic trend of improving thermodynamic efficiency leads to taller and more slender columns.

If the feed is introduced into a column too cold, then a large amount of heat exchange is required below the feed stage to strip the light components. When the heat transfer rises considerably around the feed tray location, precooling or preheating the feed can be useful to unload the top or bottom sections of columns. Preconditioning the feed is less expensive compared with the inter heating or inter cooling. Heat profiles and heat transfer lost work plots can be used together to determine if the feed preconditioning is necessary.

The exergy balance obtained from a stage-to-stage enthalpy and entropy balances shows the exergy loss profile due to the driving forces of mass and heat transfer. Such a loss profile identifies the sections, for example feed plate section that operates under large thermodynamics inefficiency. Analysis of the exergy profiles shows that columns with a more uniform driving force distribution have a higher thermodynamic efficiency than that of columns with unevenly distributed driving forces.

In many multicomponent mixture separations by distillation, components can display large concentration changes within the column, which cause considerable lost work. One of the ways of overcoming this obstacle is to remove the key components from the feed. As seen in Fig 14, light-nonkey components can be removed by using an absorber, and the bottom products of the absorber provide the feed to the main distillation column. The heavy-nonkey components are removed by using a prestripper, and the over products of the stripper becomes the feed of the main distillation column. These modifications will reduce the load of the column for debottlenecking and can reduce the required number of stages.

As actual distillation column operates under irreversible conditions, it destroys exergy. The exergy analysis determines the thermodynamic efficiency of distillation columns. Traditionally the analysis is based on the overall thermodynamic efficiency of separation and the ratio of the lost work to the ideal work required for the separation.

For a sieve tray distillation column, entropy generation rate can be calculated for heat, mass and momentum transfer accounting the movement of a bubble through a moving liquid pool. Some variables are orifice diameter and weir height for the required separation characteristics of components in vapor-liquid phases and the analysis yields the following results [34,35]:

- Mass transfer and work done against liquid during bubble growth and drag on bubble are the major causes of entropy generation on a tray.

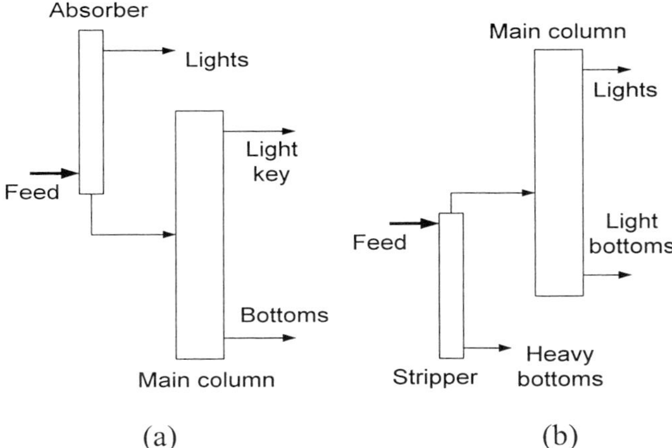

Fig. 14 Prefractionation arrangements: (a) removing light keys with absorber; (b) removing heavy keys with stripper.

- The major contribution to irreversibility on a sieve tray used to separate light hydrocarbons is the bubble-liquid interactions, while the effect of interaction of the flowing liquid with the tray internals are negligible.

- The contribution from heat transfer is small.

- When the bubbles forming at the sieve tray holes, most of entropy generation is high as most of heat and mass transfer occurs before the bubble detachment. The bubble growth after detachment is small, while viscous drag on bubble also contributes to irreversibility.

- The effect of weir height is dominant compared to the effect of sieve hole diameter.

- Increasing weir height shows a monotonic increase in the entropy generation, while increase in sieve hole diameter is associated with a maxima. This diameter range depends on the properties of the mixture on the tray.

REFERENCES

[1] A. Bejan, Entropy Generation Through Heat and Fluid Flow, Wiley, New York, 1982.
[2] A. Bejan, Advanced Engineering Thermodynamics, Wiley, New York, 1988.

[3] A. Bejan, Entropy Generation Minimization, CRC Press, Boca Raton, 1996.
[4] R.B. Bird, W.E. Stewart, E.N. Lightfoot, Transport phenomena, 2nd ed., Wiley, New York, 2002.
[5] C.G. Carrington and Z.F. Sun, Int. J. Heat Mass Transfer, 34 (1991) 2767.
[6] C.G. Carrington and Z.F. Sun, Int. J. Heat Fluid Flow, 13 (1992) 65.
[7] T.O. Ognisty, Chem. Eng. Prog., Feb. (1995) 40.
[8] B.A. Mansson, In Finite-Time Thermodynamics and Thermodynamics, Ed. S. Sienutycz, P. Salamon, Taylor & Francis, New York, 1990.
[9] Y. Demirel, Int. J. Heat Mass Transfer, 43 (2000) 4205.
[10] Y. Demirel and H.H. Al-Ali, Int. J. Heat Mass Transfer, 40 (1997) 1145.
[11] Y. Demirel and R. Kahraman, Int. J. Heat Mass Transfer, 42 (1999) 2337.
[12] Y. Demirel and R. Kahraman, Int. J. Heat Fluid Flow, 21 (2000) 442.
[13] Y. Demirel, Int. Comm. Heat Mass Transfer, 26 (1999) 75-83.
[14] Y. Demirel and S.I. Sandler, Int. J. Heat Mass Transfer, 44 (2001) 2439.
[15] W.R. Dunbar, N. Lior and R.A. Gaggioli, J. Energy Resources Tech., 114 (1992) 75.
[16] M.A. Gallis and J.K. Harvey, J. Thermophysics Heat Transfer, 10 (1996) 217.
[17] E.P. Gyftopoulos and G.P. Beretta, J. Energy Resource Tech., 115 (1993) 208.
[18] S. Kjelstrup and B. Hafskjold, Ind. Eng. Chem. Res., 35 (1996) 4203.
[19] S. Kjelstrup, E. Sauar, D. Bedeaux and H. van der Kooi, Ind. Eng. Chem. Res., 38 (1999) 3046.
[20] J. Kohler, T. Kuen and E. Blass, Chem. Eng. Sci., 49 (1994) 3325.
[21] J. Koehler, P. Poellmann and E. Blass, Ind. Eng. Chem. Res., 34 (1995) 1003.
[22] D. Kondepudi and I. Prigogine, Modern Thermodynamics, From Heat Engines to Dissipative Structures, Wiley, New York, 1999.
[23] D.A. Kouremenos, E.D. Rogdakis and G.E. Houzouris, Energy, 21 (1996) 263.
[24] G.C. Majumdar and S.P. Sengupta, Transactions ASAE, 40 (1997) 1141.
[25] B. Mansson and B. Andresen, Ind. Eng. Chem. Process Des. Dev., 25 (1986) 59.
[26] N.V. Menshutina and I.N. Dorokhov, Drying Tech., 14 (1996) 915.
[27] M.D. Mikhailov and M.N. Ozisik, Unified Analysis and Solutions of Heat and Mass Diffusion, Wiley, New York, 1984.
[28] K. Miyazaki, K. Kitahara and D. Bedeaux, Physica A 230 (1996) 600.
[29] P.K. Nag and P. Mukherjee, Int. J. Heat Mass Transfer, 30 (1987) 401.
[30] D. Poulikakos and J.M. Johnson, Energy, 14 (1989) 67.
[31] R.C. Prasad and J. Shen, Int. J. Heat Mass Transfer, 37 (1994) 2297.
[32] I.K. Puri, Int. J. Heat Mass Transfer, 35 (1992) 2571.
[33] S.K. Ratkje, E. Sauar, E.M. Hansen, K.M. Lien and B. Hafskjold, Ind. Eng. Chem. Res., 34 (1995) 3001.
[34] S. Ray, A.K. Panja and S.P. Sengupta, Chem. Eng. Sci., 49 (1994) 1472.
[35] S. Ray and S.P. Sengupta, Int. J. Heat Mass Transfer, 39 (1996) 1535.
[36] D.A. Sama, J. Energy Resources Technology, 117 (1995) 179.
[37] J.Y. San, W.M. Worek and Z. Lavan, J. Heat Transfer, 109 (1987) 647.
[38] J.Y. San, W.M. Worek and Z. Lavan, Int. J. Heat Mass Transfer, 30 (1987) 1359.
[39] E. Sauar, K.S. Ratkje and K.M. Lien, Ind. Eng. Chem. Res., 35 (1996) 41.
[40] E. Sauar, K.S. Ratkje and K.M. Lien, Computers Chem. Engng., 21, Suppl. (1997) 29.
[41] N.J. Scene and P.A. Aguirre, Ind. Eng. Chem. Res., 28 (1989) 873.
[42] J.C. Schon and B.Andresen, J. Phys. Chem., 100 (1996) 8843.
[43] D.P. Sekulic, J. Heat Transfer, 112 (1990a) 295.

[44] D.P. Sekulic, Int. J. Heat Mass Transfer, 33 (1990) 2748.
[45] S. Sieniutycz, Int. J. Heat Mass Transfer, 20 (1977) 1221.
[46] Z. Szwast and S. Sieniutycz In Eff., Costs, Opt., Sim. And Env. Impact of Energy systems, Vol. 1. pp. 207-212 (1995).
[47] S. Semiotics and J. S. Shiner, J. Non-Equilib. Thermodyn., 19 (1994) 303.
[48] S.K. Som, A.K. Mitra and S.P. Sengupta, J. Energy Resources Tech., 112 (1990) 130.
[49] M. Orin and J. Paris, Comp. Chem. Engng., 21, Suppl. (1997) 23.
[50] M.V. Orin and V.M. Brodyansky, Energy, 11 (1992) 1019.
[51] Z.F. Sun and C.G. Carrington, J. Energy Resources Tech., 113 (1991) 33.
[52] T. Takamatsu, M. Nakaiwa, K. Huang, T. Akiya, H. Noda, T. Nakanishi and K. Aso, Comp. Chem. Engng., 21, Suppl. (1997) 243.
[53] R. Taprap and M. Ishida, AIChE J, 42 (1996) 1633.
[54] D. Tondeur and E. Kvaalen, Ind. Eng. Chem. Res., 26 (1987) 50.
[55] A.M. Tsirlin, V.A. Kazakov and R. S. Berry, J. Phys. Chem., 98 (1994) 3330.
[56] R.J. Zemp, S.H.B. de Faria and M.L.O. Maia, Computers. Chem. Engng., 21 (1997) S523.

Chapter 7

Thermoeconomics

INTRODUCTION

Thermodynamics principles describe the flow, conservation, conversion and quality of energy, and hence have implications for energy management and economics. As thermodynamic imposes directions and limits on the probability of processes, it also implies the use of scarce resources. Economy of processes always involve, matter, energy, entropy and information. The following question has been historically related economics with thermodynamics: How much mechanical work can we attain from a given quantity of coal? The exergy is a physical measure of value of energy, and nonequilibrium thermodynamics has a part in economics. Economic activity aims to achieve a certain structure with minimum overall cost, and the formulation of nonequilibrium thermodynamics introduces entropy and exergy in the structure formation and energy utilization.

Thermodynamics identifies and describes relationships between process performance, and the first and second laws of thermodynamics; it defines the efficiency, which can be used to evaluate losses due to irreversibilities or costs in economic terms. Thermodynamic formulations are necessary for comparison and conversion between different kinds of energies, which may be a necessary step in the net energy analysis and the energy policy discussions. The modern theory of thermodynamics and information theory study the concepts of synergetics, structure formation, evolution, and may bring some fundamental changes in economic evolution, design and maintenance of the physical processes.

Thermal systems involve significant work and/or heat interactions with their surroundings, and appear in almost every industrial plant. Design of the thermal systems requires the application of principles from thermodynamics, fluid mechanics, heat transfer, and engineering economics. Thermoeconomics usually relates exergy and economics for optimizing the design and operating conditions of thermal systems. Optimization of subsystems individually does not guarantee an optimum for the overall system, and often various design variables must be considered and optimized simultaneously. In the optimization, the cost of thermal energy source plays an important role; changing the fuel cost from one year to another and from one place to another will eventually affect the overall design and hence the economical considerations.

1. THERMODYNAMIC ANALYSIS

Thermodynamic analysis can formulate the various performance criteria such as thermal equilibria, balance equations, kinetic relations and equations of state. Optimization seeks the best solution under specific constraints. Some methods are the linear and nonlinear programming, dynamic programming, and maximum principle. The complexity of the optimization problem depends on the complexity of the constraints, which may be in algebraic, difference, differential, and integral forms. For the static optimization problems the constraints are algebraic, for variational problems or dynamic optimization the constraints are differential equations.

We may use the *second law analysis* and *exergy analysis* to evaluate the systems. The second law analysis is based on the determination of entropy generation due to the irreversibilities in the system. The analysis can determine whether an existing design is thermodynamically sound or not for a certain operating condition. By modifying the operating conditions, sometimes, it would be desirable to minimize or to distribute the entropy generation evenly along the space and/or time variables in the system to attain a thermodynamic optimum. The entropy generation distribution ratios are especially useful in exhibiting the trade off between the sources of irreversibilities within the system.

Exergy analysis determines quantitatively the deviation from a reference (dead) state; it shows the level of irreversibility of the processes considered or the *exergy destruction* in the system. Most of the thermal systems have the exergy inputs resulting from the consumption of fossil fuel, and losses of exergy are related to the waste of these resources. The exergy balance can help to identify the locations, types and magnitudes of energy resource waste, and hence to develop designs for more efficient use of resources. Based on the exergy destruction we may decide whether the modifications in the process will lead significant energy saving. Exergy analysis completes energy analysis towards obtaining optimal design or optimal operation in the thermodynamic sense. Exergetic efficiencies can be used to evaluate the performance by comparing the efficiencies before and after the modifications. The limit of 100% exergetic efficiency, where no exergy destruction takes place, should not be regarded as a practical goal. In practical applications, decisions are usually made regarding the total costs.

Thermodynamic analysis can lead to a better understanding of the overall performance, and eventually to identifying the sources of losses due to irreversibilities in each process in the system. This will not guarantee that useful process modifications or operational changes would be generated; the relations between the energy efficiency and capital cost must be evaluated from the analysis of the overall plant system, and sometimes an improved energy efficiency will require more investment. Thermodynamics analysis, however has been shown to be of considerable value when applied to systems where an efficient energy conversion is important; exergy analysis also evaluates the merit of an energy conversion in a fuel cells or gas turbines.

The potential work of any system is given by

$$E_{es} = E - T_o S + P_o V - \sum_i \mu_{io} N_i \tag{1}$$

where μ and N are the chemical potential and number of moles of substance i, and E is the total energy including all kinetic and potential energy in addition to internal energy, the indices o denote the reference state representing the environment of the system. The term E_{es} is the essential energy in the form essential for work (power) production, so that E_{es} shows the *ess*ergy (= <u>ess</u>ential en<u>ergy</u>). The corresponding flow of essergy ψ_{es} excluding kinetic and potential energy for any uniform mixture of substances is given by

$$_{es} = H - T_o S - \sum_i \mu_{io} N_i \tag{2}$$

A refrigeration heat exchanger provides an opportunity to study the trade-off between the cost of availability destroyed and the capital cost of the exchanger. We may need to supply a refrigeration to the condenser of a distillation column that returns reflux as a condensate at a certain temperature. The refrigeration temperature must be less than the condensing temperature, and the temperature difference of the refrigerant and the condenser ΔT is an important parameter. The larger difference results in a smaller and hence less expensive condenser, however, at low temperature the power required for a unit of refrigeration increases due to the higher fuel costs to operate the refrigeration compressor. This creates a typical optimization problem between the value of ΔT and the annual cost of fuel and the condenser.

From the second law consideration, the annual cost of fuel C_f is given by

$$C_f = (c_f F_w F_b H_y T_o Q \Delta T) / T^2 \tag{3}$$

where c_f is the cost of unit of fuel, F_w is the units of shaft work required by refrigeration system to deliver unit of availability, F_b is the units of fuel fired in plant boiler per unit of shaft work produced, H_y is the operating time per year, T_o, is the absolute temperature of refrigeration system condenser, or of ambient, and Q is the heat, or refrigeration duty per unit time. The annual capital cost C_a of heat exchanger, assuming that the exchanger is large and the cost is directly proportional to its area, is given by

$$C_a = \frac{c_e F_i Q}{P_t U \Delta T} \tag{4}$$

where c_e is the purchase cost per unit of heat exchanger area, F_i is the installation cost factor, P_t is the allowable payout time in years, and U is the overall heat transfer coefficient. The differential of total annual cost of fuel and capital with respect to ΔT is set equal to zero, and is used to determine the optimum ΔT

$$\Delta T_{\text{opt}} = T\left(\frac{c_e F_i}{c_f F_w F_b H_y T_o P_t U}\right)^{1/2} \tag{5}$$

Eq. (5) shows that the optimum temperature ΔT_{opt} is proportional to the temperature level at which the heat transfer occurs, which is well known for the refrigeration systems. The value of refrigeration increases as the temperature decreases; hence the smaller ΔTs are used as the refrigeration temperature decrease. In above-ambient systems, the larger ΔTs should be used at higher temperatures, even though the value of heat increases with increasing temperature. Care should be exercised in using Eqs. (3) and (5), as the availability is not a conserved quantity. In some cases, availability saved at a heat exchanger does not lead to a fuel saving, but may be accompanied by further availability savings elsewhere in the process. Beside the cost of fuel and the capital changes from time to time and from one location to another, therefore we cannot conclude that there is a global optimum temperature ΔT for a heat exchanger.

2. THERMODYNAMIC OPTIMUM

Thermodynamics can deal with the economic balance through the availability analysis and exergy optimization. Minimization of entropy generation plays only a secondary role in thermoeconomics, mainly because the economic performance is always formulated in economic values of money and prices. Therefore the thermodynamic optimization problem may not be broken down to the problem of minimization of irreversibility. For example, the minimum overall exergy consumption may not be equivalent to the problem of minimum dissipation because of the disregarded exergy of the outgoing flows and changing prices of exergy unit. This problem mainly belongs to energy management and cost of the energy.

The industrial systems consist of various resource consumption processes and the supporting processes associated with the supply and removal of resources. The supporting processes may involve exergy loss and exergy transfer between resources, new resource upgrading, post-consumption recovery, and dispersion and degradation of resources released to the environment.

Contemporary theory of optimization can be used for analyzing the systems. The first approach is to optimize the system by adjusting the design and operating parameters through the governing equations that describe the internal changes, and by imposing a control through the system boundaries. The second approach aims to predict the system behavior under the specified external conditions by deriving its governing equations from certain variational or extremum principles.

2.1. Exergy analysis

Exergy analysis detects and evaluates the level of irreversibility of the processes considered, and hence it indicates the possibilities of improvement. However exergy analysis cannot represent the improvements identified by means of economic indices. Although exergy is applied to solve economic problems, it is however, a thermodynamic approach, not an economic one. The analysis facilitate some approximate solutions; for example, the partition of production costs between the useful products of a complex processes, which can be managed more reasonably by means of exergy than with energy. On the other hand, the exergy analysis may play a primary role in minimizing the consumption of natural resources within the context of ecological economy.

All useful products out of an industrial production line are the results of complicated network of interconnected processes, which need the supply of raw materials, fuels and other energies extracted from natural resources. The quality of the natural resources can be evaluated and expressed by means of exergy. The analysis of cumulative exergy consumption provides an insight about the possibilities of improving the technological network of production.

The following derivations are from Szargut [4]. The total consumption of natural resources involved in the production of a product can be expressed by the overall *index of cumulative exergy consumption* r_j,

$$r_j = \frac{\sum_k \Phi_{kj}}{P_j} \tag{6}$$

where Φ_{kj} is the exergy consumption, expressed as $\Phi_j = T_o \sum_k \Delta S_{kj}$, of the n natural resources (k=1,2…,n) for the product of j, and P_j is the final product from an industrial plant. The terms T_o and $\sum_k \Delta S_{kj}$ indicate the environmental temperature, and sum of entropy changes over the consumption of sources for the product j, respectively. As the value of r_j is related to a unit of the product leaving the considered system, it depends on the assumption of the system boundary. Usually the system of production processes is analyzed without considering employees and local level of consumptions, and hence r_j can be used for comparison of production processes in various countries. The exergy consumption index for a certain resource k can be determined separately for final product P_j

$$r_{kj} = \frac{\Phi_k}{P_j} \tag{7}$$

Exergy consumption of sources that are renewable should also be considered. The overall cumulative exergy consumption index can also be used to solve the various energy utilization problems for a specified product, such as the relationships between the amounts of raw materials and the products, or the cost of raw materials and the alternative production technologies available.

When analyzing the production of materials and energy flows, the values of r_j can be compared with the exergy of the product. The ratio of the specific exergy of the product x to r_j is called the *cumulative degree of thermodynamic perfection* η (CDP) for a certain production network

$$\eta = \frac{x}{r} \tag{8}$$

Here r is the cumulative exergy consumption index for a specified product. For the fabrication of major products $\eta < 1$ due to the second law of thermodynamics; sometimes for a by-product we may have $\eta > 1$, if the exergy of the product is greater than the exergy of the substituted product of a certain process. It would not be useful to calculate η for certain products, such as cars, airplanes, because their usefulness results mainly from their system features, not from the chemical composition of the components. However the calculation of r may be beneficial for all kinds of products, because the values of r can be used to compare various design variables and production technologies.

Cumulative exergy consumption can be calculated by the balance equations; the r_{kj} for the useful products equals to the sum of cumulative exergy consumptions of all raw materials and semi-finished products in the production network. For the link j of the network and for the natural resources k, the balance equations are given by

$$r_{kj}^{(l)} = \sum_i (a_{ij}^{(l)} - f_{ij}^{(l)}) \sum_t w_{it} r_{ki}^{(t)} + \Phi_{kj}^{(l)} \tag{9}$$

where k is the index of the natural resource, i and j are the indexes of the technological network, l and t are the indexes of the production technologies of the products i and j, a_{ij} is the coefficient of the gross consumption of the semi-finished product i per unit of the complex useful product containing a unit of the major product j, f_{ij} is the coefficient of the by-production of the useful product i, w_{it} is the fraction of the technology t of the fabrication of the product i in the analyzed system, and Φ_{kj} is the immediate gross exergy consumption of the neutral resources k per unit of the complex useful product.

Eq. (9) is formulated for a complex process producing more than one useful product. Mostly a complex process is related to a major product, which determines the capacity and location of the process. A useful product substituting the major product of a special process is called the by-product. The coefficient of by-production is expressed in terms of the substitution ratio z_{iu} that is the ratio of unit of the major product i substituted by the unit of the by-product u, and given by

$$\frac{f_{ij}^{(i)}}{f_{uj}^{(l)}} = z_{iu} \tag{10}$$

where f_{uj} is the coefficient of production of by-product u per unit of the major product j. Cumulative exergy consumption for the by-product u is given by

$$f_{ij}^{(i)} r_i = f_{uj}^{(l)} r_u \tag{11}$$

From Eqs. (10) and (11) we obtain

$$r_u = r_i z_{iu} \tag{12}$$

The cumulative degree of thermodynamic perfection for the real by-products u can be expressed by

$$\eta_u = \frac{x_u}{r_u} = \frac{x_u}{r_i z_{iu}} = \frac{x_u}{b_i z_{iu}} \eta_i = \frac{\eta_i}{\eta_{z,iu}} \tag{13}$$

where η_i is the cumulative *degree of thermodynamic perfection* for the major product i substituted by the product u, and $\eta_{z,iu}$ is the exergetic substitution efficiency defined by

$$\eta_{z,iu} = \frac{x_i z_{iu}}{x_u} \tag{14}$$

If $\eta_{z,iu} < \eta_i$, the value $\eta_u > 1$ results from Eq. (13). Some typical values of cumulative degree of thermodynamic perfection η are given in Table 1. For small values of r_u as a result of the substitution ratio z_{iu}, the cumulative exergy consumption will be large for the major product j of the specified process technology.

Eq. (9) can be transformed as follows

$$r_{km} = \sum_n (a_{nm} - f_{nm}) r_{kn} + \Phi_{km} \tag{15}$$

where $a_{nm} = w_{it} a_{ij}^{(l)}$, and $f_{nm} = f_{ij}^{(l)}$

Table 1

Cumulative degree of thermodynamic perfection η values for the production of some materials (Szargut [4])

Product	Specific exergy, MJ/kg	η %	Production technology
Aluminum	32.9	9.6	Bayer process and Hall cell 50 % bauxite ore
		13.2	Electrolytic method from Al_2O_3
Iron	8.2	44.0	From hematite ore in the earth
Cement	0.635	10.3	From raw materials with dry method
		6.2	Medium rotary kiln with wet method
Copper	2.11	3.2	From ore containing Cu_2S, smelting and refining
		2.6	Hydrometallurgical method
		1.5	Electrolytic
Glass	0.174	0.8	From raw material
		0.5	From panels
Ammonia gas	20.03	45.4	Steam reforming of naphtha
		64.8	Steam reforming of natural gas
		41.5	Semi-combustion of natural gas
Paper	16.5	18.7	From timber
		27.5	Integrated plant with fuels from waste products
		74.3	From waste paper
Zinc	5.19	7.6	From ore with vertical retort
		6.7	Electrothermic method
		2.7	From ZnS, metallurgical method
		4.2	From ZnS, electrolytic method
Sulphuric acid	1.66	18.3	Frasch process with sulphur combustion

For technology l every subscript j corresponds to some subscripts m, while for technology t every subscript i corresponds to some subscripts n. Eq. (15) can also be formulated for semi-finished products, consumed in other links of the technological network. In matrix form Eq. (15) is expressed by

$$\mathbf{r}(\mathbf{I}-\mathbf{A}+\mathbf{F})=\mathbf{\Phi} \tag{16}$$

where $\mathbf{I}$ is the diagonal unit matrix, $\mathbf{A}$, $\mathbf{F}$ are the square matrix with elements of MxM, $M = m_{\max}$, $\mathbf{r}$ and $\mathbf{B}$ are the rectangular matrix with elements of KxM, $K = k_{\max}$. From Eq. (16), we obtain

$$\mathbf{r}=\mathbf{\Phi}(\mathbf{I}-\mathbf{A}+\mathbf{F})^{-1}=\mathbf{\Phi}\mathbf{S}^{*} \tag{17}$$

The elements of the inverse matrix $\mathbf{S}^*$ represent the cumulative net consumption of intermediate products per unit of the major product leaving the system

$$r_{kn}=\Phi_{k1}S_{1n}^{*}+\Phi_{k2}S_{2n}^{*}+\ldots \tag{18}$$

Here S_{1n}^{*} shows the net amount of the product 1 per unit of the product n.

The difference of cumulative exergy consumption r and exergy consumption of a natural resource represents the *cumulative exergy loss* $\delta\phi$ involved in all the parts of the technological network for the production of a certain material

$$\delta\phi=r-x \tag{19}$$

The components of $\delta\phi$ provide information about possibilities to improve the technological network. The difference $(r-x)_n$ defines the constituent exergy loss or a particular semi-finished product n, and results from the thermodynamic imperfection of the constituent technological network.

In complex processes, the raw materials and semi-finished products are partially used for the fabrication of by-products. Hence the *coefficient of net consumption* A_{nm} of semi-finished products and raw materials per unit of the major product should be determined by

$$A_{nm}=a_{nm}-\sum_{u}f_{um}A_{nu}=a_{nm}-\sum_{p}f_{um}z_{pu}A_{np} \tag{20}$$

where p is the index of the major product substituted by the by-product u, and z_{pu} is the substitution ratio of the product p by the product u. The coefficient A_{nm} can be negative if the consumption of the semi-finished product n is greater, in a process substituted by the utilization of the by-product, than in the principal process considered.

The *constituent exergy losses* are calculated from the coefficients A_{nm}

$$\delta\phi_{nm} = A_{nm}(r_n - x_n), \qquad n \neq m \tag{21}$$

Some losses can be negative due to the elimination of the constituent exergy losses in the substituted process.

Local gross exergy loss $\delta\phi_m$ represents the sum of internal and external exergy losses in the used technology for the major product and by-product, and it can be calculated from the following steady-state exergy balance

$$\sum_n a_{nm} x_n = x_m + \sum_u f_{um} x_u + d\phi_m \tag{22}$$

Here the energy exchanged with surroundings should be treated as one of the useful product exergies represented by x_m, x_n, *or* x_u. The *local net exergy loss* refers to the complex of useful products containing a unit of the major product, and results from the following difference

$$\delta\phi_{nm} = \delta\phi_m - \sum_u f_{um}\delta\phi_u \tag{23}$$

where $\delta\phi_u$ is the local exergy loss due to the by-product u. The local exergy loss due to the by-product results not only from the local net exergy loss in the substituted process but also from the difference of exergy of the by-product and substituted major product

$$\delta\phi_u = z_{pu}\delta\phi_{pp} - (x_u - z_{pu}x_p) = z_{pu}\delta\phi_{pp} \text{-} (1\text{-}\eta_{z,pu})x_u \tag{24}$$

where $\delta\phi_{pp}$ is the local net exergy loss due to the major product in the substituted process, and $\eta_{z,pu}$ is the exergetic substitution efficiency (Eq. 14). Combining Eqs. (23) and Eq. (24) we obtain

$$\delta\phi_{mm} = \delta\phi_m - \sum_p f_{um} z_{pu}\delta\phi_{pp} + \sum_u f_{um}(1 - \eta_{z,pu})x_u \tag{25}$$

The local net exergy losses for all major products can be calculated by Eq. (25).

Partial exergy loss $\delta\overline{\phi}_{km}$ expresses the local net exergy loss in the link k of the technological network for the product m

$$\delta\overline{\phi}_{km} = S^*_{km}\delta\phi_{kk} \tag{26}$$

where S^*_{km} is the coefficients of cumulative net consumption shown in Eq. (18). If the product of the link m is consumed in the precedent links of the technological network, then $S^*_{mm} > 1$ and the partial exergy loss $\delta\bar{\phi}_{mm}$ is greater then the local net exergy loss $\delta\phi_{mm}$. A partial exergy loss in a complex process can be negative, if the production of the intermediate product k is smaller in the main network then in the network substituted by the utilization of by-products.

The sum of partial exergy losses equals the cumulative exergy loss

$$\delta\phi_m = \sum_k \delta\bar{\phi}_{km} \tag{27}$$

Relative constituents ϕ_c, local exergy loss ϕ_l, and partial exergy loss ϕ_p can be defined by

$$\phi_{c,nm} = \frac{\delta\phi_{nm}}{r_m}; \quad \phi_{l,mn} = \frac{\delta\phi_{mm}}{r_m}; \quad \phi_{p,km} = \frac{\delta\bar{\phi}_{km}}{r_m}; \tag{28}$$

From Eqs. (26) and (28), we obtain

$$\eta_m + \sum_{n \neq m} \phi_{c,nm} + \phi_{l,mm} = 1 \tag{29}$$

$$\eta_m + \sum_k \phi_{p,km} = 1 \tag{30}$$

The analysis of constituent and partial exergy losses may improve the processes of technological network. If some constituent exergy loss is very large, the possibility of changing the production technology or the substitution of the considered semi-finished product by another more convenient one should be considered. This would be especially suitable for a new design and technological network. If a partial exergy loss is considerably large, the reduction of thermodynamic imperfections of the technological network should be investigated. This may be achieved by a decrease of internal irreversibilities and by a better utilization of waste products or by changing the technology. A more suitable operational conditions and reduction in the consumption of some expensive intermediate products should also be considered.

2.2. Exhaustion of nonrenewable resources

Utilization of domestic natural nonrenewable resources is inevitable, and the analysis of these resources helps to assess the profitability of the import of raw materials, fuels, and semi-finished products as well as the utilization of secondary raw materials. In the analysis of *exhaustion of nonrenewable natural resources*

(ENR) the balance equations of Eq. (15) should be modified if domestic nonrenewable resources are of interest. In this case, imported raw materials, fuels and semi-finished products should be separately taken into account

$$e_{km} = \sum_n (a_{nn} - f_{um})\, e_{kn} + \sum_r a_{rm} e_{kr} + \Phi_{km} \tag{31}$$

where e_{km}, e_{kn}, and e_{kr} are the exhaustion of the domestic nonrenewable natural resources k per unit of the products m, n, and r, respectively, a_{rm} is the coefficient of gross consumption of imported raw material, fuel or semi-finished product r, and Φ_{km} is the gross consumption of the domestic nonrenewable natural resources k within the link m. The index e_{kr} should be determined by assuming that the import is economical, and the unit value of the exported and imported products is considered with the same ENR

$$e_{kr} = e_k^{(d)} D_r = \frac{\sum_n E_n e_{kn}}{\sum_n E_n D_n} D_r \tag{32}$$

where $e^{(d)}{}_k$ is the ENR per unit of the monetary values of exported products, D_r and D_n are the specified monetary value of the imported product r and exported product n, respectively, and E_n is the export of the product n.

Introducing Eq. (32) into Eq. (31), the balance equations become

$$e_{km} = \sum_n (a_{nn} - f_{um})\, e_{kn} + \sum_r a_{rm} e_{kr} + \Phi_{km} \tag{33}$$

where

$$d_{nm} = E_n \frac{\sum_r D_r a_{rm}}{\sum_n E_n D_n}$$

In matrix form Eq. (33) becomes

$$\mathbf{e}\,(\mathbf{E} - \mathbf{A} + \mathbf{F} - \mathbf{D}) = \mathbf{\Phi} \tag{34}$$

and

$$\mathbf{e} = \mathbf{\Phi}\,(\mathbf{E} - \mathbf{A} + \mathbf{F} - \mathbf{D})^{-1} \tag{35}$$

where **A**, **F**, **D** are square matrices with *M*x*M* elements, and **e** is the rectangular matrix with *K*x*M* elements.

The set of Eqs. (34) may be large as it comprises all the domestic semi-finished products and the exported products. An approximate solution is

$$e_k^{(d)} = \frac{\Sigma\Phi_k}{(DN)} \tag{36}$$

where $\Sigma\Phi_k$ is the total exhaustion of the domestic natural resources k, and (DN) is the economic value of all domestic products and exported products. Eq. (36) enables one to formulate the balance equations of Eq. (34) for the semi-finished products. The first part of Eq. (32) defines the exhaustion of nonrenewable natural resources for the imported products so Eq. (31) is fully known.

In the analysis of a material production, the utilization efficiency of nonrenewable domestic natural resources (UEN) can be defined as

$$\eta' = \frac{x}{e} \tag{37}$$

Usually $e > x$ and $\eta' < 1$, but for secondary raw materials $\eta' >> 1$, and for imported raw materials and fuels, we usually have $\eta > 1$.

For the secondary raw materials, exhaustion of nonrenewable natural resources is related to the consumption of exergy for the processing and the transportation, and usually it is much smaller than the exergy of the materials under consideration. The inequality $\eta' >> 1$ suggests that the utilization of secondary raw materials may be beneficial, since they substitute the semi-finished products requiring a large amount of exergy for their production.

We may have $x > e$ for imported raw materials, fuels, and semi-finished products if the exported goods are more advanced than the imported ones.

2.3. Ecological cost

Many environmental problems are related to the production, conversion, and utilization of energy. Some of the energy management related environmental problems are air and water pollution, impact on the use of land and rivers, thermal pollution due to mismanagement of waste heat, and global climate change. As an energy conservation equation, the first law of thermodynamics is directly related to the energy management impact on the environment.

One of the links between the second law of thermodynamics and the environment is the exergy, because it is mainly a measure of the departure of the state of a system from that of the equilibrium state of the environment; the magnitude of the exergy of a system is determined with relative to the environment.

Performing the exergy analysis on the earth's natural processes may display the disturbance due to large-scale changes, and could form a sound ecological planning and sustainable development. Some of the major disturbances are:

(i) Chaos due to destruction of order, which is a form of environmental damage. Entropy is fundamentally a measure of chaos, while exergy is a measure of order; the exergy of an ordered system is greater than that of a chaotic system.

(ii) Resource degradation, and hence the destruction of exergy. A natural resource with exergy is in nonequilibrium state compared with the environment.

(iii) Uncontrollable waste exergy emission can cause a change in the environment.

Through the exergy analysis, thermodynamics may be an important tool to interrelate energy management, environment and sustainable development in order to improve economical and environmental assessments.

The ecological cost analysis may facilitate the minimization of the consumption of nonrenewable natural resources. To understand the ecological impact of production processes, it is not sufficient to determine ENR connected with the extraction of raw materials and fuels from the natural resources; influence of the waste product discharge into the environment should also be considered. The waste products may be harmful for agriculture, for plant life and human health, and for industrial activity due to corrosion. The exhaustion of nonrenewable natural resources has been called the *index of ecological cost*. To determine the domestic ecological cost, the impact of imported materials and fuels should also be taken into account. The ecological cost φ is expressed by

$$\varphi_m = \sum_n (a_{nm} - f_{nm} + d_{nm} + \sum_s \Phi_{sm} x_{ns})\varphi_n + \sum_s \Phi_{sm}(\sum_k y_{ks} + z_s) + \sum_k \Phi_{km} \tag{38}$$

where s is the index of harmful waste product, d_{nm} is defined in Eq. (33), Φ_{km} is the immediate gross consumption of the nonrenewable domestic natural resource k per unit of complex useful products containing a unit of the major product m, Φ_{sm} is the exergy of harmful waste product s, x_{ns} is the destruction coefficient of the product n per unit of the exergy of waste product s, y_{ks} is the destruction coefficient of the nonrenewable natural resources k per unit of the exergy of waste product s, and z_s is the multiplier of exergy consumption to eliminate the results of human health deterioration per unit of exergy of the waste product s. The destruction coefficients x and y can be defined by

$$x_{ns} = \frac{\delta m_n}{\Phi_s}; \quad y_{ks} = \frac{\delta \Phi_k}{\Phi_s} \tag{39}$$

where δm_n is the amount of units of the destructed useful product, δB_k is the exergy decrease of the damaged natural resources. The coefficient x_{ns} should also take into account the reduction of agricultural and forest production. Similarly, the global ecological cost can be calculated.

The degree of negative impact of the process on natural resources can be characterized by means of the *ecological efficiency* η_e, and from Eqs. (13) and (38) we have

$$\eta_e = \frac{b}{\varphi} \tag{40}$$

Usually $\eta_e < 1$, sometimes values of $\eta_e > 1$ can appear, if the restorable natural resources are used for the considered process.

By means of the ecological cost, the ecological economy can be established to minimize the exhaustion of nonrenewable natural resources by limiting the consumption level in the optimization problem.

The transition from one form of exergy into another, for example from chemical to structural may be interpreted as creating a value. Self-organization is a production process, and exergy is necessary to build a structure with a value, which may not be measured and described by exergy. For the economical aspect of exergy, we have to replace the exergy destroyed in every process.

However, exergy is not generally and widely accepted by industry at present. One reason may be the ambiguity in defining the dead state. Some find that the exergy methods are complex, and the results are difficult to interpret, while some practicing engineers have found no direct applications of exergy. In order to correct this trend, the benefits of exergy methods should be underlined in college and university programs, continuing-education courses and on-the-job training in industry. Most of the methods and related calculations must be further simplified. Further work is also necessary to expand and improve the exergetic-costing database, which may have a major impact on the practical usefulness of the method. Exergy evaluations of emissions should be standardized, impact of the emissions on the environment should be integrated into the analysis, and user-friendly computer programs should be developed. Compatibility of various exergy analysis methods based on reliable data should be ensured.

3. AVAILABILITY

One of the important definitions in finite-time thermodynamics is the definition of a finite-time availability A given by

$$A = W_{\max} = \max\left[A(t_i) - A(t_f) - T_o \int_{t_i}^{t_f} S_{\text{tot}} dt \right] \tag{41}$$

Here t_i and t_f are the initial and final times of the irreversible process, T_o is the environmental temperature. The maximization is carried out with the constraints imposed on the process. Eq. (41) puts the second law of thermodynamics into equality form by subtracting the work equivalent of the entropy produced that is the decrease in availability in the process. Availability depends on the variables of the system as well as the variables of the environment

$$A = U + P_o V - T_o S - \sum \mu_{oi} N_i \tag{42}$$

where the temperature, pressure and chemical potential are at ambient conditions. For the optimal control problem one must specify: (i) control variables, volume, rate, voltage, and limits on the variables, (ii) equations that show the time evolution of the system usually differential equations describing heat transfer and chemical reactions, (iii) the constraints imposed on the system such as conservation equations, and (iv) the objective function, which is usually in integral form for the required quantity to be optimized. Time of the process may be fixed or may be part of the optimization.

4. EXERGY DESTRUCTION NUMBER

The use of an augmentation device results in an improved heat transfer coefficient, thus reducing exergy destruction due to convection heat transfer; however exergy destruction due to frictional effects increases. The exergy destruction number N_x is defined as the ratio of the nondimensional exergy destruction number of the augmented system to that of the unaugmented one

$$N_x = \frac{X_a^*}{X_s^*} \tag{43}$$

where subscripts a and s denote the augmented and nonaugmented cases respectively, and X^* is the nondimensional exergy destruction number, which is defined by

$$X^* = \frac{x_{fd}}{\dot{m} T_o C_p} \tag{44}$$

Here x_{fd} is the flow-exergy destruction, or irreversibility, and T_o is the reference temperature. The system will be thermodynamically advantageous only if the N_x is less than unity. The exergy destruction number is widely used in the second-law based thermoeconomic analysis of thermal processes such as heat exchangers.

5. EQUIPARTITION AND OPTIMIZATION

In the linear nonequilibrium thermodynamics theory the entropy production Φ due to a given transfer process is expressed as the product of flow J and force X

$$\Phi = JX \tag{45}$$

The flows are expressed as a linear function of forces

$$J = LX \tag{46}$$

Here L is the phenomenological coefficient.

For a heat exchanger operating at steady state, the total entropy generation P is obtained by integrating over the surface area

$$P = L \int_A X^2 dA \tag{47}$$

We consider that the duty of the exchanger is specified as J_s

$$J_s = L \int_A X dA \tag{48}$$

An average driving force over the surface area is obtained as

$$X_{\text{av}} = \frac{1}{A} \int_A X dA \tag{49}$$

So that Eq. (48) and Eq. (49) yield the specified duty J_s

$$J_s = LAX_{\text{av}} \tag{50}$$

Eq. (50) shows that for a given surface area A and constant L, specification of the duty leads to the average driving force.

The following equations are from Tondeur [7]. Minimizing integral in Eq. (47) subject to a constraint given in Eq. (48) is a variational problem, and the solution by the Euler equation in terms of the force is given by

$$\frac{\partial}{\partial X}(X^2 + \lambda X) = 0 \tag{51}$$

where λ is a Lagrange multiplier (a constant). Eq. (51) is satisfied by $X = -\lambda/2$, that is by a constant value of X. The second derivative yields

$$\frac{\partial^2}{\partial X^2}(X^2 + \lambda X) > 0 \tag{52}$$

Eq. (52) implies that the extremum is a minimum. Thus with a constant transfer coefficient, the distribution of driving force which minimizes the entropy generation under the constraint of a specified duty, is a uniform distribution. The minimal dissipation for a specified duty implies the equipartition of the driving force and entropy generation along the time and space variables of the process.

When the linear phenomenological equations do not hold, we have

$$P = \int_V jXdV; \quad J = J_o = Vj_{\text{av}} \text{ with } X = X(j) \tag{53}$$

where j is the specific flow per unit volume. The constraints on these relations are

$$jX > 0; \quad X' = \frac{dX}{dj} > 0; \quad j(X = 0) = 0 \tag{54}$$

The Lagrangian expression is given by

$$F(j) = P + \lambda(J - J_o) = \int(jX + \lambda j - \lambda j_{\text{av}})dV \tag{55}$$

The Euler equation corresponding to an extremum of P is given by

$$\frac{\partial F}{\partial j} = \int(jX' + X + \lambda)dV = 0 \tag{56}$$

which yields

$$jX' + X + \lambda = \frac{\partial}{\partial j}[(X + \lambda)j] = 0 \tag{57}$$

Eq. (57) shows that $(X+\lambda)j=$ constant, and hence the solution yields that $j=$ constant and $X=$ constant. Therefore P is stationary when the flow and forces, and the entropy generation are uniformly distributed. The sign of the second derivative reveals whether this stationary value is a minimum or not

$$\frac{\partial^2}{\partial j^2}[(X+\lambda)j]=X''j+X' \tag{58}$$

Since j and X' are positive, the quantity in Eq. (58) is always positive when $X''>0$; that means when X is a convex function of j. When the flow j is a linear or a concave function of the driving force X $(\frac{\partial^2 j}{\partial X^2}<0)$ (Fig.1a), then equipartition of entropy generation corresponds to minimal dissipation. On the other hand, when X'' is negative in Eq. (58), the sign of $\frac{\partial^2}{\partial j^2}[(X+j)j]$ may be positive or negative, and may change along the process. If $\frac{\partial^2 F}{\partial j^2}<0$, the entropy generation is maximum; and when the flow versus force curve is sufficiently convex (Fig 1b), the most dissipative configuration is the uniform one, and hence nonuniformity is then the economic trend. Such a situation may arise in an electrochemical cell, which does not obey the Ohm law.

Strongly convex flow-force curve corresponds to the ordered structures; they are dissipative, and constantly need supply of energy. For example, the Benards cells occur during a natural convection in a fluid system heated from the bottom; after the difference between the surface temperature and fluid bulk temperature exceeds a certain limit system moves into the nonlinear region in the thermodynamic branch, and the fluid shows a structured state as long as the temperature difference is maintained.

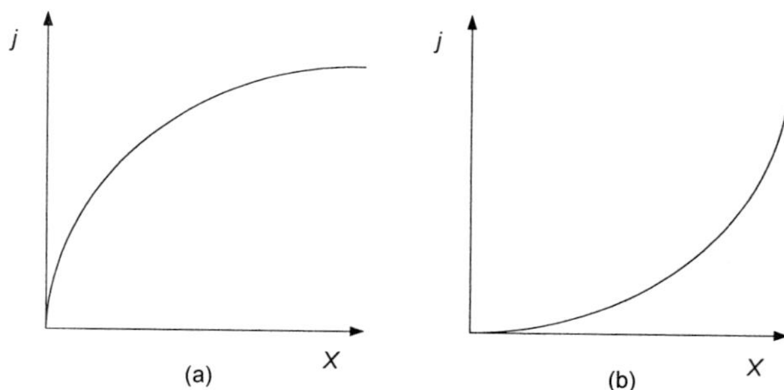

Fig. 1. Concave and convex relationships between the flow and force.

The generality of the equipartition principle should be investigated within the flow-force relationships. In the near equilibrium phenomena linear flow-force relations are valid, and optimization criteria for example for a coupled heat and mass transfer generally leads to a constant level of the entropy generation along an optimal path provided that there is no constraints imposed on the parameters controlling the system. For systems far from equilibrium, the most stable configurations may correspond to unsteady, dissipative structures. Therefore equipartition or stability should be considered as a part of economy for industrial systems or evolution for natural systems. Equipartition may also help to improve the existing design and avoid misconception in the new design of processes.

REFERENCES

[1] D. Kondepudi and I. Prigogine, Modern Thermodynamics, From Heat Engines to Dissipative Structures, Wiley, New York, 1999.

[2] S.K. Ratkje, E. Sauar, E.M. Hansen, K.M. Lien and B. Hafskjold, Ind. Eng. Chem. Res., 34 (1995) 3001.

[3] B.A. Mansson, In Finite-Time Thermodynamics and Thermodynamics, Eds. S. Sienutycz, P. Salamon, Taylor & Francis, New York, 1990.

[4] J. Szargut, In Finite-Time Thermodynamics and Thermoeconomics, Eds. S. Sieniutcyz, P. Salamon, Taylor & Francis, New York, 1990.

[5] Z. Szwast and S. Sieniutycz, In Eff. Costs Opt. Sim. and Env. Impact of Energy systems, Vol. 1, (1995).

[6] L. Connely and C.P. Koshland, Exergy Int. J., 1 (2001) 234.

[7] D. Tondeur, In Finite-Time Thermodynamics and Thermoeconomics, Eds. S. Sieniutcyz, P. Salamon, Taylor & Francis, New York, 1990.

[8] M. Yilmaz, O.N. Sara and S. Karsli, Exergy Int. J., 1 (2001) 278.

Chapter 8

Diffusion

INTRODUCTION

There are three main ways to describe multicomponent diffusion: (1) the Maxwell-Stefan diffusion where fluxes and forces are mixed, (2) Chapman-Cowling and Hirschfelder-Curtis-Bird approach where the diffusion of all the components are treated in a similar way, and (3) description with a reference to a particular component, for example solvent or mass average (barycentric) definition. In the following sections, diffusion in nonelectrolyte and electrolyte systems is presented as the linear phenomenological laws. Some keywords and concepts are mobility, electrochemical potential, electrochemical affinity, electrical conductance, and Hittorf transference number.

Under the mechanical equilibrium on a molecular scale, the exchange of momentum proceeds faster than the exchange of mass and heat, which may be the case for liquids with reasonable accuracy. On the other hand, the molecular exchange of momentum, matter and heat as determined for gases is of the same order by the Schmidt number *Sc* and the Prandtl number *Pr*, and the assumption of mechanical equilibrium in gases for heat and mass transfer may not be reliable.

For a ternary mixture of components 1, 2, and 3, the flux of component 1 in the z direction only can be expressed as

$$J_1 = D_{11}\frac{dc_1}{dz} + D_{12}\frac{dc_2}{dz} \tag{1}$$

Similar relations can be written for components J_2 and J_3. The coefficients D_{11} and D_{22} are the main coefficients; they are not self-diffusion coefficients. D_{12} and D_{21} are the cross-coefficients, and relate the flux of a component 1 to a gradient in component 2. D_{12} is normally not equal to D_{21} for multicomponent system. Frames of reference in multicomponent system must be clearly defined. One of the problems with diffusion in liquids is that even the binary diffusion coefficients are often composition dependent, and it is difficult to obtain the numerical values of coefficients relating fluxes to concentration gradients. In gas mixtures, only the binary diffusivities D_{ij} are normally assumed independent of composition, and multicomponent diffusion by the Stefan-Maxwell equation given by

$$\frac{dx_i}{dz} = \sum_{j=1}^{n} \frac{c_i c_j}{c^2 D'_{ij}} \left(\frac{J_j}{c_j} - \frac{J_i}{c_i} \right) \tag{2}$$

where c_i is the concentration of i, c is the total mixture concentration, J_i is the flux of component i, D'_{ij} is the Maxwell-Stefan diffusivity.

1. MAXWELL-STEFAN DIFFUSIVITY

Maxwell and Stefan proposed a method to describe the diffusion in multicomponent gas and liquid mixtures of isotropic systems

$$j_k = -\sum_{l=1}^{n} L_{kl} P v_l \tag{3}$$

where v_l is the velocity of component l, P is the pressure, and L_{kl} is the Onsager reciprocal relations. Eq. (3) shows the phenomenological equations to describe diffusion where the Onsgaer reciprocal relations can be expressed in terms of the diffusion coefficients. Maxwell suggested an equation for dilute gases and Stefan suggested for liquids, hence Maxwell-Stefan equations cover both phases. Here it is assumed that the diffusion results from the equal and opposite forces that are proportional to the velocity differences of the components, and the formulation is independent of the chosen reference velocity.

For an isothermal mixture the dissipation function can be expressed in terms of Eq. (3), and we have

$$\Psi = -P^2 \sum_{k=1}^{l-1} \sum_{l=2}^{n} L_{kl} (v_l - v_k)^2 \geq 0 \tag{4}$$

where $v_l - v_k$ is the velocity difference, For an isothermal system of three components Eq. (4) yields

$$\Psi = -P^2 (L_{12} v_{21}^2 + L_{13} v_{31}^2 + L_{23} v_{32}^2) \geq 0 \tag{5}$$

where $v_{kl}^2 = (v_k - v_l)^2$. Elimination of v_{32} with v_{31} - v_{21}, and dividing the result by v_{31}^2 yields the following quadratic constraint

$$-(L_{12} + L_{23}) \left(\frac{v_{21}}{v_{31}} \right)^2 + 2L_{23} \left(\frac{v_{21}}{v_{31}} \right) - (L_{13} + L_{23}) \geq 0 \tag{6}$$

The necessary conditions for satisfying this inequality are

$$-L_{21}-L_{23}=L_{22}\geq 0;\ -L_{31}-L_{32}=L_{33}\geq 0;\ -L_{12}-L_{13}=L_{11}\geq 0 \tag{7}$$

If Eq. (7) is satisfied, the constraint for the three-component system is given by

$$L_{21}L_{13}+L_{23}(L_{12}+L_{13})\geq 0 \tag{8}$$

Eq. (8) is the only sufficient constraint irrespective of the choice of independent velocity difference with $-L_{kl}\geq 0$ for $k\neq l$, and shows that it is possible to have negative phenomenological coefficients without violating the condition of positive dissipation. Condition in Eq (8) shows that $-L_{23}$ does not need to be positive subject to the following constraint

$$-L_{23}\geq \frac{L_{12}L_{13}}{L_{12}+L_{13}}<0 \tag{9}$$

by assuming that the other coefficients satisfy that $-L_{12}>0$, and $-L_{13}>0$. The constraints for n-component system is generalized as

$$-L_{kk}\geq 0 \text{ and } L_{kk}L_{ll}-L_{kl}^2\geq 0 \quad \text{for } n\geq 3 \tag{10}$$

Maxwell described the diffusion by the velocity differences, which yield forces from the friction between the molecules of different species. He considered a chemical potential gradient caused by friction, and the friction is proportional to the concentration. The diffusion coefficient of Maxwell-Stefan can be defined as

$$D'_{kl}=-\frac{x_k x_l}{PL_{kl}} \tag{11}$$

where the $x_k = c_k/c$ are the mole fractions. The following relations hold for the diffusion coefficients

$$D'_{kl}\geq 0\ ;\ D'_{kl}=D'_{lk}, \text{ and } D'_{kk}\geq 0 \tag{12}$$

Thus the Maxwell-Stefan diffusion coefficients satisfy simple symmetry relations. Since the diffusion coefficients have to be determined experimentally in a phenomenological approach, Onsager reciprocal relations reduce the number of coefficients to be determined, and are of great advantage in determining the transport coefficients, which may be difficult to measure.

Satisfying all the inequalities in Eq. (12) yields a sufficient condition for the dissipation function to be positive definite. For binary mixtures the Maxwell-Stefan diffusivity has to be positive, but for multicomponent system, negative diffusivities are possible, for example negative diffusivities have been reported in electrolyte solutions. From Eq. (12) the Maxwell-Stefan diffusivities in an n-component system satisfy the following inequality as

$$\sum_{\substack{l=1\\k\neq l}}^{n} \frac{x_l}{D'_{kl}} \geq 0 \tag{13}$$

Possible negative Maxwell-Stefan diffusivities are allowed if they satisfy

$$D'^2_{kl} \geq D'_{kk} D'_{ll} \tag{14}$$

It is useful to express Eq. (14) in terms of D'_{kl} as follows

$$\sum_{k=1}^{n} \frac{x_k}{D'_{kl}} \sum_{k=1}^{n} \frac{x_l}{D'_{kl}} - \frac{x_k x_l}{D'^2_{kl}} \geq 0 \tag{15}$$

For nonisothermal conditions the diffusion equation becomes

$$J_k = \sum_{l=1}^{n} \frac{x_k x_l}{D'_{kl}} (v_l - v_k) - L_{kq} T^{-1} \text{grad} T \tag{16}$$

The term L_{kq} shows that for all $J_k = 0$, diffusion velocities can arise due to temperature gradient, and called the thermal diffusion or the Soret effect. The thermal diffusion is important in engineering practice whenever large temperature gradients coexist with large molecular weight difference. Thermal diffusion appears if $J_k = 0$ and grad $T \neq 0$, and mass flux is given as

$$\rho_k v_k = -\rho D_k^T \frac{\text{grad} T}{T} \tag{17}$$

Here the coefficient D_k^T is defined as the barycentric coefficients of thermal diffusion. Due to

$$\sum \rho_k v_k = 0 \tag{18}$$

thermal diffusion coefficients satisfy the following constraint

$$\sum D_k^T = 0 \tag{19}$$

From Eq. (16), we have the phenomenological coefficients expressed by

$$-L_{kq} = L_{qk} = \sum \frac{x_k x_l}{D_{kl}^{'}} \left(\frac{D_l^T}{w_l} - \frac{D_k^T}{w_k} \right) \tag{20}$$

Using Eq. (20), The Maxwell-Stefan diffusion can be written as

$$J_k = \sum \frac{x_k x_l}{D_{kl}^{'}} (v_l^o - v_k^o) \tag{21}$$

where v_k^o is defined as

$$v_k^o = v_k + \frac{D_k^T}{w_k} \frac{\mathrm{grad}T}{T} \tag{22}$$

Eq. (21) is considered as a generalization of the Maxwell diffusion equations. Curtiss and Hirschfelder have also derived similar equations for dilute gases by using the kinetic theory of gases.

The Maxwell-Stefan equations do not depend on a special choice of the reference velocity, and therefore they are a proper starting point for other descriptions of multicomponent diffusion. For n-component ideal gases the diffusivities $D_{kl}^{'}$ are independent of the composition of ideal gas mixtures, and are equal to the diffusivity $D_{lk}^{'}$ of the binary pair of kl. In an n-component system only $n(n\text{-}1)/2$ different Maxwell-Stefan diffusivities are required as a result of the simple symmetry relations. Some advantages of the Maxwell-Stefan description of diffusion are:

1. The diffusion is independent of the choice of the reference velocity.
2. Diffusion of all the components is treated equally.
3. Diffusion is in agreement with the results of the kinetic theory of dilute monatomic gases.
4. The Maxwell-Stefan binary diffusivities $D_{kl}^{'}$ are independent of the concentration of the components in the multicomponent system of ideal and nonideal mixtures.

The mass and molar diffusion are important in practice, and can be derived from the Maxwell-Stefan description of diffusion. The Maxwell-Stefan multicomponent diffusivities are obtained from the binary diffusivities, which are easy to measure.

For the application of Maxwell-Stefan description of diffusion to binary isotropic systems we consider components 1 and 2. To solve the mass balance equations, the diffusion flux has to be known. The diffusion flow without electromagnetic field and external forces is given by

$$J_1 = -J_2 = \frac{\rho_1}{P}[(\text{grad}\,\mu_1)_{T,P} + (v_1 - v)\text{grad}\,P] \tag{23}$$

where

$$v_1 - v = (w_1 + w_2)v_1 - w_1 v_1 - w_2 v_2 = w_2(v_1 - v_2)$$

The thermodynamic correction factor Γ is defined using the Gibbs-Duhem relation

$$\Gamma = \frac{\rho_1}{P}\left(\frac{\partial \mu_1}{\partial x_1}\right)_{T,P} = \frac{\rho_2}{P}\left(\frac{\partial \mu_2}{\partial x_2}\right)_{T,P} \tag{24}$$

The thermodynamic factor is a measure of deviation from the ideal behavior, and $\Gamma = 1$ for an ideal system. Eq. (24) can be rearranged as

$$\frac{\rho_1}{P}(\text{grad}\mu_1)_{T,P} = \frac{\rho_1}{P}\left(\frac{\partial \mu_1}{\partial x_1}\right)_{T,P} \text{grad}\,x_1 = \Gamma \text{grad}\,x_1 \tag{25}$$

Diffusion flow in Eq. (23) can be expressed as in terms of Γ

$$J_1 = -J_2 = \Gamma(\text{grad}\,\text{x}_1) + \rho w_1 w_2 (v_1 - v_2)\frac{\text{grad}\,P}{P} \tag{26}$$

For a binary system, grad w_1 is expressed in terms of grad x_1 by using the summation relation $x_1 + x_2 = 1$ and $M = M_1 x_1 + M_2 x_2$, and we obtain

$$dw_1 = d\left(\frac{M_1}{M}\right)x_1 = \frac{M_1}{M}\left(dx_1 - x_1\frac{dM}{M}\right); \text{ and } dw_1 = \frac{M_1 M_2}{M^2}dx_1 \tag{27}$$

where M_1, M_2 are the molar masses of components 1 and 2, respectively. From Eq. (27) we have

$$J_1 = \frac{M^2}{M_1 M_2}\Gamma \text{grad}\,w_1 + \rho w_1 w_2 (v_1 - v_2)\left(\frac{\text{grad P}}{P}\right) \tag{28}$$

For a binary system from Eq. (21), we have

$$v_2^o - v_1^o = \frac{D_{12}^{'}}{x_1 x_2} J_1 \tag{29}$$

Finally, the diffusion flux J_1 becomes

$$J_1 = \rho_1 v_1 = -\rho D_{12} \text{ grad } w_1 - \rho D_1^P \left(\frac{\text{grad } P}{P} \right) - \rho D_1^T \frac{\text{grad } T}{T} \tag{30}$$

where the binary diffusivity D_{12} is called the Fick diffusivity, which can be used in the barycentric (mass average) as well as in the molar description. The binary pressure diffusivity D_1^P is defined by

$$D_1^P = \rho D_{12}^{'} \frac{M_1 M_2}{M^2} w_1 w_2 (v_1 - v_2) \tag{31}$$

where $D_{12}^{'} = \dfrac{D_{12}}{\Gamma}$

Eq. (30) shows that in a binary mixture, diffusion occurs due to concentration difference (ordinary diffusion), pressure difference (pressure diffusion), and temperature difference (thermal diffusion) without external forces and electromagnetic field.

In nonideal binary system, the Fick diffusivity varies considerably with the concentration. As seen in Fig. 1, water-acetone and water- ethanol systems exhibit a minimum at intermediate concentrations. As the nonideality increases, the minimum may approach to zero or even become negative causing the mixture to be split into two liquid phases, as is the case for benzene-water system.

The behavior of the Fick diffusion coefficient in nonideal system may be complicated, while the Maxwell-Stefan diffusion coefficients behave quite well, and are always positive for binary systems. A solution is a condensed phase of several components, which may be subject to strong intermolecular forces. This may not be the case in a gas. The molar concentration in a solution is much larger than that in a gas at the same temperature. Despite the fundamental differences between the solutions and gases, some laws for solutions are analogous to those for gases. If the solution is sufficiently dilute, the osmotic pressure is described by an equation similar to that for an ideal gas, and ideal solutions are treated as a special case of ideal gas.

The chemical potential of an ideal solution may be expressed by

$$\mu_k = \mu_k^o(T,P) + R_k T \ln x_k \tag{32}$$

where $R_k = R/M_k$, R is the universal gas constant, and M_k is the molar mass. Dilute solutions, in which the molar concentrations are sufficiently close to zero, behave like ideal solutions. The level of low concentrations for a solution to be ideal depends to a large extent on the nature of the solvent and the dissolved substances. On the other hand, in electrolytes, deviations from the ideal behavior can occur even in very dilute solutions. This can be linked to the spatial range of the electromagnetic forces.

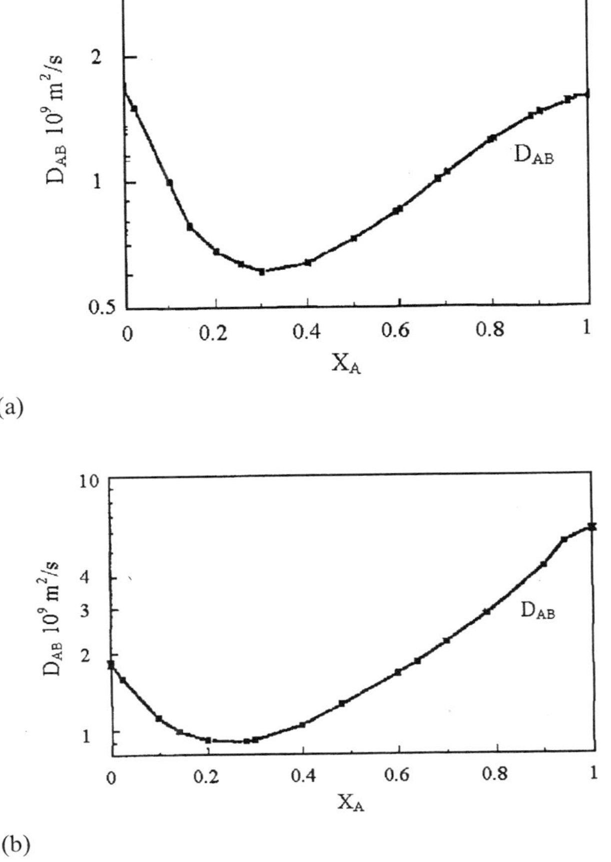

Fig. 1. Concentration dependence of the Fick diffusivity for the binary mixtures: (a) water(A)-acetone(B) at 45 °C, (b) water(A)-ethanol(B) at 40 °C in terms of water mole fraction, Tyn and Calus [10].

Solutions of molecules with normal size and similar chemical structures can behave ideally over a large region of concentrations. The principal difference between ideal gases and ideal solutions is caused by the short time of influence between the molecules in gases and the large intermolecular times of influence in liquids. Therefore the pressure effect on $\mu_k^o(T,P)$ for ideal gases is different from the corresponding quantities for ideal solutions. This can be shown through the isothermal compressibility coefficient κ_k given by

$$\kappa_k = -\frac{1}{v_k}\left(\frac{\partial v_k}{\partial P}\right)_T \tag{33}$$

The compressibility for a liquid is practically is independent of the pressure, and we have

$$v_k(T,P) = v_k(T,0)(1-\kappa_k P) \tag{34}$$

The specific volume v_k does not depend on the concentration in ideal system.

For all ideal systems (both gases and liquids) we have

$$\left(\frac{\partial \mu_k}{\partial P}\right)_T = \left(\frac{\partial \mu_k^o(T,P)}{\partial P}\right)_T = v_k \tag{35}$$

From Eqs (34) and (35), we get

$$\mu_k(T) = \mu_k^o(T,P) = P(1-\frac{1}{2}\kappa_k P)v_k(T,0) \tag{36}$$

The isothermal compressibility for an ideal gas mixture κ is given by $1/P$, whereas for liquids the compressibility is negligible. The activities can be introduced to describe the deviations from the ideal behavior of solutions; the activities are expressed in terms of the activity coefficients.

For nonideal systems the interactions between the mass fluxes may be simplified if the activities are introduced for the diffusion potentials. The deviations from ideal systems can be expressed by using the relations between the diffusivities of Fick and of Maxwell-Stefan. The thermodynamic factor in this relation can be expressed in terms of the activities, and the Maxwell-Stefan diffusivities can be calculated from those of the Fick diffusivities.

The chemical potential μ_k can be expressed in terms of temperature, pressure and activity a_k

$$\mu_k = \mu_k^o(T, P) + R_k T \ln a_k \tag{37}$$

and the total differential is given by

$$d\mu_k = \left(\frac{\partial \mu_k}{\partial a_k}\right)_{T,P} da_k + \left(\frac{\partial \mu_k}{\partial P}\right)_{a_k,T} dP + \left(\frac{\partial \mu_k}{\partial T}\right)_{P,a_k} dT \tag{38}$$

The partial differentiation of chemical potential with respect to the activity is obtained from Eq. (37), and given by

$$\left(\frac{\partial \mu_k}{\partial a_k}\right)_{T,p} = \frac{R_k T}{a_k} \tag{39}$$

For nonideal systems Eq. (39) may be rearranged as

$$\frac{\rho_k}{P}(\text{grad } \mu_k)_{T,P} = \frac{\rho_k R_k T}{P a_k}(\text{grad } a_k)_{T,P} = \frac{x_k}{a_k}(\text{grad } a_k)_{T,P} \tag{40}$$

where $P_k = \rho_k R_k T$, and $x_k = P_k / P$. The diffusion flux for nonideal systems, with the gradient of activity as the driving force can be written as

$$J_k = \frac{x_k}{a_k}(\text{grad } a_k)_{T,P} + (\rho_k v_k - w_k)\frac{\text{grad } P}{P} \tag{41}$$

Description of diffusion may be complex in mixtures with more than two components; the diffusion coefficients are generally unknown, although sufficient experimental and theoretical information on binary systems is available. The Maxwell-Stefan diffusivities can be estimated for dilute monatomic gases from $D'_{lk} \approx D_{kl}$ when the Fick diffusivity D_{kl} is available. The Maxwell diffusivity is independent of the concentration for ideal gases, and almost independent of the concentration for ideal liquids mixtures. Since the Fick diffusivities are commonly measured, the Maxwell-Stefan diffusivities can be calculated from

$$D'_{kl} = \frac{D_{kl}}{\Gamma} \tag{42}$$

The thermodynamic correction factor Γ can be expressed in terms of the activity using Eq. (39)

$$\Gamma = x_k \frac{\partial \ln a_k}{\partial x_k} = \frac{\partial \ln a_k}{\partial \ln x_k} \tag{43}$$

Using the definition of activity as $a_k = x_k \gamma_k$, Eq. (43) becomes

$$\Gamma = 1 + \frac{\partial \ln \gamma_k}{\partial \ln x_k} \tag{44}$$

If the activity coefficients are known the Maxwell-Stefan diffusivities can be calculated from the Fick diffusivities by means of Eqs. (42) and (44). The activity coefficients of nonideal mixtures can be calculated using the molecular models of NRTL, UNIQUAC or the group contribution method of UNIFAC equation with temperature dependent parameters, since nonideality may be strong function of temperature and composition. The Maxwell-Stefan diffusivity for binary mixture of water-ethanol can be considered to be independent of the concentration of the mixture at around 40 °C. However, for temperatures above 60 °C the deviations from the ideal behavior increase, and the Maxwell-Stefan diffusivities can no longer be approximated as concentration independent constants. For highly nonideal mixtures, empirical correlations for estimating the concentration dependence of the diffusivities have to be used.

Binary systems

For a binary mixture, if experimental diffusivities do not exist over the whole range of concentration, one has to rely on interpolation of the diffusivities at infinite dilution $D_{kl}^{'\infty}$. In calculating the diffusivities at infinite dilution by the Stokes-Einstein relation, we consider small isolated hard spheres submerged in a liquid that are subjected to Brownian motion. The friction of the spheres in the liquid is given by the Stokes law; Einstein had used this law to calculate the mean-square displacement of the particle; the displacement increases linearly with time, and the proportionality constant is the Stokes-Einstein diffusivity $k_B T / 6\pi\eta r$, where r is the radius of the particle, k_B is the Boltzmann constant, and η is the viscosity of solvent. The generalized Stokes-Einstein diffusivity is modified to account the particle sizes of solute and solvent, and given by

$$D_{kl}^{'\infty} = \frac{k_B T}{n_c \pi \eta_l r_k} \left(\frac{r_l}{r_k} \right) \tag{45}$$

where n_c is the number replacing 6, and is around 3.5; 3.54 and 3.53 for nonassociating and associating binary organic mixtures, respectively; it is 3.65 for self-diffusivities of organic species; and 3.47 for polar organic species in water, where the hydrogen bonding occurs. The radius is calculated from the Van der Waals volumes. The estimates of the diffusivities with the modified Stokes-Einstein relation are comparable or more accurate than those estimated with the Wilke-Chang relation. A modified version of the Vignes interpolation for concentrated binary solutions is adequate for nearly ideal systems

$$D_{kl}^{'\infty} = \frac{1}{\eta}(D_{kl}^{'k\to 1}\eta_k)^{x_k}(D_{kl}^{'l\to 1}\eta_l)^{x_l} \tag{46}$$

where η is the viscosity of the mixture. The diffusivity of moderately ideal mixtures is estimated with an accuracy of the order of 10%. The simple linear interpolation with the viscosity correction correlates the diffusivities within the experimental error

$$D_{kl}^{'\infty} = \frac{1}{\eta}(D_{kl}^{'k\to 1}\eta_k x_k + D_{kl}^{'l\to 1}\eta_l x_l) \tag{47}$$

Interpolations of Eqs. (46) and (47) are satisfactory for nonideal nonassociating systems with accuracy of about 10%; however for associating mixtures the expressions yield relatively large errors.

Ternary systems

Ternary diffusion is more complicated than binary diffusion, and there is lack of experimental data on ternary diffusion. To obtain estimates for ternary mixtures the interpolation relations, given in Eqs. (46) and (47), are extended as

$$D_{kl}^{'\infty} = \frac{1}{\eta}(D_{kl}^{'k\to 1}\eta_k)^{x_k}(D_{kl}^{'l\to 1}\eta_l)^{x_l}(D_{kl}^{'i\to 1}\eta_i)^{x_i} \tag{48}$$

$$D_{kl}^{'\infty} = \frac{1}{\eta}(D_{kl}^{'k\to 1}\eta_k x_k + D_{kl}^{'l\to 1}\eta_l x_l + D_{kl}^{'i\to 1}\eta_i x_i) \tag{49}$$

For the complete estimate of the ternary diffusion for nonideal mixtures, six diffusivities at infinite dilution and three diffusivities of the type $D_{kl}^{'k\to 1}$ are needed. The nonelectrolyte multicomponent diffusion coefficients are reduced to binary diffusivities, while the multicomponent electrolyte systems cannot be reduced to a set of binary systems; but to a set of ternary systems. Negative diffusion coefficients can exist in ternary systems and are consistent with the nonequilibrium thermodynamics approach. More experimental and theoretical studies are necessary in the field of multicomponent diffusion.

Some of the molecular theories of multicomponent diffusion in mixtures led to expressions for the mass fluxes of the Maxwell-Stefan form, and predicted the mass fluxes dependent on the velocity gradients in the system. Such dependencies are not allowed in linear nonequilibrium thermodynamics. Mass-flux contains concentration rather than activity as driving forces. In order to overcome this inconsistency, the starting point is Jaumann's entropy balance equation

$$\rho \frac{DS}{Dt} = -(\nabla \cdot \mathbf{s}) + \Phi \tag{50}$$

where ρ is the density of the fluid mixture, S is the entropy per unit mass, $\mathbf{s}$ is the entropy-flux vector, and Φ is the rate of entropy production per unit volume. The operator

$$\frac{D}{Dt} = \frac{\partial}{\partial t} + (\mathbf{v} \cdot \nabla) \tag{51}$$

is the substantial derivative. From the balance equations of mass, momentum, energy, and the Gibbs relation, one obtains explicit expressions for $\mathbf{s}$ and Φ, which were derived in chapter 4.

For multicomponent diffusion, and the mass-flux expressions, we mainly use the Fick and the Maxwell-Stefan forms. Using the symmetric-diffusivity, in length^2/time, we have definition

$$D_{ij}^{'} = \frac{cRTL_{ij}}{\rho_i \rho_j} \tag{52}$$

where ρ_i is the density of component i, the L_{ij} are the phenomenological coefficients, and c is the total molar concentration

$$c = \sum c_i = \sum \frac{\rho_i}{M_i} \tag{53}$$

where M_i shows the molecular weight of component i. The diffusivity coefficients have the following properties

$$D_{ij}^{'} = D_{ji}^{'} \qquad i,j = 1,2,..,n \tag{54}$$

$$\sum_{i=1}^{n} w_i D_{ij}^{'} = 0 \qquad j = 1,2,..,n \tag{55}$$

In the diffusivity matrix, and there are ½$n(n-1)$ independent diffusivities D'_{ij}, which are also the coefficients in a positive definite quadratic form, since according to the second law of thermodynamics, the internal entropy for an uncoupled process never decreases. In terms of these symmetric diffusivities, the mass-flux expression becomes

$$\mathbf{j}_i = -D_i^T \nabla \ln T - \rho_i \sum_{j=1}^{n} D'_{ij} X_j \qquad i = 1,2,..,n \tag{56}$$

where D_i^T is the generalized thermal diffusion coefficient in mass/(length)(time). The generalized driving force X_i is given by

$$cRTX_i = \nabla P_i - w_i \nabla P - \rho_i F_i + w_i \sum_{j=1}^{n} \rho_j F_j \tag{57}$$

where F_i is the force per unit mass acting on the ith species. Using Eqs (56) and (57) we can express the mass-flux in terms of general driving force that is the Fick form

$$\mathbf{j}_i = -D_i^T \nabla \ln T - \frac{\rho_i}{cRT} \sum_{j=1}^{n} D'_{ij} (\nabla P_j - w_j \nabla P - \rho_j F_i + w_j \sum_{k=1}^{n} \rho_k F_k) \tag{58}$$

Instead of expressing the mass-flow vector $\mathbf{j}_i$ in terms of the driving forces X_i, it is sometimes convenient to express the X_i as a linear function of the $\mathbf{j}_i$ that is in the Maxwell-Stefan form

$$\begin{aligned} cRT \sum_{k \neq i}^{n} C'_{ik} \left(\frac{j_k}{\rho_k} - \frac{j_i}{\rho_i} \right) &= \nabla P_i - w_i \nabla P - \rho_i F_i + w_i \sum_{j=1}^{n} \rho_j F_j \\ &- cRT \sum_{k \neq i}^{n} C'_{ik} \left(\frac{D_k^T}{\rho_k} - \frac{D_i^T}{\rho_i} \right) (\nabla \ln T) \end{aligned} \tag{59}$$

where C'_{ij} is the inverse diffusivity, and sometimes is expressed as $C'_{ij} = x_i x_k / D'_{ik}$, and D'_{ik} are the Maxwell-Stefan diffusivities. The expressions in Eqs (58) and (59) contain the same information and are interrelated through the connection between the multicomponent diffusivities D'_{ij} and the multicomponent inverse diffusivities C'_{ij}. For low density gases $\nabla P_i = \nabla(c_i RT)$, and standard results are obtained. For polymeric liquids, a similar form to Eq. (59) can be found from a molecular theory [3] by replacing the pressure P and the partial pressure P_i with the total stress tensor and the partial stress tensor. The mass flux is related to the velocity gradient via the stress tensor, temperature, and concentration gradients. The linear nonequilibrium thermodynamics is able to generalize these expressions to include thermal, pressure, and forced diffusion.

2. DIFFUSION IN NONELECTROLYTE SYSTEMS

The linear phenomenological law of diffusion for a binary system is given by

$$\mathbf{j}_1 = \rho_1(v_1 - v_2) = \frac{L}{T}\left(\mathbf{F}_1 - \frac{\partial \mu_1}{\partial x_1}\right) \tag{60}$$

For a perfect gas or an ideal solution

$$\mu_j = \mu_j^o(T, P) + RT \ln c_1 \tag{61}$$

Inserting Eq. (61) into Eq. (60) yields

$$c_1(v_1 - v_2) = \frac{L}{T}\left(\mathbf{F}_1 M_1 - \frac{\partial \mu_1}{\partial x_1}\right) = -\frac{L}{T}\frac{RT}{c_1}\left(\frac{\partial c_1}{\partial x_1} - \frac{\mathbf{F}_1 M_1 c_1}{RT}\right) \tag{62}$$

Here, we can distinguish the two separate systems:
(i) For a uniform system, where ($\partial c_1 / \partial x_1 = 0$), we have

$$(v_1 - v_2) = \frac{L}{Tc_1}\mathbf{F}_1 M_1 \tag{63}$$

The coefficient of proportionality between the relative velocity $(v_1 - v_2)$ and the force $F_1 M_1$ is called the *mobility* of component 1 B^*

$$B^* = \frac{L}{Tc_1} \tag{64}$$

(ii) For a system without the external forces ($\boldsymbol{F}_1 = 0$), we have

$$c_1(v_1 - v_2) = -\frac{L}{T}\frac{RT}{c_1}\frac{\partial c_1}{\partial x_1} \tag{65}$$

The coefficient of proportionality between the flux of diffusion $c_1(v_1 - v_2)$ and the concentration gradient is the diffusion coefficient

$$D = \frac{L}{T}\frac{RT}{c_1} \tag{66}$$

Comparing Eqs. (64) and (66) yields the Einstein relation between the mobility and diffusion coefficient

$$D = RTB^* \tag{67}$$

For a system without the external force, Eq. (62) can be written as

$$c_1(v_1 - v_2) = -\frac{L}{T}\frac{\partial \mu_1}{\partial N_1}\frac{\partial N_1}{\partial x_1} \tag{68}$$

The phenomenological law defines the diffusion coefficient D as

$$c_1(v_1 - v_2) = -Dc\frac{\partial N_1}{\partial x_1} \tag{69}$$

so that

$$D = \frac{1}{Tc}L\frac{\partial \mu_1}{\partial N_1} \tag{70}$$

This definition is equivalent to Eq. (66) for a perfect gas or for an ideal solution.

Eq. (70) shows that the diffusion coefficient is the product of the phenomenological coefficient L and the thermodynamic quantity $\frac{1}{Tc}(\partial \mu_1 / \partial N_1)$. The coefficient L is positive, and so is the quantity $\partial \mu_1 / \partial N_1$ for all ideal systems. This means that the diffusion coefficient is positive, and according to Eq. (69), the diffusion flow has the direction imposed by the existing concentration gradient. In some highly nonideal systems of partially or completely immiscible mixtures, such as water-butane and water-benzene, the quantity $\partial \mu_1 / \partial N_1$ may be negative corresponding to the thermodynamic instability. Such systems may split into two liquid phases, and may have negative diffusion coefficients in the immiscible region. This is in contrast to the thermal conductivity, which is always positive, and results from the fact that the diffusion coefficient is a product of two quantities with only one of them, which is L, has a definite sign.

3. DIFFUSION IN ELECTROLYTE SYSTEMS

For the treatment of electrostatic potentials and electric current of charged ionic species, we start with the fundamental Gibbs equation

$$dU = TdS - \delta W \tag{71}$$

and reconsider the work term δW. Usually δW means the work of compression, PdV, and the work involved in changing the number of moles of the components (chemical work), $-\sum \mu_i dN_i$. When we have a region with an electrostatic potential ψ, a change in the charge de results the electrostatic work. Hence Eq. (71) may be extended

$$dU = TdS - PdV + \sum \mu_i dN_i + \psi de \tag{72}$$

When de is due to changes in the concentration of ionic species, we get

$$de = \sum z_i F dN_i \tag{73}$$

then Eq. (72) becomes

$$dU = TdS - PdV + \sum_i \mu_i dN_i + \sum_i z_i F \psi dN_i \tag{74}$$

combining the last two summations, we have

$$dU = TdS - PdV + \sum_i (\mu_i + z_i F \psi) dN_i \tag{75}$$

Eq. (75) indicates that the chemical work in electrolytes contains a chemical term μdN_i and an electrical term $z_i F \psi dN_i$. In practice the sum can be measured. This fact was recognized by Gibbs, and Guggenheim defined the sum $\mu_i + z_i F \psi$ as the *electrochemical potential* $\tilde{\mu}_i$ of the ionic substance *i*

$$\tilde{\mu}_i = \mu_i + z_i F \psi \tag{76}$$

If we have a phase in which the composition is identical at points 1 and 2 but $\psi_1 \neq \psi_2$, then we have

$$\mu_i^1 = \mu_i^2 \tag{77}$$

$$\tilde{\mu}_i^1 - \tilde{\mu}_i^2 = z_i F (\psi_1 - \psi_2) \tag{78}$$

Using Eq. (76) in Eq. (75) we get

$$dU = TdS - PdV + \sum \tilde{\mu}_i dN_i \tag{79}$$

We may also use the change in the Gibbs free energy in terms of the chemical potential

$$dG = -SdT + VdP + \sum \tilde{\mu}_i dN_i \tag{80}$$

or the Gibbs-Duhem relation,

$$SdT - VdP + \sum N_i d\tilde{\mu}_i = 0 \tag{81}$$

which under isothermal, isobaric conditions reduces to

$$\sum_i N_i d\tilde{\mu}_i = 0 \tag{82}$$

It is often useful to express the electrochemical potential as a sum of explicit terms of activity and electrostatic as follows

$$\tilde{\mu}_i = \mu_i^o(T) + VP + RT \ln a_i + z_i F\psi \tag{83}$$

Thus, in the case of an ion distributed between two phases of α and β, the condition of thermodynamic equilibrium is

$$\tilde{\mu}_i^\alpha = \tilde{\mu}_i^\beta \tag{84}$$

Introducing Eq. (83) into Eq. (84) yields

$$\Delta\mu_i^o + (V_i^\alpha P^\alpha - V_i^\beta P^\beta) + RT \ln \Delta a_i + z_i F\Delta\psi = 0 \tag{85}$$

If, for example, α and β are aqueous phases separated by a membrane, then

$$(\mu_i^o)^\alpha = (\mu_i^o)^\beta \quad \text{and} \quad V_i^\alpha = V_i^\beta$$

so that Eq. (85) becomes

$$V_i \Delta P + RT \ln \Delta a_i + z_i F\Delta\psi = 0 \tag{86}$$

In most cases of interest $V_i\Delta P$ may be neglected in comparison with the other terms, so that the condition of phase equilibrium across a membrane becomes

$$RT \ln \frac{a_i^\alpha}{a_i^\beta} = -z_i F \Delta \psi \tag{87}$$

For an ideal solution the condition of phase equilibrium becomes

$$RT \ln \frac{c_i^\alpha}{c_i^\beta} = -z_i F \Delta \psi \tag{88}$$

For practical calculations, Eq. (88) is written with base 10 logarithms

$$-\Delta \psi = \frac{2.3RT}{z_i F} \log \frac{c_i^\alpha}{c_i^\beta} = \frac{58}{z_i} \log \frac{c_i^\alpha}{c_i^\beta} \qquad \text{(mV at 20 °C)} \tag{89}$$

As a practical example let us consider the aqueous solution with N_1 moles of sodium chloride and N_2 moles of calcium chloride. An increase in the concentrations of both salts by amounts dN_1 and dN_2 will cause the following changes in the ionic concentrations

$$dN_{\mathrm{Na}} = dN_1, \quad dN_{\mathrm{Ca}} = dN_2, \quad dN_{\mathrm{Cl}} = dN_1 + 2dN_2 \tag{90}$$

By introducing these expressions into Eq. (79), we obtain

$$dU = TdS - PdV + \tilde{\mu}_{\mathrm{Na}} dN_{\mathrm{Na}} + \tilde{\mu}_{\mathrm{Ca}} dN_{\mathrm{Ca}} + \tilde{\mu}_{\mathrm{Cl}} dN_{\mathrm{Cl}} \tag{91}$$

or

$$dU = TdS - PdV + (\tilde{\mu}_{\mathrm{Na}} + \tilde{\mu}_{\mathrm{Cl}}) dN_1 + (\tilde{\mu}_{\mathrm{Ca}} + 2\tilde{\mu}_{\mathrm{Cl}}) dN_2 \tag{92}$$

The corresponding chemical potentials of the electroneutral combinations are

$$\tilde{\mu}_{\mathrm{Na}} + \tilde{\mu}_{\mathrm{Cl}} = \mu_{\mathrm{Na}} + F\psi + \mu_{\mathrm{Cl}} - F\psi = \mu_{Na} + \mu_{Cl} \tag{93}$$

and

$$\tilde{\mu}_{\mathrm{Ca}} + 2\tilde{\mu}_{\mathrm{Cl}} = \mu_{\mathrm{Ca}} + 2F\psi + 2\mu_{\mathrm{Cl}} - 2F\psi = \mu_{\mathrm{Ca}} + 2\mu_{\mathrm{Cl}} \tag{94}$$

The physical significance of these combinations is readily seen by consideration of the dissociation equilibria

$$\mathrm{NaCl} \Leftrightarrow \mathrm{Na}^+ + \mathrm{Cl}^-$$

$$\mathrm{CaCl_2} \Leftrightarrow \mathrm{Ca}^{++} + 2\mathrm{Cl}^-$$

which are characterized thermodynamically by the following equations

$$\mu_{\text{NaCi}} = \widetilde{\mu}_{\text{Na}} + \widetilde{\mu}_{\text{Cl}} = \mu^{\text{o}}_{\text{NaCl}} + PV_{\text{NaCl}} + RT\ln a_{\text{Na}} a_{\text{Cl}} \tag{95}$$

$$\mu_{\text{CaCl}_2} = \widetilde{\mu}_{\text{Ca}} + 2\widetilde{\mu}_{\text{Cl}} = \mu^{\text{o}}_{\text{CaCl}_2} + PV_{\text{CaCl}_2} + RT\ln a_{\text{Ca}} a^2_{\text{Cl}} \tag{96}$$

Therefore Eq. (92) may be written in terms of electrically neutral species

$$dU = TdS - PdV + (\mu_{\text{NaCl}})dN_1 + (\mu_{\text{CaCl}_2})dN_2 \tag{97}$$

Conditions of phase equilibria of the salts across a membrane separating two solutions of α and β are

$$\mu^{\alpha}_{\text{NaCl}} = \mu^{\beta}_{\text{Na}Cl} \tag{98}$$

$$\mu^{\alpha}_{\text{CaCl}_2} = \mu^{\beta}_{\text{CaCl}_2} \tag{99}$$

Let us assume that one side of the membrane contains a chloride salt of a macromolecule to which the membrane is impermeable. Other side contains a solution of CaCl_2 alone. The concentration of the macromolecule is c_m and the number of charged groups per molecule is v. The concentration of CaCl_2 in the solution containing the macromolecule is c^{α}_s, and the concentration in the other phase is c^{β}_s. For the equilibrium case we have

$$\begin{aligned} &\mu^{o}_{\text{CaCl}_2} + V_{\text{CaCl}_2} P^{\alpha} + RT\ln c^{\alpha}_{\text{Ca}} (c^{\alpha}_{\text{Cl}})^2 \\ &= \mu^{o}_{\text{CaCl}_2} + V_{\text{CaCl}_2} P^{\beta} + RT\ln c^{\beta}_{\text{Ca}} (c^{\beta}_{\text{Cl}})^2 \end{aligned} \tag{100}$$

Since the pressure terms are negligible, this expression reduces to

$$c^{\alpha}_{\text{Ca}} (c^{\alpha}_{\text{Cl}})^2 = c^{\beta}_{\text{Ca}} (c^{\beta}_{\text{Cl}})^2 \tag{101}$$

We know that

$$c^{\alpha}_{\text{Ca}} = c^{\alpha}_s \text{ and } c^{\alpha}_{\text{Cl}} = vc_m + 2c^{\alpha}_s \tag{102}$$

$$c^{\beta}_{\text{Ca}} = c^{\beta}_s \text{ and } c^{\beta}_{\text{Ca}} = 2c^{\beta}_s \tag{103}$$

and hence

$$c_s^\alpha(\nu c_m + 2c_s^\alpha)^2 = c_s^\beta(2c_s^\beta)^2 = 4(c_s^\beta)^3 \tag{104}$$

Eq. (104) shows the well-known *Donnan equilibrium* of salt across a membrane in the presence of a polyelectrolyte, to which the membrane is permeable. It demonstrates the characteristic properties of chemical potentials of neutral salts.

4. IRREVERSIBLE PROCESSES IN ELECTROLYTE SYSTEMS

The local dissipation function for the systems with charged components is

$$\Psi = -J_s \cdot \nabla T - \sum_i J_i \nabla \tilde{\mu} + J_r \tilde{A} \tag{105}$$

where $\tilde{A}$ is the *electrochemical affinity*, and given by

$$\tilde{A} = -\sum \nu_i \tilde{\mu}_i = -\sum \nu_i(\mu_i + z_i F\psi) = -\sum \nu_i \mu_i - F\psi \sum \nu_i z_i \tag{106}$$

Since the charge is conserved in the reaction, $\sum \nu_i z_i = 0$, so that

$$\tilde{A} = -\sum_i \nu_i \mu_i = A \tag{107}$$

For an isothermal system, in which the diffusing component does not react, Eq. (105) reduces to

$$\Psi = -\sum_i J_i \nabla \tilde{\mu}_i \tag{108}$$

electrochemical potentials also obey the Gibbs-Duhem equation

$$\sum_i c_i \nabla \tilde{\mu}_i = 0 \tag{109}$$

In an n-component system there are n-1 independent forces $-\nabla\tilde{\mu}_i$. Eq. (109) is used to eliminate the force on the solvent from Eq. (108)

$$\Psi = \sum_i^{n-1}\left(J_i - \frac{c_i}{c_w}J_w\right)\nabla(-\tilde{\mu}_i) = \sum_i^{n-1} J_i^d \nabla(-\tilde{\mu}_i) \tag{110}$$

where J_i^d is the flow of solute relative to that of the solvent.

For solution of a single electrolyte dissociating into two ions, the dissipation function can be expressed by

$$\Psi = -J_1^d \nabla\tilde{\mu}_1 - J_2^d \nabla\tilde{\mu}_2 \tag{111}$$

The phenomenological equations relating the flows and forces defined by Eq. (111) are

$$J_1^d = -L_{11}\nabla\tilde{\mu}_1 - L_{12}\nabla\tilde{\mu}_2 \tag{112}$$

$$J_2^d = -L_{21}\nabla\tilde{\mu}_1 - L_{22}\nabla\tilde{\mu}_2 \tag{113}$$

The physical meaning of Eqs. (112) and (113) can be clarified by making one of the forces vanish; if $\nabla\tilde{\mu}_2 = 0$ then $J_1^d = -L_{11}\nabla\tilde{\mu}_1$, which indicates that L_{11} is the generalized mobility of the cation, since it is the proportionality coefficient relating the flow to its conjugate force. In this case J_2^d is not zero, but is given by $J_2^d = -L_{21}\nabla\tilde{\mu}_1$, indicating that the diffusion of the cation causes a drag effect on the anions, and such interactions are determined by the coefficient L_{12} or L_{21}.

Eqs. (112) and (113) can be used in a special case of an electrical-conductance measurement. This analysis is usually carried out under isothermal, isobaric, and uniform concentration ($\nabla\mu_i = 0$) for all components in the cell. The electric current I is driven by a potential difference between two nonpolarizable electrodes, and the local field intensity ε is defined by

$$\varepsilon = -\nabla\psi \tag{114}$$

Then the forces acting on a z_1-valence cation and a z_2-valence anion become

$$\nabla\tilde{\mu}_1 = \nabla\mu_1 + z_1 F\nabla\psi = -z_1 F\varepsilon \tag{115}$$

$$\nabla\tilde{\mu}_2 = \nabla\mu_2 + z_2 F\nabla\psi = -z_2 F\varepsilon \tag{116}$$

So that Eqs. (112) and (113) become

$$J_1^d = (z_1 L_{11} + z_2 L_{12})F\varepsilon \tag{117}$$

$$J_2^d = (z_1 L_{12} + z_2 L_{22})F\varepsilon \tag{118}$$

For a monomonovalent salt such as NaCl or KCl, for which $z_1 = - z_2 = 1$, we have

$$J_1^d = (L_{11} - L_{12})F\varepsilon \tag{119}$$

$$J_2^d = (L_{12} - L_{22})F\varepsilon \tag{120}$$

The electric current due to the transport of all ionic species is given by the sum over all the charges carried by the ionic flows

$$I = \sum_{i=1}^{n} z_i F J_i^d \tag{121}$$

For a single electrolyte, we obtain

$$I = z_1 F J_1^d + z_2 F J_2^d = (z_1^2 L_{11} + 2 z_1 z_2 L_{12} + z_2^2 L_{22}) F^2 \varepsilon \tag{122}$$

Due to the condition of electroneutrality, diffusion flows can be used in Eq. (122). The flows are given with relative to the water velocity

$$J_1^d = c_1 (v_1 - v_w)$$

$$J_2^d = c_2 (v_2 - v_w)$$

We also have $c_1 = \nu_1 c_s$ and $c_2 = \nu_2 c_s$, so that Eq. (122) becomes

$$I = z_1 F \nu_1 c_s v_1 + z_2 F \nu_2 c_s v_2 - v_w F c_s (\nu_1 z_1 + \nu_2 z_2) \tag{123}$$

However electroneutrality implies that $_1 z_1 + \nu_2 z_2 = 0$, so that

$$I = z_1 F J_1 + z_2 F J_2 \tag{124}$$

where J_1 and J_2 are the absolute flows $J_1 = c_1 v_1$, $J_2 = c_2 v_2$, respectively. Ohm's law holds for homogeneous, isothermal salt solutions, therefore the relation between the current and the electric field intensity may also be written as

$$I = \kappa \varepsilon \tag{125}$$

where κ is the *electrical conductance* of the solution.

Comparison of Eqs. (124) and (125) indicates that

$$\kappa = L' F^2 \tag{126}$$

where L' is given by

$$L' = z_1^2 L_{11} + 2 z_1 z_2 L_{12} + z_2^2 L_{22} \tag{127}$$

Eq. (127) shows the direct relation between the electrical conductance of the solution and the phenomenological coefficient. Similar relations are obtained by measuring the fraction of the total current that is carried by each ion, also under conditions of $\nabla \mu_i = 0$. This fraction is called the *Hittorf transference number* t_i, and is expressed by

$$t_i = \left(\frac{z_i F J_i^d}{I} \right)_{\nabla \mu_i = 0} \tag{128}$$

For the case of a single electrolyte, t_1 and t_2 may be evaluated by introducing Eqs. (117) and (118) into Eq. (128), and we have

$$t_1 = \frac{z_1 F J_1^d}{I} = \frac{z_1 (z_1 L_{11} + z_2 L_{12})}{L'} \tag{129}$$

$$t_2 = \frac{z_2 F J_2^d}{I} = \frac{z_2 (z_1 L_{12} + z_2 L_{22})}{L'} \tag{130}$$

It is apparent that the two transference numbers are not independent since $t_1 + t_2 = 1$. Therefore an additional expression is required to evaluate the three coefficients L_{11}, L_{12}, and L_{22}. Such a relation may be obtained from the diffusion of the electrolyte. In this case, there is no electric current in the system and the total transport of charge must vanish

$$z_1 J_1^d + z_2 J_2^d = \frac{I}{F} = 0 \tag{131}$$

By introducing Eqs. (112) and (113) for J_1^d and J_2^d, we obtain a relation between the forces acting on the two ions

$$- z_1 (L_{11} \nabla \tilde{\mu}_1 + L_{12} \nabla \tilde{\mu}_2) - z_2 (L_{12} \nabla \tilde{\mu}_1 + L_{22} \nabla \tilde{\mu}_2) = 0 \tag{132}$$

or using the general form of Eq. (95) $_1 \nabla \tilde{\mu}_1 + \nu_2 \nabla \tilde{\mu}_2 = \nabla \mu_s$, and from the neutrality condition $_1 z_1 + \nu_2 z_2 = 0$, we have

$$\nabla \tilde{\mu}_1 = \frac{z_2}{\nu_1} \left(\frac{z_1 L_{12} + z_2 L_{22}}{L'} \right) \nabla \mu_s \tag{133}$$

$$\nabla \tilde{\mu}_2 = \frac{z_1}{\nu_2} \left(\frac{z_1 L_{11} + z_2 L_{12}}{L'} \right) \nabla \mu_s \tag{134}$$

Introducing Eqs. (133) and (134) into (112) and (113), and rearranging yields

$$J_1^d = \frac{z_1 z_2}{v_2}\left(\frac{L_{11}L_{22} - L_{12}^2}{L'}\right)\nabla\mu_s \tag{135}$$

$$J_2^d = \frac{z_1 z_2}{v_1}\left(\frac{L_{11}L_{22} - L_{12}^2}{L'}\right)\nabla\mu_s \tag{136}$$

The flows of J_s^d of the neutral salt is given by

$$J_s^d = \frac{J_1^d}{v_1} = \frac{J_2^d}{v_2} = \frac{z_1 z_2}{v_1 v_2}\left(\frac{L_{11}L_{22} - L_{12}^2}{L'}\right)\nabla\mu_s \tag{137}$$

Fick's law expresses the diffusion of a neutral salt in a binary solution as

$$J_1^d = -D\nabla c_s \tag{138}$$

and comparing the resulting Eqs. (137) and (138), we may express the diffusion coefficient in terms of the L_{ij} as

$$D = -\frac{z_1 z_2}{v_1 v_2}\left(\frac{L_{11}L_{22} - L_{12}^2}{L'}\right) \tag{139}$$

The relations for the electrical conductance, transference number, and diffusion coefficient provide the three relations from which the phenomenological coefficients can be determined, and for a monomonovalent salt we have

$$L_{11} = \frac{D}{\mu_{ss}} + \frac{\kappa t_1^2}{F^2} \tag{140}$$

$$L_{22} = \frac{D}{\mu_{ss}} + \frac{\kappa t_2^2}{F^2} \tag{141}$$

$$L_{12} = \frac{D}{\mu_{ss}} + \frac{\kappa t_1 t_2}{F^2} \tag{142}$$

Calculations for the NaCl solution show that the straight coefficients, L_{11} and L_{22}, are nearly linear functions of concentrations, while the cross coefficient L_{12} is highly dependent on concentration and becomes quite small at high dilution, where the interactions between the ions are minimal.

For the properties of the phenomenological coefficients it may be advantageous to consider the mobilities, which express the behavior of ions in the solutions similar to those gained by considering the of frictional coefficients in the case of membrane permeability. The mobility may be defined by using the explicit expressions for the flows under uniform chemical potentials

$$J_1^d = c_1(v_1 - v_w) = \nu_1 c_s \omega_1 z_1 F\varepsilon \qquad (143)$$

where ${}_1c_s = c_1$ is the concentration of ion 1, and ω_1 is the ionic mobility. Eq. (143) shows that ω_1 is the relative velocity of the ion per unit electrical force that is the velocity acquired as a result of the operation of a force of 1 dyne. In practice, the mobilities u_i are defined as the velocity of the ions acquired in a field of $\varepsilon = 1$ volt/cm, and mobilities u_1 and u_2 of the cation and anion respectively are

$$u_1 = z_1\omega_1 F \text{ and} - u_2 = z_2\omega_2 F \qquad (144)$$

Therefore from Eq. (143) we may write

$$J_1^d = \nu_1 c_s u_1 \varepsilon \text{ , and } J_2^d = -\nu_2 c_s u_2 \varepsilon \qquad (145)$$

and the total electric current becomes

$$I = z_1 F J_1^d + z_2 F J_2^d = \nu_1 z_1 c_s F(u_1 + u_2)\varepsilon \qquad (146)$$

therefore, in terms of the mobilities, the electrical conductance is given by

$$\kappa = \nu_1 z_1 c_s F(u_1 + u_2) \qquad (147)$$

It is often convenient to consider the equivalent conductance λ_{eq} instead of κ

$$\lambda_{eq} = \frac{\kappa}{\nu_1 z_1 c_s} = F(u_1 + u_2) \qquad (148)$$

similarly conductance of a single ion can be defined as

$$\lambda_1 = Fu_1 \text{ and } \lambda_2 = Fu_2 \qquad (149)$$

So that $\lambda_{eq} = \lambda_1 + \lambda_2$, which is the well-known *expression of Kholrausch.*

For the cation, we can express the diffusion in terms of the mobility

$$J_1^d = \nu_1 c_s u_1 \varepsilon = (z_1 L_{11} + z_2 L_{22})F\varepsilon \qquad (150)$$

Therefore

$$u_1 = \frac{(z_1 L_{11} + z_2 L_{12})F}{c_1} = \frac{z_1^2 L_{11} F}{c_1 z_1} + \frac{z_1 z_2 L_{12} F}{c_1 z_1} \tag{151}$$

We now define the *reduced phenomenological mobility* u_{ij}

$$u_{11} = \frac{z_1^2 L_{11} F}{v_1 z_1 c_s} \text{ and } -u_{12} = \frac{z_1 z_2 L_{12} F}{v_1 z_1 c_s} \tag{152}$$

where u_{11} is the reduced phenomenological mobility of ion 1, and u_{12} is a measure of the interaction of ions 1 and 2, so that

$$u_1 = u_{11} - u_{12}$$

$$u_2 = u_{22} - u_{12} \tag{153}$$

Introducing Eq. (153) into Eq. (147), we have

$$\kappa = c_1 z_1 F(u_{11} - 2u_{12} + u_{22}) \tag{154}$$

The equivalent conductances become

$$\lambda_{eq} = F(u_{11} - 2u_{12} + u_{22}) \tag{155}$$

$$\lambda_1 = F(u_{11} + u_{12}) \tag{156}$$

$$\lambda_2 = F(u_{22} + u_{12}) \tag{157}$$

The transference number can be obtained as

$$t_1 = \frac{u_{11} - u_{12}}{u_{11} - 2u_{12} + u_{22}} \tag{158}$$

$$t_2 = \frac{u_{22} - u_{12}}{u_{11} - 2u_{12} + u_{22}} \tag{159}$$

Finally, the diffusion coefficient of the salt can be expressed in terms of the mobilities

$$D = \frac{c_s \mu_{ss}}{v_1 z_1} \left(\frac{u_{11} u_{22} - u_{12}^2}{\lambda_{eq}} \right) \tag{160}$$

From Eqs. (155) to (160), the reduced phenomenological mobilities are obtained as

$$u_{11} = F\frac{\nu_1 z_1 D}{c_s \mu_{ss}} + \frac{\lambda t_1^2}{\lambda_{eq} F} \tag{161}$$

$$u_{22} = F\frac{\nu_1 z_1 D}{c_s \mu_{ss}} + \frac{\lambda t_2^2}{\lambda_{eq} F} \tag{162}$$

$$u_{12} = F\frac{\nu_1 z_1 D}{c_s \mu_{ss}} - \frac{\lambda t_1 \lambda_2}{\lambda_{eq} F} \tag{163}$$

These expressions can be used to calculate u_{ij} from known values of the other parameters. Calculations for NaCl show that u_{11} and u_{22} remain approximately constant over a relatively wide range of concentrations, while u_{12} changes considerably.

REFERENCES

[1] A. Bouddour, J.L. Auriault, M. Mhamdi-Alaoui and J.F. Bloch, Int. J. Heat Mass Transfer, 41 (1998) 2263.
[2] M. Braun and U. Renz, Int. J. Heat Mass Transfer, 40 (1997) 131.
[3] C.F. Curtiss and R.B. Bird, Ind. & Eng. Chem. Res., 38 (1999) 2515.
[4] E.L. Cussler, R. Aris and A. Bhown, J. Memb. Sci., 43 (1989) 149.
[5] P.R. Danesi and L.R. Yinger, J. Memb. Sci., 29 (1986) 195.
[6] J.A. Daoud, S.A. El-Reefy and H.F. Aly, Sep. Sci. Tech., 33 (1998) 537.
[7] Y. Demirel and S.I. Sandler, Int. J. Heat Mass Transfer, 44 (2001) 2439.
[8] B.C. Eu, Kinetic Theory and Irreversible Thermodynamics, John Wiley, New York, 1992.
[9] D. Kondepudi and I. Prigogine, Modern Thermodynamics, From Heat Engines to Dissipative Structures, Wiley, New York, 1999.
[10] M.T. Tyn and W.F. Calus, J. Chem. Eng. Data, 20 (1975) 310.

Chapter 9

Heat and mass transfer

INTRODUCTION

Simultaneous heat and mass transfer plays an important role in various physical, chemical and biological processes, hence a vast amount of work on the subject is available in the literature; heat and mass transfer occurs in absorption, distillation, and extraction, drying, melting and crystallization, and evaporation-condensation, such as thermal diffusion effects in partial film condensation. Various formulations and methodologies have been suggested for describing combined heat and mass transfer problems. Mikhailov and Ozisik [5] used the integral transform technique in the development of the general solutions. In this chapter the cross phenomena or the coupled heat and mass transfer is discussed using the linear nonequilibrium thermodynamics theory.

The heat fluxes $\mathbf{J}_q^{'}$ and $\mathbf{J}_q^{''}$ are related through the internal energy flux $\mathbf{J}_u$

$$\mathbf{J}_u = \mathbf{J}_q^{''} + \sum_{i=1}^{n} \bar{h}_i \mathbf{j}_i = \mathbf{J}_q^{'} + \sum_{i=1}^{n} \bar{u}_i \mathbf{j}_i$$

Similarly the entropy flux is expressed by [Chapter 4, Eq. (59)]

$$\mathbf{J}_s = \frac{\mathbf{J}_q^{''}}{T} + \sum_{i=1}^{n} \bar{s}_i \mathbf{j}_i$$

where $\bar{s}_i$ is the partial specific entropy and, $\bar{h}_i, \bar{u}_i$ are the partial specific enthalpy and partial specific internal energy, respectively.

Consider the heat and diffusion flows in a fluid in mechanical equilibrium without chemical reaction; the dissipation function is given by

$$\Psi = -\mathbf{J}_q^{''} \nabla \ln T - \sum_{i,k=1}^{n-1} \mathbf{j}_i \cdot a_{ik} \left[\sum_{j=1}^{n-1} \left(\frac{\partial \mu_k}{\partial w_j} \right)_{T,P,w_{i \neq j}} \nabla w_j \right] \geq 0 \quad (1)$$

where $a_{ik} = \delta_{ik} + w_k / w_n$, and δ_{ik} is the unit tensor, and $\mathbf{j}_i$ is the diffusion flow of component i.

1. HEAT AND MASS TRANSFER

Using the dissipation function, the conjugate flows and forces are identified and used in the phenomenoligical equations for simultaneous heat and mass transfer.

1.1. Binary systems

For a binary liquid mixture, the independent forces identified from the dissipation function of Eq. (1) for the heat and mass flows are

$$X_q = -\nabla \ln T \tag{2}$$

$$X_1 = -\frac{1}{w_2}\left(\frac{\partial \mu_1}{\partial w_1}\right)_{T,P} \nabla w_1 \tag{3}$$

and the phenomenological equations are given by

$$-\mathbf{J}_q'' = L_{qq} \nabla \ln T + L_{q1} \frac{1}{w_2}\left(\frac{\partial \mu_1}{\partial w_1}\right)_{T,P} \nabla w_1 \tag{4}$$

$$-\mathbf{j}_1 = \mathbf{j}_2 = L_{1q} \nabla \ln T + L_{11} \frac{1}{w_2}\left(\frac{\partial \mu_1}{\partial w_1}\right)_{T,P} \nabla w_1 \tag{5}$$

By the Onsager reciprocal relations, the matrix of phenomenological coefficients is symmetric $L_{1q} = L_{q1}$. Since the dissipation function is positive, the phenomenological coefficients must satisfy the inequalities

$$L_{qq} > 0,\ L_{11} > 0;\quad L_{qq} L_{11} - L_{q1}^2 > 0 \tag{6}$$

Fourier's law describes heat conduction caused only by the temperature gradient

$$\mathbf{J}_q'' = -k \nabla T \tag{7}$$

where k is the thermal conductivity in the absence of concentration gradient. Comparison of Eqs. (4) and (7) yields the relationship between the phenomenological coefficient L_{qq} and the thermal conductivity coefficient

$$L_{qq} = kT \tag{8}$$

Fick's law describes the diffusion flow caused only by the concentration gradient for an isothermal fluid

$$-\mathbf{j}_1 = \mathbf{j}_2 = -\rho D \nabla w_1 \tag{9}$$

which contains the diffusion coefficient D given by

$$D = D_{11} = L_{11} \frac{1}{\rho w_2} \left(\frac{\partial \mu_1}{\partial w_1} \right)_{T,P} \tag{10}$$

The diffusion caused only by the temperature gradient is called the *thermal diffusion* (*Soret effect*). When concentration gradient vanishes, Eq. (5) reduces to

$$\mathbf{j}_1 = -L_{1q} \nabla \ln T = -\frac{\rho}{T} D_{T1} \nabla T \tag{11}$$

The *thermal diffusion coefficient* of component 1, ($D_{T1} = L_{1q} / \rho$) is given by

$$D^{'} = \frac{D_{T1}}{w_1 w_2 T} = \frac{L_{1q}}{\rho w_1 w_2 T} \tag{12}$$

The Dufour effect is the flow of heat arising only from a concentration gradient, and expressed by

$$\mathbf{J}_q^{''} = -L_{q1} \frac{1}{w_2} \left(\frac{\partial \mu_1}{\partial w_1} \right)_{T,P} \nabla w_1 = -\rho_1 \left(\frac{\partial \mu_1}{\partial w_1} \right)_{T,P} T D^{''} \nabla w_1 \tag{13}$$

The *Dufour coefficient* D'' is related to the phenomenological coefficient L_{q1}

$$D^{''} = \frac{L_{q1}}{\rho w_1 w_2 T} \tag{14}$$

and the Onsager reciprocal relations yield

$$D'' = D' \tag{15}$$

So that Eqs. (4) and (5) can be expressed in terms of the transport coefficients of k and D

$$-\mathbf{J}_q'' = k\nabla T + \rho_1 \left(\frac{\partial \mu_1}{\partial w_1} \right)_{T,P} TD'\nabla w_1 \tag{16}$$

$$-\mathbf{j}_1 = \mathbf{j}_2 = \rho(w_1 w_2 D'\nabla T + D\nabla w_1) \tag{17}$$

The *thermal diffusion ratio* K_T is defined as

$$K_T = \frac{D_{T1}}{D} = w_1 w_2 T \left(\frac{D'}{D} \right) = w_1 w_2 T s_T \tag{18}$$

where s_T is called the *Soret coefficient*, and given by

$$s_T = \left(\frac{D'}{D} \right) = \frac{K_T}{w_1 w_2 T} = \left(\frac{L_{1q}}{L_{11}} \right) \left[w_1 T \left(\frac{\partial \mu_1}{\partial w_1} \right) \right]^{-1} \tag{19}$$

If K_T is positive, component 1 diffuses to cooler region, otherwise it diffuses to a hotter region. The *thermal diffusion factor* α, which is mainly independent of concentration for gases, is given by

$$\alpha = T \left(\frac{D'}{D} \right) = T s_T \tag{20}$$

Inequalities in Eq. (6) can now be written in terms of the transport coefficients of the thermal conductivity and mass diffusivity by using the *thermodynamic stability condition* $(\partial \mu_1 / \partial w_1)_{T,P} \geq 0$

$$k > 0, \; D > 0, \; (D')^2 < \frac{kD}{\rho w_1^2 w_2 T \left(\frac{\partial \mu_1}{\partial w_1} \right)_{T,P}} \tag{21}$$

For a binary mixture of ideal gases the molar fraction x_1 and the chemical potential μ_1 of the first component are given by

$$x_1 = \frac{M_2 w_1}{M_2 w_1 + M_1 w_2} \tag{22}$$

$$\mu_1 = \mu^o (T, P) + RT \ln P x_1 \tag{23}$$

The partial derivative of chemical potential of the mixture at constant temperature and pressure is given by

$$\left(\frac{\partial \mu_1}{\partial w_1}\right)_{T,P} = \frac{RT}{w_1 (M_2 w_1 + M_1 w_2)} \tag{24}$$

So that the diffusion coefficient, given in Eq. (10), becomes

$$D = L_{11} \frac{RT}{\rho} \frac{1}{w_1 w_2 (M_2 w_1 + M_1 w_2)} \tag{25}$$

and the phenomenological equation of the heat flow is given by

$$-\mathbf{J}_q'' = k \nabla T + D' \frac{RT^2 \rho}{M_2 w_1 + M_1 w_2} \nabla w_1 \tag{26}$$

With the thermal diffusion ratio K_T, Eq. (17) becomes

$$-\mathbf{j}_1 = \mathbf{j}_2 = \rho (w_1 w_2 D' \nabla T + D \nabla w_1) = \rho D (K_T \nabla \ln T + \nabla w_1) \tag{27}$$

1.2. Multicomponent systems

The dissipation function $\Psi = T\Phi$, resulting from the heat and mass transfer is expressed by

$$\Psi = -\mathbf{J}_u \cdot \nabla \ln T - \sum_{i=1}^{n} \mathbf{j}_i \cdot \left[T \nabla \left(\frac{\mu_i}{T} \right) - \mathbf{F}_i \right] \tag{28}$$

A transformation of Eq. (28) may be useful by using the *total potential* μ', which is the summation of the chemical potential and the potential energy

$$\mu' = \mu + e_p \tag{29}$$

so that we have

$$T\nabla\left(\frac{\mu_i}{T}\right) - F_i = \nabla\mu_i + \bar{s}_i\nabla T + \nabla e_{pi} - \frac{h_i\nabla T}{T} = \nabla_T\mu_i^{'} - \frac{h_i\nabla T}{T} \tag{30}$$

where $\nabla_T\mu_i^{'}$ is the isothermal gradient of the total potential, and given by

$$\nabla_T\mu_i^{'} = \nabla\mu_i + \bar{s}_i\nabla T + \nabla e_{pi} = \nabla\mu_i^{'} + \bar{s}_i\nabla T \tag{31}$$

Using the total potential together with the internal energy flux $\mathbf{J}_u$

$$\mathbf{J}_u = \mathbf{J}_q^{''} + \sum_{i=1}^{n} \bar{h}_i\mathbf{j}_i \tag{32}$$

Dissipation function, given by Eq. (28), becomes

$$\Psi = -\mathbf{J}_q^{''}\cdot\nabla\ln T - \sum_{i=1}^{n} \mathbf{j}_i\cdot\nabla_T\mu_i^{'} \tag{33}$$

Since only n-1 diffusion fluxes $\mathbf{j}_i$ are independent, we have

$$\sum_{i=1}^{n} \mathbf{j}_i\cdot\nabla_T\mu_i^{'} = \sum_{i=1}^{n-1} \mathbf{j}_i\cdot\nabla_T(\mu_i^{'} - \mu_n^{'}) \tag{34}$$

So that we transform the dissipation function, [Eq. (33)], into the form that has all the forces independent

$$\Psi = -\mathbf{J}_q^{''}\cdot\nabla\ln T - \sum_{i=1}^{n-1} \mathbf{j}_i\cdot\nabla_T(\mu_i^{'} - \mu_n^{'}) \tag{35}$$

We may consider the phenomenological equations for the n+1 vector flows of $\mathbf{J}_q^{''}$ and $\mathbf{j}_i$, and n+1 forces of $\nabla(1/T)$ and $\nabla_T\mu'$. Assuming the linear relations between the forces and the flows, we have the following phenomenological equations

$$-\mathbf{J}_q^{''} = L_{qq}\nabla\ln T + \sum_{j=1}^{n} L_{qj}\nabla_T\mu_j^{'} \tag{36}$$

$$-\mathbf{j}_i = L_{iq}\nabla\ln T + \sum_{j=1}^{n} L_{ij}\nabla_T\mu_j^{'} \qquad (i=1,2,..,n) \tag{37}$$

More general relations would contain inertial and viscous terms.

For a system in mechanical equilibrium in which the pressure gradient is balanced by the mass forces, the Gibbs-Duhem relation is given by

$$\sum_{j=1}^{n} \rho_j \nabla_T \mu_j' = 0 \tag{38}$$

therefore, the forces $-\nabla_T \mu_j'$ are not all independent. Summation of the diffusion flows $\mathbf{j}_i$ yield

$$\sum_{i=1}^{n} \mathbf{j}_i = \sum_{i=1}^{n} \rho_i (\mathbf{v}_i - \mathbf{v}) = 0 \tag{39}$$

where $\mathbf{v}$ refers to the mass average velocity. Eqs. (38) and (39) show that the diffusion flows $\mathbf{j}_i$ are not all independent. For a system which are not in mechanical equilibrium, the chemical potential gradients $-\nabla_T \mu_j'$ are independent. Thus all the forces in Eqs. (36) and (37) may be varied independently. The coefficient L_{qq} is related to the thermal conductivity, while L_{iq} and L_{qi} define the thermal diffusion and heat transferred by mass diffussion (Dufour effect) of component i, respectively. The coefficient L_{ii} determines that part of the diffusion current $\mathbf{j}_i$ arising from its own chemical potential gradient of component i, while the codiffusion coefficient L_{ij} defines that part of $\mathbf{j}_i$ arising from the chemical potential gradients of component j. The codiffusion coefficients L_{ij} are affected by the forces acting between the dissimilar molecules. If the average intermolecular force between i and j is repulsive, the diffusion of j induces a diffusion current of i in the opposite direction, and L_{ij} is negative. Otherwise, L_{ij} is positive and the diffusion of component j induces a diffusion current of component i in the same direction.

In principle, each phenomenological coefficient may be measured by a suitable experimental procedure. The Onsager reciprocal relations reduces the number of unknown coefficients to be determined. If we substitute Eq. (37) into Eq. (39), we find that the coefficients L_{iq} and L_{ij} obey the following relations

$$\sum_{i=1}^{n} L_{iq} = 0 \ ; \sum_{i=1}^{n} L_{ij} = 0 \qquad (j=1,2,..,n) \tag{40}$$

For a system, which is not in mechanical equilibrium, the forces $-\nabla_T \mu_j'$ are not all independent. From the solution of Eq. (38) for $-\nabla_T \mu_j'$, and substuting the resulting expression into Eq. (37), we obtain

$$-\mathbf{j}_i = L_{iq}\nabla \ln T + \sum_{j=1}^{n}\left(L_{ij} - \frac{\rho_j L_{in}}{\rho_n}\right)\nabla_T \mu'_j \qquad (i=1,2,..,n) \tag{41}$$

Since the (n-1) chemical potential gradients are independent, the coefficients $\left(L_{ij} - \rho_j L_{in} / \rho_n\right)$ are determined, however the individual phenomenological coefficients L_{ij} cannot be measured experimentally.

In order to find the independent forces for the heat and diffusion flows in a system with mechanical equilibrium, we express the dissipation function due to heat and diffusion in the form given in Eq. (35). Later, we establish the linear relations for the flows and the forces, in which all the forces are independent

$$-\mathbf{J}''_q = L_{qq}\nabla \ln T + \sum_{j=1}^{n-1} L_{qj}\nabla_T(\mu'_j - \mu'_n) \tag{42}$$

$$-\mathbf{j}_i = L_{iq}\nabla \ln T + \sum_{j=1}^{n-1} L_{ij}\nabla_T(\mu'_j - \mu'_n) \qquad (i=1,2,..,n\text{-}1) \tag{43}$$

These linear relations are necessary to define uniquely the following phenomenological coefficients for the system in mechanical equilibrium

$$L_{qn} = -\sum_{j=1}^{n-1} L_{qj} \;\; ; \;\; L_{nq} = -\sum_{j=1}^{n-1} L_{jq} \;\; ; \;\; L_{nn} = \sum_{i=1}^{n-1}\sum_{j=1}^{n-1} L_{ij}$$

$$L_{in} = -\sum_{j=1}^{n-1} L_{ij} \;\; ; \;\; L_{ni} = -\sum_{j=1}^{n-1} L_{ji} \qquad (i=1,2,..,n\text{-}1) \tag{44b}$$

2. HEAT OF TRANSPORT

The 'transportation quantities' are useful to describe the transport phenomena in a multicomponent fluid. In order to understand the concept, consider rewriting Eq. (43) as

$$-(\mathbf{j}_i + L_{iq}\nabla \ln T) = \sum_{j=1}^{n-1} L_{ij}\nabla_T(\mu'_j - \mu'_n) \qquad (i=1,2,..,n\text{-}1) \tag{45}$$

then the thermodynamic forces can be determined by

$$\nabla_T(\mu'_j - \mu'_n) = -\sum_{i=1}^{n-1} K_{ji}(\mathbf{j}_i + L_{iq}\nabla \ln T) \qquad (j=1,2,..,n\text{-}1) \tag{46}$$

Since the matrix of resistance coefficients K_{ji} is the inverse of the matrix of coefficients L_{ij}, we have

$$K_{ji} = L_{ij}^{-1}$$

Combining Eq. (46) with Eq. (42) yields

$$-\mathbf{J}_q'' = (L_{qq} - \sum_{i,j=1}^{n-1} L_{iq} L_{qj} K_{ji}) \nabla \ln T - \sum_{i,j=1}^{n-1} L_{qj} K_{ji} \mathbf{j}_i \tag{47}$$

From Eq. (47) the *heat of transport*, Q_i^* of component i is defined by

$$Q_i^* = \sum_{j=1}^{n-1} L_{qj} K_{ji} = \sum_{j=1}^{n-1} L_{qj} L_{ij}^{-1} \qquad (i=1,2,..,n\text{-}1) \tag{48}$$

From the Onsager reciprocal relations, Eq. (48) can be rewritten as

$$Q_i^* = \sum_{j=1}^{n-1} L_{jq} K_{ij} = \sum_{j=1}^{n-1} L_{jq} L_{ji}^{-1} \qquad (i=1,2,..,n\text{-}1) \tag{49}$$

The heat of transport can be used in the phenomenological equations in order to eliminate the coefficients L_{qj} or L_{jq}. After introducing Eq. (49) into Eq. (47), we obtain the expression for heat flow in terms of the heat of transport

$$-\mathbf{J}_q'' = (L_{qq} - \sum_{i,j=1}^{n-1} L_{iq} Q_i^*) \nabla \ln T - \sum_{i,j=1}^{n-1} Q_i^* \mathbf{j}_i \tag{50}$$

For an isothermal system where $\nabla \ln T = 0$, we have an interpretetaion of the physical meaning of the heat of transport expressed by

$$\mathbf{J}_q'' = \sum_{i=1}^{n-1} Q_i^* \mathbf{j}_i \text{ and } Q_i^* = \left(\frac{\mathbf{J}_q''}{\mathbf{j}_i} \right)_T \tag{51}$$

Eq. (51) shows that the heat of transport Q_i^* is the heat carried by a unit diffusion flow of component i when there is no temperature gradient and no diffusion of other components. For a binary fluid, the heat of transport is expressed by

$$Q_1^* = \frac{L_{q1}}{L_{11}} = \frac{L_{1q}}{L_{11}} = U_1^* - (\bar{h}_1 - \bar{h}_2) \tag{52}$$

where U_1^* is called the *energy of transport*, which is the internal energy carried by a unit diffusion of component when there is no temperature gradient and no diffusion of the other components. The heat of transport is the flow of heat entering through the surface of contact to maintain isothermal conditions if a unit of mass leaves the region in equilibrium. The value of Q_i^* can be calculated analytically when the energy field of molecules crossing the surface is known.

$$\mathbf{J}_u = \sum_{i=1}^{n-1} U_i^* \mathbf{j}_i \tag{53}$$

Similarly we may also define the *entropy of transport* S_i^* by

$$\mathbf{J}_s = \sum_{i=1}^{n-1} S_i^* \mathbf{j}_i \tag{54}$$

We can relate the heat of transport and energy of transport by using the relation [Chapter 4, Eq. (68)]

$$\mathbf{J}_q' = \mathbf{J}_u - \sum_{i=1}^{n-1} (\bar{h}_i - \bar{h}_n)\mathbf{j}_i$$

and obtain

$$Q_i^* = U_i^* - (\bar{h}_i - \bar{h}_n) \qquad (\nabla T = 0) \qquad (i=1,2,..,n\text{-}1) \tag{55}$$

Likewise we can relate the heat of transport and entropy of transport by using the relation [Chapter 4, Eq. (66)]

$$\mathbf{J}_q'' = T[\mathbf{J}_s - \sum_{i=1}^{n-1} (\bar{s}_i - \bar{s}_n)\mathbf{j}_i]$$

and find

$$S_i^* = \frac{Q_i^*}{T} + (\bar{s}_i - \bar{s}_n) = \frac{1}{T}(U_i^* - \mu_i + \mu_n) \tag{56}$$

In order to eliminate the cross-effect coefficient L_{q1} and L_{1q} from the phenomenological Eqs. (4) and (5), we can use the heat of transport given in Eq. (52) and obtain

$$-\mathbf{J}_q^{''} = L_{qq}\nabla \ln T + L_{11}Q_1^* \frac{1}{w_2}\left(\frac{\partial \mu_1}{\partial w_1}\right)_{T,P} \nabla w_1 \tag{57}$$

$$-\mathbf{j}_1 = \mathbf{j}_2 = L_{11}\left[Q_1^*\nabla \ln T + \frac{1}{w_2}\left(\frac{\partial \mu_1}{\partial w_1}\right)_{T,P} \nabla w_1\right] \tag{58}$$

3. THE DEGREE OF COUPLING

The *degree of the coupling q* results from the relationships between the forces and the flows. For a binary system, using Eqs. (42) and (43) we can define the ratio of forces [$\lambda = \nabla \ln T / \nabla_T(\mu_1^{'} - \mu_2^{'})$], and the ratio of flows ($\eta = \mathbf{J}_q^{''} / \mathbf{j}_1$) for a binary mixture. Dividing Eq. (42) with Eq. (43), and further dividing both the numerator and denominator by $(L_{qq}L_{11})^{1/2}[\nabla_T(\mu_1^{'} - \mu_2^{'})]$, we obtain

$$\eta = \frac{(L_{qq}/L_{11})^{1/2}\lambda + L_{1q}/(L_{qq}L_{i11})^{1/2}}{L_{1q}/(L_{qq}L_{11})^{1/2}\lambda + (L_{11}/L_{qq})^{1/2}} \tag{59}$$

Eq. (59) shows that the ratio of flows η varies with the ratio of forces, λ. As the quantity $L_{1q}/(L_{qq}L_{11})^{1/2}$ approaches zero, each flow becomes independent, and we have the ratio of flows approaching $\eta \to L_{qq}/L_{11}\lambda$. If $L_{1q}/(L_{qq}L_{11})^{1/2}$ approaches ± 1, then the two flows are not associated with the forces, and the ratio of flows approaches a fixed value $\eta \to \pm(L_{qq}/L_{11})^{1/2}$. This is the case, where the matrix of the phenomenological coefficients becomes singular.

The ratio

$$q = \frac{L_{1q}}{(L_{qq}L_{11})^{1/2}} \tag{60}$$

is called the degree of coupling for a two-flow system. The degree of coupling is a dimensionless parameters quantifying the coupling of the energy conversion in a process.

As the heat and mass flows are both the vectors, the sign of q indicates the direction of forces on the substances. If $L_{iq} > 0$ hence $q > 0$, substance may drag another substance in the same direction, however the substance diffuses in the opposite direction if $L_{iq} < 0$ and $q < 0$. For the heat and mass flows the dissipation function, given in Eq. (35), defines the two limiting values of q between +1 and –1. An incomplete coupling takes a value between these two limits.

The coupling may be further clarified by the ratio

$$z^2 = \frac{L_{qq}}{L_{11}} \tag{61}$$

and Eq. (59) becomes

$$\eta = \frac{q + \lambda z}{\lambda q + (1/z)} \tag{62}$$

Eq. (62) indicates that for the known values of z and λ, the ratio of flows is determined by the degree of coupling q. With complete coupling, λ is equal to z and q becomes +1 or –1. The minus sign of λ stems from the situations where the chemical potential may have negative values due to nonideality of the mixture. In nonideal systems the change of the Fick diffusivity with concentration may be zero, and sometimes negative, and phase splitting may occur in the liquid flows. This complex behavior needs to be determined from the thermodynamic models.

The reduced force $z\lambda$, and the reduced flow η/z can be related by

$$\frac{\eta}{z} = \frac{q + z\lambda}{qz\lambda + 1} \tag{63}$$

Two reference stationary states in the coupled processes can be defined as the level flow where $X_i = 0$, and the static head where $\mathbf{J}_i = 0$. Examples of the static head states are the open-circuited fuels cell and active transport in a cell membrane, whereas the examples of the level flow states are the short-circuited fuel cells and the salt and water transport in kidneys. In incompletely coupled systems, a constant supply of energy is necessary to maintain these reference states without output or work.

4. COUPLING IN LIQUID MIXTURES

Examples for some coupled processes include separations by thermal diffusion (or thermal osmosis), thermoelectric phenomena, and the active transport of a substance coupled to a chemical reaction. Since many chemical reactions within a biological cell produce or consume heat, local temperature gradients may contribute in the transport of materials across membranes. We can identify the independent driving forces that couple to produce various flows by using the dissipation function obtained from the theory of linear nonequilibrium thermodynamics based on the Onsager relations. This is called the Dissipation-Phenomenological Equation (DPE) approach in which the conjugate forces and flows are used to express the transport and rate processes [3]. The phenomenological equations are capable of displaying the interactions between the various processes through the coupling or the cross-phenomenological coefficients (L_{ij} with $i \neq j$) that are closely related to the transport coefficients. The Onsager reciprocal relations state that the cross coefficients are equal to each other for the proper conjugate forces and flows. Once the values for cross coefficients are known, we can calculate the degree of coupling and the rate of entropy production or the dissipation for the system.

When we consider a diffusion flow $\mathbf{j}_i$, and a temperature gradient X_q, which are both the vectors, the cross coefficients L_{qi} are scalar quantities, and hence consistent with the isotropic character of the mixture; the coefficients L_{iq} do not need to vanish, and not only a temperature gradient cause a heat flux in fluid mixtures, but also isothermal chemical potential gradient. This latter effect is known as the diffusion thermoeffect or the Dufour effect, and is characterized by a transport property called the heat of transport Q_1^*, which represents the heat flow due to the diffusion of component i under isothermal conditions. Rowley and Horne [9] measured the heat of transport in carbon tetrachloride-cyclohexane mixture, and showed that even though the temperature gradients in initially isothermal liquid mixtures are small, the heat of transport itself may be large. The heat of transport contains information on the molecular interactions; hence it can be used for the development of the generalized molecular transport models, and for the theories on intermolecular potential wells and mixing rules.

4.1. Coupling in binary liquid mixtures

For a binary fluid, where $-\mathbf{j}_1 = \mathbf{j}_2$, at mechanical equilibrium with a negligible body force, and for diffusion based on the mass average velocity, we can now establish a set of phenomenological equations representing the coupling between heat and mass transfer [Eqs. (4) and (5)]

$$-\mathbf{J}_q'' = L_{qq} \nabla \ln T + L_{1q} \frac{1}{w_2} \left(\frac{\partial \mu_1}{\partial w_1} \right)_{T,P} \nabla w_1$$

$$-\mathbf{j}_1 = L_{q1} \nabla \ln T + L_{11} \frac{1}{w_2} \left(\frac{\partial \mu_1}{\partial w_1} \right)_{T,P} \nabla w_1$$

These equations obey the Onsager reciprocal relations, which states that the phenomenological coefficient matrix is symmetric. The coefficients L_{qq} and L_{11} are associated to the thermal conductivity k and the mutual diffusivity D, respectively, while the cross coefficient L_{1q} or L_{q1} defines the coupling phenomena, namely the thermal diffusion (Soret effect) and the heat flow due to the diffusion of substance i (Dufour effect). These effects are referred to as the forces, however the roles played by the forces and flows are symmetric.

The heat of transport, Q_1^* of component 1 is defined by Eq. (52)

$$Q_i^* = \frac{L_{1q}}{L_{11}}$$

and it can be used in the phenomenological equations in order to eliminate the coefficients L_{q1} or L_{1q}. If we express L_{1q} in terms of the heat of transport Q_1^* ($L_{1q} = Q_1^* L_{11}$), Eq. (59) becomes

$$\eta = \frac{(L_{qq}/L_{11})^{1/2}\lambda + Q_1^*(L_{11}/L_{qq})^{1/2}}{Q_1^*(L_{11}/L_{qq})^{1/2}\lambda + (L_{11}/L_{qq})^{1/2}} \tag{64}$$

The degree of coupling q can be expressed in terms of the heat of transport

$$q = Q_1^*\left(\frac{L_{11}}{L_{qq}}\right)^{1/2} \tag{65}$$

The degree of coupling may be used as a basis for comparison of systems with various coupled forces. The phenomenological coefficients are expressed in terms of the transport coefficients, k, D and Q_1^*

$$L_{1q} = Q_1^* \frac{\rho D M_1 M_2 w_1 w_2}{[MRT(1+\Gamma_{11})]} \tag{66}$$

$$L_{qq} = kT \tag{67}$$

$$L_{11} = \rho w_2 D\left(\frac{\partial \mu}{\partial w_1}\right)^{-1} \tag{68}$$

where $\Gamma_{11} = (\partial \ln \gamma_1 / \partial \ln x_1)_{T,P}$ is known as the thermodynamic factor, and can be determined from experimental data or an activity coefficient model such as NRTL or UNIFAC. The *phenomenological stoichiometry* z is defined in Eq. (61)

$$z = \left(\frac{L_{qq}}{L_{11}}\right)^{1/2} \tag{69}$$

With the definitions of degree of coupling q and the phenomenological stoichiometry z, Eq. (64) can be written as

$$\eta = \frac{z\lambda + Q_1^* / z}{Q_1^* \lambda / z + 1/z} \tag{70}$$

Eq. (70) shows that, as the degree of coupling approaches zero, each flow becomes independent, and we have the ratio of flows approaching $\eta \to z^2\lambda$. If q approaches ± 1, then the two flows are no longer associated with the forces, and the ratio of flows approaches a fixed ratio $\lambda \to \pm z$. This case is complete coupling. Negative values of η stem from the situations where the differentiation of chemical potential with respect to concentration is negative due to nonideality of the mixture. The degree of coupling is not a unique characteristic of the system since there may be various ways of describing flows and forces consistent with a given dissipation rate. For complete coupling $q = \pm 1$ for any choice of flows and forces, and z has a unique value.

The degree of coupling q and the thermal diffusion ratio of component 1 $K_{T,1}$ can be expressed in terms of the transport coefficients and Γ by using Eq. (65)

$$q = \frac{Q_1^*}{T}\left(\frac{\rho D M_1 M_2 w_1 w_2}{kMR(1+\Gamma_{11})}\right)^{1/2} \tag{71}$$

$$K_{T,1} = q\left(\frac{k M_1 M_2 w_1 w_2}{\rho DMR(1+\Gamma_{11})}\right)^{1/2} \tag{72}$$

Eq. (71) shows that q is directly proportional to the heat of transport, and inversely proportional to temperature. The degree of coupling is a function of the heat of transport, the thermal conductivity, and the diffusion coefficient, while the L_{1q} is independent of the thermal conductivity. As the heat and diffusion flows are both vectors, the sign of q is related to the direction of flows of a substance. If $q >$ 0, the flow of a substance may drag another substance in the same direction, however it may push the other substance in the opposite direction if $q < 0$. For heat and mass flows the dissipation function of Eq. (64) defines the two limiting values of q as +1 and −1.

Based on the dissipation function, given in Eq. (1), we can define the ratio of dissipations due to the heat and mass flows in terms of the reduced force ratio and the degree of coupling q

$$\eta\lambda = -\frac{\mathbf{J}_q'' X_q}{\mathbf{j}_1 X_1} \tag{73}$$

Eq. (73) is known as the *efficiency of energy conversion* in physical and biological systems, and $\mathbf{j}_1 X_1$ shows the input, while $\mathbf{J}_q'' X_q$ is the output power. Therefore, diffusion drives the heat flow, and the term $(z\lambda/q)$ changes between 0 and −1. Since $\eta\lambda$ is zero when either $\mathbf{J}_q''$ or X_q is zero, then it must pass through a maximum at intermediate values. The values of $\eta\lambda$ are often small in regions of physical interest, and the maximum depends on the degree of coupling only

$$(\eta\lambda)_{\max} = \frac{q^2}{(1+\sqrt{1-q^2})^2} \tag{74}$$

Inserting Eqs. (4) and (5) into the dissipation function of Eq. (1) we obtain the three contributions due to heat, mass, and the coupled transport, respectively

$$\begin{aligned}\Psi = L_{qq}\left(\frac{1}{T}\nabla T\right)^2 + L_{11}\left[\frac{1}{w_2}\left(\frac{\partial \mu_1}{\partial w_1}\right)\nabla w_1\right]^2 \\ + 2L_{1q}\left(\frac{1}{T}\nabla T\right)\left[\frac{1}{w_2}\left(\frac{\partial \mu_1}{\partial w_1}\right)\nabla w_1\right]\end{aligned} \tag{75}$$

We can also use the transport coefficients, given in Eqs. (66)-(68), and the degree of coupling in the dissipation function, and we obtain

$$\begin{aligned}\Psi = kT\left(\frac{1}{T}\nabla T\right)^2 + \rho D\left[\frac{1}{w_2}\left(\frac{\partial \mu_1}{\partial w_1}\right)(\nabla w_1)^2\right] \\ + 2\rho D Q_1^*\left(\frac{1}{T}\nabla T\right)\left[\frac{1}{w_2}\left(\frac{\partial \mu_1}{\partial w_1}\right)\nabla w_1\right]\end{aligned} \tag{76}$$

By differentiating the chemical potential in terms Γ

$$\left(\frac{\partial \mu_1}{\partial w_1}\right) = \frac{MRT(1+\Gamma_{11})}{w_1 M_1 M_2} \tag{77}$$

and using the definition of q [(Eq. 71)], we can express the dissipation function in terms of the degree of coupling q, the thermodynamic factor Γ, and the transport coefficients of thermal conductivity k and diffusion coefficient D

$$\Psi = kT\left(\frac{1}{T}\nabla T\right)^2 + \frac{\rho DMRT(1+\Gamma_{11})}{w_1 w_2 M_1 M_2}(\nabla w_1)^2 + 2q\left(\frac{\rho DkMR(1+\Gamma_{11})}{w_1 w_2 M_1 M_2}\right)^{1/2}(\nabla T)(\nabla w_1) \tag{78}$$

The terms on the right-hand side of Eq. (78) show that the dissipation due to the heat flow, mass flow and coupling between the heat and mass flows, respectively.

Rowley et al. [11,12] measured the heat of transport, the thermal conductivity, and the mutual diffusivity coefficients at 30 °C and ambient pressure for a number of binary systems. The data has error levels of 2 % for diffusivities and thermal conductivities, and about 4% for heat of transport measurements. The degree of coupling is directly proportional to the product $Q_1^*(D/k)^{1/2}$, and hence the error level of the predictions of q is mainly related to the reported error levels of Q_1^* values. The polynomial fits to the thermal conductivity, mass diffusivity, and heat of transport for the alkanes in chloroform and in carbon tetrachloride are given in Appendix B, Tables B1 to B3. Thermal conductivity for the hexane-carbon tetrachloride mixture has been predicted by the local composition model [10]. The NRTL and UNIFAC models with the data given in DECHEMA [2] series were used to calculate the thermodynamic factors. However, it should be noted that, the thermodynamic factors obtained from the various molecular models as well as from two sets of parameters for the same model might be different.

The liquid mixtures consist of six-to-eight carbon alkanes of n-hexane, n-heptane, n-octane, 3-methylpentane, 2,3-dimethylpentane, and 2,2,4-trimethylpentane in chloroform and in carbon tetrachloride. These systems represent straight and branched chains of the alkanes in two solvents. As the degree of coupling and the thermal diffusion ratio depend on the heat and mass transfer coefficients, the plots of q and $K_{T,1}$ versus the alkane compositions x_1 show the combined effect of the transport coefficients on q and $K_{T,1}$ in various solvents, as seen in Figs 1 and 2, respectively.

The degree of coupling q between heat and mass flows and the thermal diffusion ratio $K_{T,1}$ are calculated for the full composition range from Eqs. (71) and (72), and are shown in Figs. 1 and 2; the plots show four important properties of coupling. The first is that the absolute values of q and $K_{T,1}$ reach the peak values at a certain concentration of the alkane, and these peak values decrease as the molecular weight increases. Secondly the solute concentrations at the peak values of coupling decrease gradually as the molecular weights increase. The third is that the behavior of alkanes are similar up to a certain concentration of solute depending upon the combined effect of branching and the solvent on q and $K_{T,1}$ (through approximately $x_1 = 0.2$), but at higher concentrations they behave differently. The fourth is that the absolute maximum extent of coupling is small as expected, and the branching of alkanes has only marginal effect on the coupling phenomena.

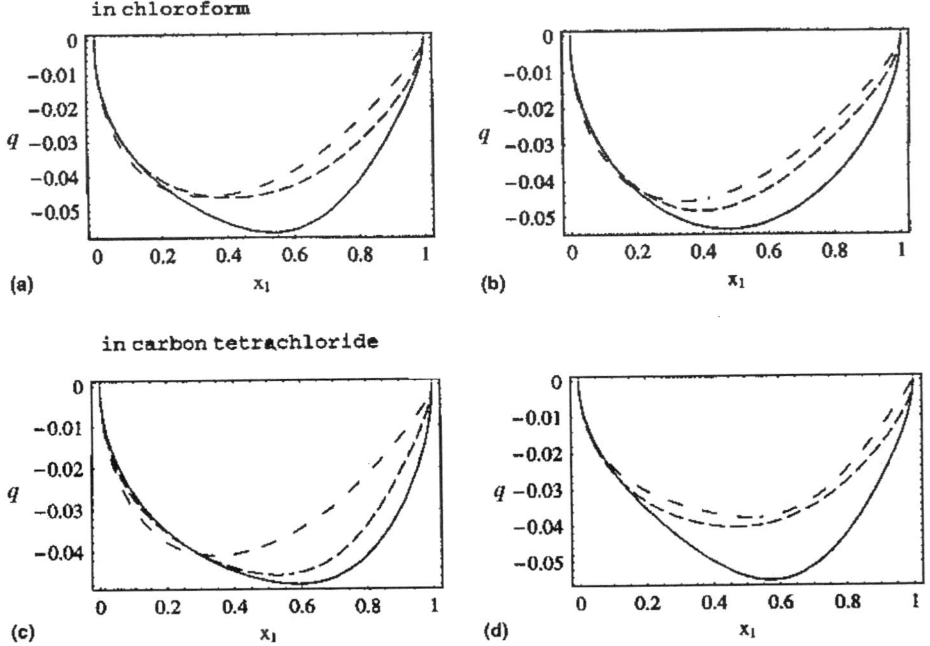

Fig.1. Change of the degree of coupling q with the alkane concentrations x_1 at 30 °C and ambient pressure: (a) and (c) straight chain alkanes; (—) n-hexane, (---), n-heptane, (- - -) n-octane; (b) and (d) branched-chain alkanes; (—) 3-methylpentane, (---) 2,2-dimethynentane, (- - -) 2,2,4-trimethylpentane. Reprinted with the permission from Elsevier, Int. J. Heat Mass Transfer, 43 (2002) 75.

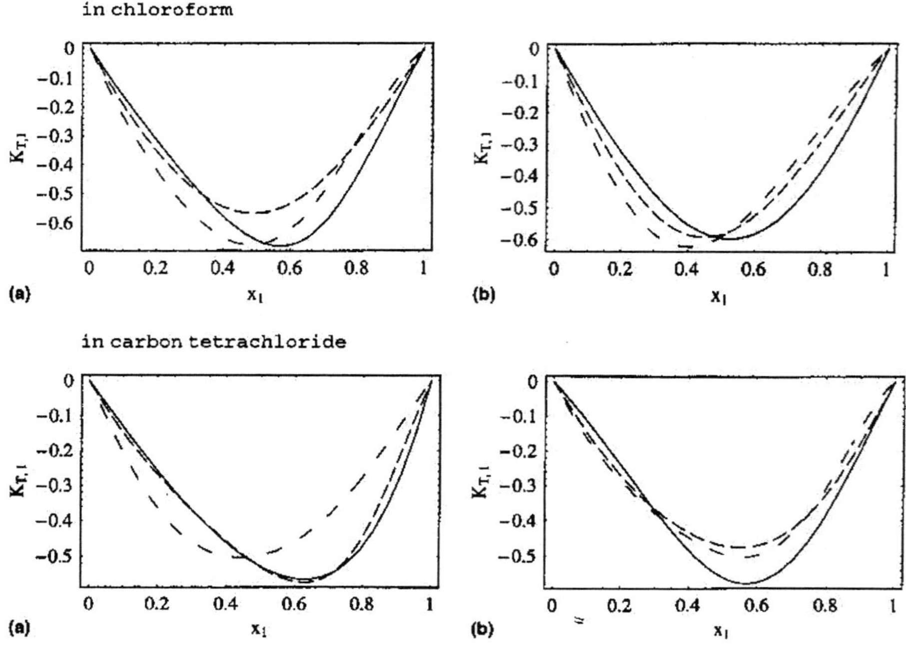

Fig.2. Change of the thermal diffusion ratio $K_{T,1}$ with the alkane concentrations x_1 at 30 °C and ambient pressure: (a) straight chain alkanes; (—) n-hexane, (---), n-heptane, (- - -) n-octane; (b) branched-chain alkanes; (—) 3-methylpentane, (---) 2,2-dimethynentane, (- - -) 2,2,4-trimethylpentane. Reprinted with the permission from Elsevier, Int. J. Heat Mass Transfer, 43 (2002) 75.

In Tables 1 and 2, the values of q and $K_{T,1}$ and the concentrations at where the peaks occur are shown for both the straight and branched alkanes separately in both the solvents chloroform and in carbon tetrachloride. General trend is that the branching of the solute molecule has a minimal effect on the coupling for the considered alkanes. The tables also show the effect of the solvent on q and $K_{T,1}$. The alkane concentrations where the peaks of q and $K_{T,1}$ occur are lower in chloroform than in carbon tetrachloride. Generally, the peak values of q_{max} are also smaller in carbon tetrachloride than in chloroform. Consequently, it appears that concentration is important in the level of coupling in fluid mixtures.

4.2. Coupling in Ternary Mixtures

Platt et al. [6-8] measured the heat of transport in binary and ternary mixtures using diffusion thermoeffect method. The heat of transport of component i is a measure of the local heat addition or removal required to maintain isothermal conditions as molecular diffusion of component i takes place from a higher chemical potential to a lower one. Since there are only $n - 1$ independent diffusion flows in an n component mixture, there are only $n - 1$ independent heats of transport.

Table 1
Degree of coupling q, and maximum ratio of dissipation $(\eta\lambda)_{max}$
(a) for alkanes in chloroform

Straight-chain alkanes				Branched-chain alkanes			
Solute	$-q_{max}$	x_1^*	$(\eta\lambda)_{max}$ 10^4	Solute	$-q_{max}$	x_1^*	$(\eta\lambda)_{max}$ 10^4
n-Hexane	0.056	0.534	7.993	3-Methylpentane	0.053	0.475	7.085
n-Heptane	0.046	0.408	5.365	2,3-Dimethylpentane	0.048	0.392	5.815
n-Octane	0.046	0.337	5.295	2,2,4-Trimethylpentane	0.046	0.340	5.226
(b) for alkanes in carbon tetrachloride							
n-Hexane	0.048	0.588	5.718	3-Methylpentane	0.055	0.570	7.574
n-Heptane	0.045	0.527	5.203	2,3-Dimethylpentane	0.040	0.467	4.083
n-Octane	0.041	0.341	4.185	2,2,4-Trimethylpentane	0.038	0.513	3.574

Table 2
Thermal diffusion ratio for solute $K_{T,1}$
(a) for alkanes in chloroform

Straight-chain alkanes			Branched-chain alkanes		
Solute	$-K_{T,1}$	x_1^*	Solute	$-K_{T,1}$	x_1^*
n-Hexane	0.679	0.569	3-Methylpentane	0.596	0.527
n-Heptane	0.565	0.488	2,3-Dimethylpentane	0.591	0.449
n-Octane	0.675	0.484	2,2,4-Trimethylpentane	0.622	0.402
(b) for alkanes in carbon tetrachloride					
n-Hexane	0.564	0.629	3-Methylpentane	0.579	0.569
n-Heptane	0.572	0.628	2,3-Dimethylpentane	0.476	0.550
n-Octane	0.503	0.441	2,2,4-Trimethylpentane	0.503	0.557

Using the forces and flows identified in Eq. (1), and the Gibbs-Duhem equation for an n component system at constant temperature and pressure, we obtain

$$\nabla_T \mu_n = -\sum_{k=1}^{n-1} (w_k / w_n) \nabla_T \mu_k \tag{79}$$

The phenomenological equations are expressed by

$$-\mathbf{J}_q'' = L_{qq} \nabla \ln T + \sum_{j=1}^{n-1} \sum_{k=1}^{n-1} \sum_{l=1}^{n-1} L_{qj} \left(\delta_{jk} + \frac{w_k}{w_n} \right) \left(\frac{\partial \mu_k}{\partial w_l} \right)_{T,P} \nabla w_l \tag{80}$$

$$-\mathbf{j}_1 = L_{iq} \nabla \ln T + \sum_{j=1}^{n-1} \sum_{k=1}^{n-1} \sum_{l=1}^{n-1} L_{ij} \left(\delta_{jk} + \frac{w_k}{w_n} \right) \left(\frac{\partial \mu_k}{\partial w_l} \right)_{T,P} \nabla w_l \tag{81}$$

The coefficients L_{qq} and L_{ij} are associated with the thermal conductivity k and the mutual diffusivity D, respectively, while the cross coefficients L_{iq} and L_{qi} define the coupling.

In ternary mixtures, there are two independent heats of transport related to the two independent cross-phenomenological coefficients L_{q1} and L_{q2}

$$L_{q1} = L_{11} Q_1^* + L_{21} Q_2^* \tag{82}$$

$$L_{q2} = L_{12} Q_1^* + L_{22} Q_2^* \tag{83}$$

We relate the thermal conductivity k and the thermal diffusion coefficient $D_{T,i}$ to the phenomenological coefficients as $L_{qq} = kT$ and $L_{iq} = \rho D_{T,i}$. Therefore Eqs. (80) and (81) can be expressed in terms of the transport coefficients

$$-\mathbf{J}_q'' = k \nabla T + \sum_{k=1}^{n-1} \sum_{l=1}^{n-1} \rho Q_k^* D_{kl} \nabla w_l \tag{84}$$

$$-\mathbf{j}_1 = \rho D_{T,i} \nabla \ln T + \sum_{l=1}^{n-1} \rho D_{il} \nabla w_l \tag{85}$$

Eqs. (84) and (85) are valid for mixtures in mechanical equilibrium, containing no external body forces, and with negligible surface effects. Also mass-average velocity is small even under an initially large concentration gradient. It can be assumed that thermal diffusion effect causes an insignificant change to the mass flow because of the small temperature difference. For a ternary mixture, Eqs. (84) and (85) become

$$-\mathbf{J}_q'' = k\nabla T + \rho(Q_1^* D_{11} + Q_2^* D_{21})\nabla w_1 + \rho(Q_1^* D_{12} + Q_2^* D_{22})\nabla w_2 \tag{86}$$

$$-\mathbf{j}_1 = \rho D_{T,1}\nabla \ln T + \rho D_{11}\nabla w_1 + \rho D_{12}\nabla w_2 \tag{87}$$

$$-\mathbf{j}_2 = \rho D_{T,2}\nabla \ln T + \rho D_{21}\nabla w_1 + \rho D_{22}\nabla w_2 \tag{88}$$

where D_{il} is the diffusion coefficient, and related to the phenomenological coefficients as

$$D_{il} = \frac{1}{\rho} + \sum_{j=1}^{n-1}\sum_{k=1}^{n-1} L_{qj}\left(\delta_{jk} + \frac{w_k}{w_n}\right)\left(\frac{\partial \mu_k}{\partial w_l}\right)_{T,P} \tag{89}$$

For a ternary mixture, there are two independent degrees of coupling between the heat and mass flows, and are given by

$$q_{q1} = \frac{L_{q1}}{(L_{qq}L_{11})^{1/2}} \tag{90}$$

$$q_{q2} = \frac{L_{q2}}{(L_{qq}L_{22})^{1/2}} \tag{91}$$

Eqs. (90) and (91) show the relationships between the degrees of coupling and the cross-phenomenological coefficients L_{qi}.

Platt et al. [8] have measured the heats of transport and phenomenological coefficients in ternary mixture of toluene(1)-chlorobenzene(2)-bromobenzene(3), and Rowley and Hall [13] reported the heats of transport in binary liquid mixtures of toluene, chlorobenzene, and bromobenzene at 25 and 35 °C. These sets of data enable one to examine the effect of concentration and temperature on the coupling for the ternary liquid mixture.

The sign of heat of transport is an artifact of the numbering the substances since $-Q_1^* = Q_2^*$ in a binary mixture of substances 1 and 2. The negative sign with the numbering system used here indicates that heat is transported down the composition gradient of the more concentrated substance.

The absolute values of the degree of coupling decrease with increasing temperature for the binary mixtures, while the effect of the composition on the degree of coupling is more complex. The degree of coupling decreases gradually with increasing concentration of toluene for toulene(1)-chlorobenzene(2), while it increases with increasing chlorobenzene concentration at 35 °C, and remains almost the same at 25 °C for chlorobenzene(1)-bromobenzene(2) [4]. The heats of transport have a complex composition dependence, and are sensible to the composition of the heavy component bromobenzene.

For the ternary mixture, the coefficients of the fitted equations for the phenomenological coefficients computed from the diffusion coefficients [17] are given in Table 3. Platt et al. [8] fitted the values of L_{q1} and L_{q2} in kg m^{-1} s^{-1}, calculated from Eqs. (82) and (83), as functions of compositions and temperature as follows

$$L_{q1} = 10^{-7}(15.61 - 0.059T - 0.0501Tw_1 + 2.687Tw_2) \tag{92}$$

$$\begin{aligned} L_{q2} = 10^{-7}\, w_1 w_3 [&-533.0 + 2.185T - 441.5w_1 + (718.1 - 4.025T)w_2 \\ &+ 1011 w_1 w_2 + (4096 - 12.53T)w_1^2 + 300.3w_2^2] \end{aligned} \tag{93}$$

Beside the cross coefficients, the straight coefficients of L_{qq}, L_{11} and L_{22} should also be calculated. The values of L_{qi}, L_{qq} and L_{ii} are used in Eqs. (90) and (91) to calculate the degrees of coupling in the ternary mixture.

Table 3
Coefficients in the smoothing equation for phenomenological coefficients for the ternary mixture of toluene (1)-chlorobenzene (2)-bromobenzene (3):

$$L_{ik} = a_o + a_1 w_1 + a_2 w_1^2 + a_3 w_2 + a_4 w_1 w_2 + a_5 w_2^2 .$$

L_{ik}	T, K	a_o	a_1	a_2	a_3	a_4	a_5
L_{qq}	298.15	67.0426	10.8125	11.9207	8.81046	17.9015	5.73382
	308.15	69.9331	11.2839	12.451	9.11118	18.7197	6.05464
L_{11}	298.15	-3.2126	101.41	93.8366	6.23984	-25.0849	-4.27267
	308.15	-3.2166	111.835	-104.64	5.10796	-24.6334	-3.37569
L_{22}	298.15	-3.37541	7.13475	-1.68174	108.765	-36.1574	-104.986
	308.15	-3.19829	7.1013	-1.69981	118.181	-36.1694	-114.52

Fig. 3 shows the degrees of coupling for the range of $0.1 \leq w_i \leq 0.6$ ($i = 1,2$) and $w_3 \geq 0.1$ at 25 and 35 °C and ambient pressure; the degree of coupling q_{q2} changes its direction with changing bromobenzene composition. At high concentration of bromobenzene q_{q2} is positive and the flows of the components are in the same direction, at lower concentration however q_{q2} becomes negative hence the components flow in the opposite directions; the cross-phenomenological coefficien L_{q2} changes its sign as a function of the mass fraction of the heavy component bromobenzene. This means that the direction of coupling due to the heat transported by the flow of chlorobenzene relative to the mass-average velocity in toluene-chlorobenzene mixture can be reversed by controlling the mass fraction of bromobenzene in the mixture. From the stand point of thermal diffusion, addition of bromobenzene to the mixture toluene-chlorobenzene can change the magnitude and direction of the separation. Mainly the effect of temperature on q_{q1} and q_{q2} is marginal. As the number of components increase the relative compositions of each component may play important role in the coupling between the two-flow systems.

REFERENCES

[1] C.F. Curtiss and R.B. Bird, Ind. & Eng. Chem. Res., 38 (1999) 2515.

[2] J. Gmehling, U. Onken and W. Arlt, Vapor-Liquid Equilibrium Data Collection; DECHEMA Chem. Data Series, Vol. 1, Part 6a, 6b, Verlag and Druckerel Friedrich Bischoff, Frankfurt, 1977.

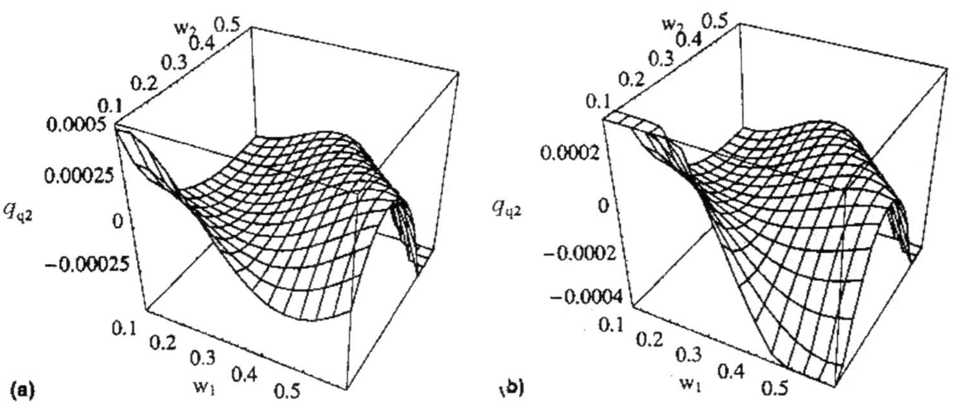

Fig. 3. Change of degree of coupling q_{q1} and q_{q2} with weight fraction of toluene w_1 and chlorobenzene w_2 at (a) 25 °C, (b) 35 °C. Reprinted with the permission from Elsevier, Int. J. Heat Mass Transfer, 43 (2002) 75.

[3] Y. Demirel and S.I. Sandler, Int. J. Heat Mass Transfer, 44 (2001) 2439.
[4] Y. Demirel and S.I. Sandler, Int. J. Heat Mass Transfer, 45 (2002) 75.
[5] M.D. Mikhailov and M.N. Ozisik, Unified Analysis and Solutions of Heat and Mass Diffusion, Wiley, New York, 1984
[6] G. Platt, T. Vongvanich and R.L. Rowley, J. Chem. Phys., 77 (1982a) 2113.
[7] G. Platt, T. Vongvanich, G.Fowler and R.L. Rowley, J. Chem. Phys., 77 (1982b) 2121.
[8] G. Platt, T, Vongvanich and R.L. Rowley, J. Non-Equilib. Thermody., 8 (1983) 1.
[9] R.L. Rowley and F.H. Horne, J. Chem. Phys., 68, 325 (1978).
[10] R.L. Rowley, Chem. Eng. Sci., 37 897 (1982).
[11] R.L. Rowley, S.C. Yi, V. Gubler and J. M. Stoker, Fluid Phase Equlib., 36 (1987) 219.
[12] R.L. Rowley, S.C. Yi, V. Gubler and J. M. Stoker, J. Chem. Eng. Data, 33 (1988) 362.
[13] R.L. Rowley and M.D. Hall, J. Chem. Phys., 85 (1986) 3550.
[14] S. Sieniutycz and A.N. Beris, Int. J. Heat Mass Transfer, 42 (1999) 2695.
[15] S. Sieniutycz, Int. J. Heat Mass Transfer, 20 (1977) 1221.
[16] S. Wisniewski, B Staniszewski and R. Szymanik, Thermodynamics of Nonequilibrium Processes, D. Reidel Publishing Company, Dordrecht, 1976.
[17] J.K. Burchard and H.I. Toor, J. Phys. Chem., 66 (19620 2015.

Chapter 10

Chemical reactions

INTRODUCTION

A chemical reaction is an irreversible process and produces entropy. In this chapter dissipation function for chemical reactions and coupled reactions are presented. The general criterion of irreversibility is $d_i S > 0$. Criteria applicable under particular conditions are readily obtained from the Gibbs equation. The changes in thermodynamic potentials for chemical reactions are given by

$$\left(\frac{\partial U}{\partial \xi}\right)_{S,V} = \left(\frac{\partial H}{\partial \xi}\right)_{S,P} = \left(\frac{\partial A}{\partial \xi}\right)_{T,V} = \left(\frac{\partial G}{\partial \xi}\right)_{T,P} = -A \tag{1}$$

All four potentials U, H, A, and G decrease as chemical reaction proceeds. Rate of reaction that is the change of extent of reaction with time has the same sign as the affinity. The reaction system is in the equilibrium state when the affinity is zero.

1. DISSIPATION FOR CHEMICAL REACTIONS

In a nonviscous fluid, where l chemical reactions take place, the dissipation functions are limited to the scalar chemical affinities A and chemical rates J_r

$$\Psi = T\Phi = -\sum_{j=1}^{l} A_j J_{r,j} \geq 0 \tag{2}$$

The phenomenological equations are

$$J_{r,i} = -\sum_{k=1}^{l} L_{jk} A_k = -\sum_{k=1}^{l} L_{jk} \sum_{i=1}^{n} \nu_{ik} M_i \mu_i \tag{3}$$

and the Onsager reciprocal relations link the coefficients $L_{jk} = L_{kj}$. The dissipation function is a quadratic function of the chemical affinities

$$\Psi = T\Phi = \sum_{j,k=1}^{l} L_{jk} A_j A_k \geq 0 \tag{4}$$

The positive values of Ψ imply the following inequalities

$$L_{kk} > 0, \; L_{kk} \, L_{jj} - L_{kj}^2 > 0 \tag{5}$$

The velocity of a chemical reaction is proportional to the product of the concentration of the reactants. For a reaction

$$A \underset{k_b}{\overset{k_f}{\rightleftarrows}} B \tag{6}$$

the reaction rates for the forward and backward directions are given by

$$r_f = k_f \prod_i c_{i,f}^{\nu_{i,f}} \tag{7}$$

$$r_b = k_b \prod_i c_{i,b}^{\nu_{i,b}} \tag{8}$$

For a reaction j the rate of reaction is given by

$$J_{r,j} = r_f - r_b = r_f \left(1 - \frac{\prod_i c_i^{\nu_i}}{K_c} \right) \tag{9}$$

where $K_c = k_f / k_b$ is a constant at equilibrium conditions.

For an ideal gas, the chemical affinity is given by

$$A_j = \sum_{i=1}^{n} \nu_{ij} M_i \mu_i = -RT \ln \frac{\prod_i c_i^{\nu_i}}{K_c} \tag{10}$$

where

$$\frac{\prod_i c_i^{\nu_i}}{K_c} = \exp\left(-\frac{A_j}{RT} \right) \tag{11}$$

so that the reaction rate becomes

$$J_{r,j} = r_f\left(1 - \exp\left(-\frac{A_j}{RT}\right)\right) \tag{12}$$

If we expand the expression in brackets, and consider the case of near equilibrium state, which may be specified by the inequality $\left|A_j / RT\right| << 1$, then we have a linear relationship between the reaction rate and the chemical affinity

$$J_{r,j} = \frac{r_f}{RT}A_j \tag{13}$$

Therefore the phenomenological coefficient is equal to

$$L_{ij} = \frac{r_f}{RT} \tag{14}$$

Very large values of the affinities may cause instability, and lead to a new steady state that is no more homogeneous in space. This causes a discontinuous decrease of entropy and symmetry breaking, and has important consequences in oscillating chemical reactions, which are of great biological interest. Such reactions are far from equilibrium, and present undamped fluctuation on a macroscopic scale. Even if it causes a negative contribution to the entropy production; oscillation around a stationary state is possible as long as the total entropy production is positive in a coupled phenomenon.

Some physical and biological structures can only originate in a dissipative (nonequilibrium) medium and be maintained by a continuous supply of energy. Such dissipative structures exist for only narrow limits due to delicate balance between reaction rates and diffusion. If one of those factors is changed, then the balance is affected and the whole organization would collapse. This is similar to a cell with the same chemical composition may be alive or dead.

In a system of two simultaneous reactions, thermodynamic coupling allows one of the reactions to progress in a direction contrary to that imposed by its own affinity provided that the total dissipation is positive. A coupled reaction is equivalent to an endothermic reaction, and measurement of heat effect during the embryonic period of a biological system may yield valuable information.

Consider a simple system 1 consisting of n components and subject to r chemical reaction mechanisms, and having specified values U of the energy, V of the volume, and values of N_1, N_2,..,N_r of the amounts of components that are obtained from given values N_{1a}, N_{2a},..,N_{ar}. Such a system admits a very large number of states. But the second law requires that among all these states the chemical equilibrium state is the only stable equilibrium state. In this state we have

$$N_{io} = N_{ia} + \sum_{j=1}^{r} \nu_i^{(j)} \varepsilon_{jo} \qquad i = 1,2,\ldots,r \tag{15}$$

The values of U, V, N_a, and ν determine uniquely the values of the entropy S, the reaction coordinate ε, and each N_{io}

$$S = S(U, V, N_a, \nu) \tag{16}$$

$$\varepsilon_{jo} = \varepsilon_{jo}(U, V, N_a, \nu) \tag{17}$$

$$N_{io} = N_{io}(U, V, N_a, \nu) \tag{18}$$

In an isolated system 1 with n components and r chemical reaction mechanisms, the system passes through a sequence of nonequilibrium states, and entropy is generated until the system reaches chemical equilibrium. In chemical equilibrium, the rate of change of each reaction coordinate is zero.

The evaluation of entropy generation as a function of time from a state that is not stable to a chemical or stable equilibrium state is not possible. This evaluation requires the solution of a general equation of motion that is still to be established. However, we can approximate the rate of entropy generation in terms of the r affinities of the surrogate system 2 of 1, and the rates of change of the reaction coordinates of the r chemical reactions. The surrogate system 2 is a simple system consisting of the same r types of reactions as system 1, but will have all the chemical reaction mechanisms inhibited. The following relations are from Gyftopoulos and Beretta [3]. At time t the amounts of n substances satisfy

$$N_i = N_{ia} + \sum_{j=1}^{r} \nu_i^{(j)} \varepsilon_j \qquad i = 1,2.,..,n \tag{19}$$

The entropy of the system 1 is a function of time. If S_{eq} is the entropy at stable equilibrium, the values of $\Phi(t)$ are smaller than the value S_{eq}. The rate of entropy generation Φ in the isolated system 1 is given by

$$\Phi = \sum_{j=1}^{r} \sum_{i=1}^{n} \left(-\frac{\nu_i^{(j)} \mu_i}{T_{\mathrm{eq}}} \right) \dot{\varepsilon}_j = \sum_{j=1}^{r} X_j J_j \tag{20}$$

Since each of the functions Φ and X_j depends solely on the reaction coordinate ε_j, we have

$$\frac{\partial \Phi_{\text{eq}}}{\partial \varepsilon_k} = -\sum_i^n \nu_i^{(k)} \frac{\mu_{i,\text{eq}}}{T_{\text{eq}}} \tag{21}$$

and

$$\frac{\partial^2 \Phi_{\text{eq}}}{\partial \varepsilon_l \partial \varepsilon_k} = \frac{\partial^2 \Phi_{\text{eq}}}{\partial \varepsilon_k \partial \varepsilon_l} \tag{22}$$

or equivalently

$$\left(\frac{\partial X_k}{\partial \varepsilon_l}\right)_{U,V,\mathbf{N_a},\mathbf{v},\varepsilon} = \left(\frac{\partial X_j}{\partial \varepsilon_k}\right)_{U,V,\mathbf{N_a},\mathbf{v},\varepsilon} \tag{23}$$

Eq. (23) indicates that the r x r matrix with elements $\left(\frac{\partial X_k}{\partial \varepsilon_l}\right)_{U,V,\mathbf{N_a},\mathbf{v},\varepsilon}$ is symmetric for all values of matrix ε. From the properties of Jacobians, the matrix with elements $\left(\frac{\partial \varepsilon_k}{\partial X_l}\right)_{U,V,\mathbf{N_a},\mathbf{v},\mathbf{X}}$ is symmetric also for both zero and nonzero values of **X**, and we have

$$\left(\frac{\partial \varepsilon_k}{\partial X_l}\right)_{U,V,\mathbf{N_a},\mathbf{v},\mathbf{X}} = \left(\frac{\partial \varepsilon_j}{\partial X_k}\right)_{U,V,\mathbf{N_a},\mathbf{v},\mathbf{X}} \tag{24}$$

Eqs. (23) and (24) are among the many Maxwell relations that can be established for the stable equilibrium states of a multicomponent system 2.

Assuming that the reacting system 1 belongs at each instant of time to the family of states 1ε, and the rate of change of each reaction coordinate is a function of the element of vector **X**, we have

$$\mathbf{J} = \mathbf{J}(\mathbf{X}) \tag{25}$$

and

$$\mathbf{J}(0) = 0 \tag{26}$$

Therefore at equilibrium $X_i = 0$. Eq. (24) shows that along the time evolution, the surrogate system 2 proceeds through stable equilibrium states, and system 1 proceeds through states X_ε. This condition is stated without any reference to microscopic reversibility, and for all values of $\mathbf{X}$ representing the nonequilibrium as well as the chemical equilibrium states. We can expand each of the r reaction into a Taylor series around the chemical equilibrium state at which $\mathbf{X} = 0$

$$\mathbf{J} = \mathbf{L} \cdot \mathbf{X} + \text{higher order terms} \tag{27}$$

where $\mathbf{L}$ is a r x r matrix of coefficients, each defined by the relation

$$L_{kl} = \left(\frac{\partial J_k}{\partial X_l} \right)_{U,V,\mathbf{N_a},\mathbf{v},\mathbf{X}=0} \qquad k = 1,2,..,r, \quad l = 1,2,..,r \tag{28}$$

Substituting of Eq. (24) into Eq. (28) indicates that the matrix $\mathbf{L}$ is symmetric and the matrix elements, L_{kl} obey the Onsager reciprocal relations.

For the small values of $\mathbf{X}$**,** (near chemical equilibrium state) the linear term predominates in Eq. (27), and the entropy generation becomes

$$\Phi = \mathbf{X}^{\mathrm{T}} \cdot \mathbf{L} \cdot \mathbf{X} \tag{29}$$

where $\mathbf{X}^{\mathrm{T}}$ is the row vector of $\mathbf{X}$. Because $\Phi \geq 0$ in general, the right hand side of Eq. (29) is a quadratic form, and the matrix $\mathbf{L}$ is be positive semi-definite.

Each X_i can be regarded as a driving force, and each rate of change of a reaction coordinate J_j as a flow that depends on all the forces; this indicates a coupled phenomenon. The reciprocal relations are valid both for the states that are in chemical equilibrium and for those that are not [3].

1.1. Michaelis-Menten kinetics

Since enzymes are proteins made of amino acid molecules, they are temperature and pH sensitive, which can change either their state of dissociation or dimensional conformation as a result of hydrogen bonds and Van der Waals forces. Only a certain, small portion of the amino acids that comprise an enzyme is involved in the catalytic reaction. This region is called the active site, and is closely involved towards the formation of products. For example the amino acid residues of proteins are greatly influenced by their local pH values, and the activity of proton acceptors and donors occurs in the active site. Enzymes are highly specific, both in the reaction catalyzed and in their choice of reactants, which are called the substrates.

Let us consider the simplest enzyme reaction of converting a substrate S into a product P by means of the following chemical reactions

$$F_o + S \underset{k_2}{\overset{k_1}{\rightleftarrows}} F_1 \tag{30}$$

$$F_1 \underset{k_4}{\overset{k_3}{\rightleftarrows}} F_o + P \tag{31}$$

where F_o and F_1 are the concentrations of enzyme and substrate-enzyme complex, respectively. The terms k_1, k_2, k_3, and k_4 represent the rate constants in the forward and backward directions. If we assume $k_4 = 0$, then the kinetic equations are

$$\dot{S} = -k_1 F_o S + k_2 F_1 \tag{32}$$

$$\dot{F}_1 = k_1 F_o S - (k_2 + k_3) F_1 \tag{33}$$

$$\dot{P} = k_3 F_1 \tag{34}$$

As the total concentration of the enzyme is constant $E = F_o + F_1 =$ constant, Eq. (33) becomes

$$\dot{F}_1 = k_1 ES - (k_2 + k_3 + k_1 S) F_1 \tag{35}$$

If $S >> E$ in the stationary state we assume that $\dot{F}_o = \dot{F}_1 = 0$. From Eq. (35) we express F_1 by

$$F_1 = \frac{k_1 ES}{k_2 + k_3 + k_1 S} \tag{36}$$

Since $P = -S$ coincides with the reaction rate J_r, we obtain

$$J_r = \frac{k_3 k_1 ES}{k_2 + k_3 + k_1 S} \tag{37}$$

Dividing the numerator and denominator of Eq. (37) by k_1, and defining the Michaelis constant k_m by

$$k_m = \frac{k_2 + k_3}{k_1} \tag{38}$$

leads to the *Michaelis-Menten equation*

$$J_r = \frac{k_3 ES}{k_m + S} \tag{39}$$

The maximum velocity $J_{r,\max}$ can be expressed by

$$J_{r,\max} = \lim_{S\to 0} \frac{k_3 ES}{k_m + S} = k_3 E \tag{40}$$

So that Eq. (39) can also be written as

$$J_r = \frac{J_{r,\max} S}{k_m + S} \tag{41}$$

Eq. (41) indicates that the relation between the reaction rate and the substrate concentration is not linear. However, if we study the reaction in certain regions, it can be approximated as linear; the curve of J_r versus S yields a rectangular hyperbola, and approaches $J_{r,\max}$ asymptotically at large values of S leading to concentration independent rate of reaction.

2. COUPLED CHEMICAL REACTIONS

Beside the transport, the most important processes in biological systems are those related to chemical reactions of metabolism. One of the typical aspects is the requirement regarding the apparent stoichiometry of two partially coupled reactions, and the study of the efficiency of such reactions as limited by the constraints of the second law of thermodynamics.

When we have number of moles of species i and reactions j, we can express the Gibbs equation in terms of the extent of reaction ξ_j and the affinity A_j

$$dS = T^{-1} dU + PT^{-1} dV - T^{-1} \sum_j A_j d\xi_j \tag{42}$$

where $d\xi_j = \dfrac{dN_{ij}}{\nu_{ij}}$, and $A_j = -\sum_j \nu_{ij} \mu_j$

In an isolated system in which U and V do not vary, $dU = dV = 0$, the dissipation function is given by the second law of thermodynamics

$$\Psi = T\frac{dS}{dt} = \sum_j A_j \left(\frac{d\xi_j}{dt}\right) \geq 0 \tag{43}$$

According to the law of mass action of chemical kinetics, the rate of reaction is proportional to the product of the concentrations. The reaction velocity is

$$J_{r,i} = \frac{d\xi_j}{dt} \tag{44}$$

With the respective affinities as the thermodynamic forces, Eq. (43) becomes

$$\Psi = \sum_j A_j J_{r,j} \geq 0 \tag{45}$$

Therefore A and J_r must have the same sign. In the chemical equilibrium state we have $A = 0$, which leads to

$$A = -\sum_i \mu_i \nu_i = 0 \tag{46}$$

2.1. Two-reaction coupling

Despite its limitations for the chemical reactions, the linear nonequilibrium thermodynamics theory has a useful conceptual base. Consider the linear phenomenological equations for two chemical reactions with flows of $J_{r,1}$ and $J_{r,2}$

$$J_{r,1} = L_{11}A_1 + L_{12}A_2 \tag{47}$$

$$J_{r,2} = L_{21}A_1 + L_{22}A_2 \tag{48}$$

These relations are based on the dissipation function

$$\Psi = J_{r,1}A_1 + J_{r,2}A_2 \geq 0 \tag{49}$$

For the sum in Eq. (49) to be positive, we can have either both contributions to be positive, or one of them ($A_1J_{r,1}$) is negative, while the other is positive and large enough to compensate the negative effect of the first term. When $A_1J_{r,1}$ is negative, for example, the reaction is carried out in the direction opposite to the direction imposed by its affinity. This is only possible if a coupling takes place between these reactions, and a sufficiently large entropy production is attained by the other reaction, which occurs in the direction predicted by its own affinity. This effect may cause one process to drive another process.

The coupled processes are of great interest in biological systems, since in many situations the synthesis of reactions or the transport of substrates takes place in the direction opposite to that predicted by its thermodynamic force.

The degree of coupling is defined by the ratio

$$q_{12} = \frac{L_{12}}{(L_{11}L_{22})^{1/2}} \tag{50}$$

The second law imposes $L_{11}L_{22} \geq (L_{12})^2$, and therefore the degree of coupling is limited between −1 and +1. When $q = \pm 1$, the system is completely coupled and the two processes become a single process. When $q = 0$, the two processes are completely uncoupled and do not have any energy-conversion interactions.

Let us consider $J_{r,1}$ as output and $J_{r,2}$ as input flows, the ratio $J_{r,1}/J_{r,2}$ is the stoichiometric ratio, which indicates the number of moles reacting in reaction 2 in order to produce a certain rate of reaction 1. From Eqs. (47) and (48), we have

$$\frac{J_{r,1}}{J_{r,2}} = \frac{L_{11}A_1 + L_{12}A_2}{L_{21}A_1 + L_{22}A_2} \tag{51}$$

Dividing the numerator and the denominator of Eq. (51) by $(L_{11}L_{22})^{1/2} A_2$, we obtain

$$\frac{J_{r,1}}{J_{r,2}} = z\frac{q + zx}{1 + qzx} \tag{52}$$

where z and x are given by

$$z = \left(\frac{L_{11}}{L_{22}}\right)^{1/2}, \text{ and } x = \frac{A_1}{A_2}$$

Eq. (52) shows that the ratio of flows (rate of reactions) depends on the ratio of forces and the degree of coupling. We observe that when the degree of coupling q goes to zero, $J_{r,1}/J_{r,2} = z^2x$, so that z is a stoichiometric parameter.

In the case of two coupled reactions, one spontaneous and the other forced, it is customery to define efficiency as the ratio of dissipation in reaction 1 to the dissipation in reaction 2, that is

$$\eta = -\frac{J_{r,1}A_1}{J_{r,2}A_2} \tag{53}$$

From Eq. (52) we can write

$$\eta = -\frac{zx(q+zx)}{1+qzx} \tag{54}$$

For the maximum efficiency of the energy transfer from reaction 2 to 1, we differentiate η with respect to x and equate to zero, then we have

$$x_{\max} = \frac{-1+(1-q^2)^{1/2}}{qz} \tag{55}$$

After mathematical arrangement, the maximum efficiency is obtained as

$$\eta_{\max} = \frac{q^2}{[1+(1-q^2)^{1/2}]^2} \tag{56}$$

The value of $\eta_{\max}$ depends on only the extent of coupling, and tends to vanish when the degree of coupling decreases. When the flows are completely uncoupled, the efficiency must be zero, when $q = 1$, we have the maximum efficiency of unity.

REFERENCES

[1] S.R. Caplan and A. Essig, Bioenergetics and Linear Nonequilibrium Thermodynamics, The Steady State, Harvard University Press, Cambridge, 1983.

[2] W. Ebel, J. Math. Biology, 21 (1985) 243.

[3] E.P. Gyftopoulos and G.P. Beretta, Thermodynamics. Foundations and Applications, Macmillan, New York, 1991.

[4] J.T. Edsall and H. Gutfreund, Biothermodynamics: To study of Biochemical Processes at Equilibrium, Wiley, Chichester, 1983.

[5] A. Katchalsky and P.F. Curran, Nonequilibrium Thermodynamics in Biophysics, Harvard University Press, Cambridge, 1967.

[6] D. Kondepudi and I. Prigogine, Modern Thermodynamics, From Heat Engines to Dissipative Structures, Wiley, New York, 1999.

[7] I. Prigogine, Introduction to Thermodynamics of Irreversible Processes, Wiley, New York, 1967.

Chapter 11

Membrane transport

INTRODUCTION

Conventional approach to analyze the membrane transport is based on the specific models of classical laws of electrostatic (Poisson's equation), hydrodynamics (Navier-Stokes equations) combined with Fick's diffusion equation; Nernst-Planck equation is used for the transport of charged particles in an electrical field. However, the specific equations of Fick or Nernst-Planck disregard the cross effects, and treat all flows as independent requiring the same mobility under gradients of electrostatic and chemical potentials, which may never be fulfilled especially in concentrated solutions. Interaction (coupling) is a common phenomenon in membrane transport. One of the trends in such analysis is the approach of linear nonequilibrium thermodynamics and phenomenological equations to incorporate the coupled phenomena into the membrane transport [1-7]. Since the interactions between the permeant and membrane may be complex, it may be useful to describe the problem by phenomenological approach without the need of detailed mechanism of transport and interactions. The nonequilibrium thermodynamics approach identifies the possible pathways and interactions, such as the coupling between flow of a substance and a reaction, or between two flows. In this chapter, we summarize the formulations for passive, facilitated and active transport problems, and the degree of coupling.

1. PASSIVE TRANSPORT

In passive transport, electrolytes and other substances are transported due to diffusion or a pressure gradient. It is always desirable to characterize a membrane with the minimum number of parameters. We consider a single solute and a single solvent and nonelectrolytes substances. Total volume flux across the membrane J_v and the velocity of the solute relative to the solvent J_d are defined respectively as

$$J_v = J_w V_w + J_s V_s \tag{1}$$

$$J_d = v_s - v_w \tag{2}$$

Here J_s is the solute flux and J_w is the solvent flux, while V_s and V_w are the partial volumes of the solute and the solvent, and v_s and v_w are their respective velocities of solute and solvent. The dissipation function for such a system is given by

$$\Psi = -J_v \Delta P - J_d \Delta \Pi \tag{3}$$

With the conjugated forces and flows identified by Eq. (3), the phenomenological equations that describe the transport through a membrane are

$$J_v = L_p \Delta P + L_{pd} \Delta \Pi \tag{4}$$

$$J_d = L_{dp} \Delta P + L_d \Delta \Pi \tag{5}$$

where $\Delta\Pi$ is the osmotic pressure difference. Due to the Onsager reciprocity relations, $L_{pd} = L_{dp}$, the membrane could be characterized by three parameters.

If we consider a situation in which the solute concentration is the same on both sides of the membrane and therefore $\Delta\Pi = 0$, but a hydrostatic pressure difference ΔP exist between the two sides, we will have a flux J_v that is a linear function of ΔP. The term L_p is called the *mechanical filtration coefficient*, which represents the velocity of the fluid per unit pressure difference between the two sides of membrane. The cross-phenomenological coefficient L_{dp} is called the *ultrafiltration coefficient*, which is related to the coupled diffusion of the solute with respect to the solvent induced by a mechanical pressure. Osmotic pressure difference produces a diffusion flux characterized by the permeability coefficient, which indicates the movement of the solute with respect to solvent due to inequality of concentrations at both sides of the membrane.

The cross-coefficient L_{pd} relates the flow J_v at $\Delta P = 0$ to $\Delta\Pi$, and is called the *coefficient of osmotic flow*. The cross-coefficients are imposed by the nature of the flow in the membrane, for example L_{pd} shows the selectivity. If $J_v = 0$, we have

$$(\Delta P)_{J_v=0} = -\frac{L_{pd}}{L_p} \Delta \Pi \tag{6}$$

Only if $-L_{pd} = L_p$, then we have

$$(\Delta P)_{J_v=0} = \Delta \Pi \tag{7}$$

This is the condition for an *ideal semipermeable membrane*, which blocks the transport of solute no matter what the value of ΔP and $\Delta\Pi$. When this is not the case, the membrane allows some solute to pass

$$-L_{pd}/L_p < 1$$

The ratio $-L_{pd}/L_p$ is called the *reflection coefficient* σ. The value of $\sigma = 1$ indicates that all solute is reflected (ideal membrane); the solute cannot cross the membrane. When $\sigma < 1$, on the other hand, some of the solute is reflected and the rest crosses the membrane; when $\sigma = 0$, the membrane is completely permeable and is not *selective*. If we introduce σ into Eq. (4), we have

$$J_v = L_p(\Delta P - \sigma\Delta\Pi) \tag{8}$$

The *solute permeability coefficient* ω is defined as

$$\omega = \left(\frac{\bar{c}_s}{L_p}\right)(L_p L_d - L_{pd}^2) \tag{9}$$

where $\bar{c}_s$ is the average solute concentration. For an ideal membrane, $\sigma = 1$, and $\omega = 0$. For *nonselective membranes* $L_{pd} = 0$, and $\omega = c_s L_d$. The permeability coefficient is a characteristic parameter both in synthetic and natural membranes.

The solute flux can be expressed in terms of the solute permeability.

$$J_s = \bar{c}_s(1-\sigma)J_v + \omega\Delta\Pi \tag{10}$$

For a simplified model of the membrane having cylindrical pores of radius r and length l of N_p pores per unit surface area, Poiseuille's law expresses the volume flow Q for a pressure difference ΔP as follows

$$Q = \frac{\pi r^4 \Delta P}{8\eta l} \tag{11}$$

where η is the viscosity of the liquid. The flux J_v is expressed by

$$J_v = \frac{N_p \pi r^4 \Delta P}{8\eta l} \tag{12}$$

Using Eqs. (4) and (11), we find the filtration coefficient L_p

$$L_p = \frac{N_p \pi r^4}{8\eta l} \tag{13}$$

Let us assume that $\Delta P = 0$, but there is a concentration difference and the solute can diffuse. If the solution is ideal van't Hoff's law states that

$$\Delta\Pi = RT\Delta c_s \tag{14}$$

Then from Eq. (10), we have

$$J_s = \omega\Delta\Pi = RT\omega\Delta\bar{c}_s \tag{15}$$

According to Fick's law of diffusion, the diffusion flux in each pore of transverse area of πr^2 and length l is given by

$$J = D(\pi r^2 / l)\Delta\bar{c}_s \tag{16}$$

where D is the diffusion coefficient. Then the total flux becomes

$$J_s = N_p D(\pi r^2 / l)\Delta\bar{c}_s \tag{17}$$

Comparison of Eqs. (15) and (17) yields

$$\omega = \frac{N_p D}{RT}\frac{\pi r^2}{l}(1-\sigma') \tag{18}$$

where σ' is the fraction of solute reflected. From a hydrodynamic point of view, the reflection coefficient σ may be defined as

$$1-\sigma = \left(1-\frac{r_s}{r}\right)^2 \tag{19}$$

where r_s and r are the radius of the solute molecule and the pore, respectively. If $r_s > r$, and we assume that $\sigma = 1$, since the solute molecules do not fit into the pores, all the solute is reflected. Eq. (19) implies the ratio of effective area of the pores, which is $\pi(r-r_s)^2$, to the total area

$$\frac{\text{effective area}}{\text{actual area}} = \frac{(r-r_s)^2}{r^2} = \left(1-\frac{r_s}{r}\right)^2 \tag{20}$$

Generally the reflection coefficient decreases as the molecular radius increases.

Consider two containers of volumes V_1 and V_2 and with the initial solute concentrations of c_1 and c_2 ($c_1 > c_2$) separated by a permeable membrane. We would be interested to know the time necessary to reach the equilibrium at which $c_1 = c_2$. The changes in the number of moles of solute in containers 1 and 2 are calculated as a function of the area A of the membrane and the solute flux J_s

$$\frac{dN_1}{dt} = -J_s A; \quad \frac{dN_2}{dt} = J_s A \tag{21}$$

We know that

$$N_i = V_i c_i; \quad J_s = \omega RT(c_1 - c_2) \tag{22}$$

then

$$V_1 \frac{dc_1}{dt} = -A\omega RT(c_1 - c_2) \tag{23}$$

$$V_2 \frac{dc_2}{dt} = A\omega RT(c_1 - c_2) \tag{24}$$

After subtracting Eq. (24) from Eq. (23), we obtain

$$\frac{d(c_1 - c_2)}{dt} = -A\omega RT\left(\frac{1}{V_1} + \frac{1}{V_2}\right)(c_1 - c_2) \tag{25}$$

The solution of Eq. (25) is

$$\Delta c(t) = (c_1 - c_2) = \Delta c(0)\exp\left(-\frac{t}{t_o}\right) \tag{26}$$

where t_o is defined using Eq. (25)

$$t_o = \frac{V_1 V_2}{(A\omega RT)(V_1 + V_2)} \tag{27}$$

If we assume that $V_1 >> V_2$, we have a simplified approximation for t_o

$$t_o \approx \frac{V_2}{A\omega RT} \tag{28}$$

In practice, after a time of $4t_o$, the concentration differences between the two containers will vanish, and t_o is considered the characteristic time. This approximation may be useful in cellular transport, artificial kidneys and lungs.

We can express the phenomenological coefficients in terms of the frictional forces; assuming that for a steady state flow, the thermodynamic forces X are counter balanced by a sum of suitable frictional forces F. Thus for a solute in an aqueous solution, we have

$$X_s = -F_{sw} - F_{sm} \tag{29}$$

where F_{sw} is the frictional force between solute and water, and F_{sm} is the corresponding friction with the membrane. Similarly the force on water is

$$X_w = -F_{ws} - F_{wm} \tag{30}$$

The terms F_{sm} and F_{wm} comprise complex hydrodynamic interactions within the membrane matrix and should be regarded as macroscopic averages. For sufficiently swollen membranes, however F_{sw}, which indicates the interactions of solute and solvent, may approach to the free diffusion.

The individual frictional forces F_{ij} are assumed to be linearly proportional to the relative velocity

$$v_{ij} = v_i - v_j \tag{31}$$

and the proportionality factor f_{ij} being the frictional coefficient per mole of *i*th component, then we have

$$F_{ij} = -f_{ij}(v_i - v_j) \tag{32}$$

The f_{ij}'s obey the reciprocity relations

$$C_i f_{ij} = C_j f_{ji} \tag{33}$$

The terms C and c denote the concentrations in the membrane and in the external solutions, respectively.

Introducing Eqs. (33) and (32) into Eqs. (29), and (30) we obtain

$$X_s = f_{sw}(v_s - v_w) + f_{sm}(v_s - v_m) \tag{34}$$

Choosing the membrane as the frame of reference and for $v_m = 0$, we have

$$X_s = v_s(f_{sw} + f_{sm}) - v_w f_{sw} \tag{35}$$

and

$$X_w = v_s \frac{C_s F_{sw}}{C_w} + v_w \left(f_{wm} + \frac{C_s}{C_w} f_{sw} \right) \tag{36}$$

Since the flows J_s and J_w are expressed as $J_s = C_s v_s$ and $J_w = C_w v_w$, we may transform Eq. (35) and (36) into expressions of the forces in terms of the flows

$$X_s = \frac{f_{sw} + f_{sm}}{C_s} J_s - \frac{f_{sw}}{C_w} J_w \tag{37}$$

$$X_w = -\frac{f_{sw}}{C_w} J_s + \frac{C_w f_{wm} + C_s f_{sw}}{(C_w)^2} J_w \tag{38}$$

Eqs. (37) and (38) are equivalent to the resistance type of formulations

$$X_s = K_s J_s + K_{sw} J_w \tag{39}$$

$$X_w = K_{sw} J_s + K_w J_w \tag{40}$$

The resistance phenomenological coefficients are given by

$$K_s = \frac{f_{sw} + f_{sm}}{C_s}; \quad K_{\mathrm{sw}} = -\frac{f_{sw}}{C_w}; \; K_w = \frac{C_w f_{wm} + C_s f_{sw}}{(C_w)^2} \tag{41}$$

In Kirkwood's procedure the differences of chemical potentials (for nonelectrolyte solutions) are expressed by

$$\Delta\mu_w = v_w \Delta P - RT\Delta \ln c_w = v_w \Delta P - \frac{RT\Delta c_s}{c_w} = v_w \Delta P - \frac{\Delta\Pi}{c_w} \tag{42}$$

$$\Delta\mu_s = v_s \Delta P + RT\Delta \ln c_s = v_s \Delta P + \frac{RT\Delta c_s}{c_s} = v_w \Delta P + \frac{\Delta\Pi}{c_s} \tag{43}$$

where $\Delta\Pi = RT\Delta c_s$ is the gradient of osmotic pressure across the membrane. If the forces are expressed in terms of the chemical potentials, we have

$$X_i = -\frac{d\mu_i}{dx} \tag{44}$$

Integration of Eq. (44) gives

$$\int_0^{\Delta x} X_s dx = \int_{\mu_s^{\mathrm{I}}}^{\mu_s^{\mathrm{II}}} -\frac{d\mu_s}{dx} dx = \Delta\mu_s = v_s \Delta P + \frac{RT\Delta c_s}{c_s} = \frac{\phi_s \Delta P + \Delta\Pi}{c_s} \tag{45}$$

$$\int_0^{\Delta x} X_w dx = \int_{\mu_s^{\mathrm{I}}}^{\mu_s^{\mathrm{II}}} -\frac{d\mu_w}{dx} dx = \Delta\mu_w = v_w \Delta P - \frac{RT\Delta c_s}{c_w} = \frac{\phi_w \Delta P - \Delta\Pi}{c_w} \tag{46}$$

where $\phi = c_i v_i$ is the mean volume fraction of component i in the external solution.

On the other hand, integration of the right-hand-sides of Eqs. (37) and (38) requires the concentration profiles in the membrane and the dependence of the f_{ij}'s on X. For a homogeneous membrane, it is possible to assume that the f_{ij}'s are constant, and instead of integrating over the concentration, we can use the mean values given by

$$\overline{C}_s = \frac{(C_s)_o + (C_s)_{\Delta x}}{2}; \quad C_w = \frac{\phi_w}{v_w} \tag{47}$$

where ϕ_w is the mean volume fraction of water in the membrane. From the assumption of the equality of the chemical potential on the surface of the membrane and in the surrounding solution, we may write

$$(C_s)_o = Kc_s^{\mathrm{I}}; \ (C_s)_{\Delta x} = Kc_s^{\mathrm{II}} \tag{48}$$

and

$$\overline{C}_s = K\frac{c_s^{\mathrm{I}} + c_s^{\mathrm{II}}}{2} = K\overline{c}_s \tag{49}$$

where K is the *distribution coefficient* of solute between the membrane and the solution in equilibrium with it. The K is taken as a concentration-independent parameter uniform through out the membrane in an ideal behavior. These assumptions permit the integration of the right-hand side of Eqs. (37) and (38). Since in the majority of all cases $\phi_w = c_w v_w \sim 1$, and ϕ_s is usually sufficiently small to make $\phi_s \Delta P < \Delta\Pi$, Eqs. (37) and (38) yield the following expressions

$$\Delta\Pi = \Delta x \frac{f_{sw} + f_{sm}}{K} J_s - \Delta x \bar{c}_s \frac{f_{sw} v_w}{\phi_w} J_w \tag{50}$$

$$\Delta P - \Delta\Pi = \Delta x \frac{f_{sw}}{\phi_w} J_s + \Delta x (\frac{f_{wm}}{\phi_w} + K \frac{c_s f_{sw}}{(\phi_w)^2}) J_w \tag{51}$$

where Δx denotes the thickness of the membrane.

The permeability coefficient of the solute flow is determined at zero volume flow

$$(J_s)_{J_v=0} = \omega \Delta\Pi \tag{52}$$

In this case the solute flow at $J_w = 0$ is practically the same as at $J_v = 0$. So that Eq. (50) at $J_w = 0$ leads to

$$\Delta\Pi = \Delta x \frac{f_{sw} + f_{sm}}{K} J_s \tag{53}$$

therefore

$$\omega = \frac{K}{\Delta x (f_{sw} + f_{sm})} \tag{54}$$

In the ideal case, K and the frictional coefficients are concentration-independent, and ω becomes a constant characterizing the mobility of the solute in membrane.

Eq. (54) shows the characteristics of the permeability coefficient ω. It is obvious that the total frictional resistance is larger than in a membrane than in free solution. Although f_{sw} is approximately equal to f_{sw}° of the solute diffusing freely in the solvent, we have to consider the additional friction f_{sm} due to interactions of the solute with the membrane. If the membrane, for example, has a porous structure and the penetrating molecules are sufficiently large, then f_{sm} may be very large and $\omega \sim 0$. The system in this case will behave as an ideal semipermeable membrane. If the solute penetrates by dissolving in the lipoid components of the membrane, then a large friction between solute and membrane will develop, and the value of f_{sm} will be large. However, if the attraction between solute and lipoid is very large, K may increase to a larger value than f_{sm}.

The reflection coefficient σ is also defined at $J_v = 0$ using the relation

$$\sigma = \left(\frac{\Delta P}{\Delta\Pi} \right)_{J_v=0} \tag{55}$$

However, the coefficient σ' at $J_w = 0$ is not exactly equal to the coefficient σ at $J_v = 0$. Relationship between σ and σ' is given by

$$\sigma = \sigma' - \frac{\omega v_s}{L_p} \tag{56}$$

The coefficient σ' is readily obtained from Eqs. (50) and (51) at $J_w = 0$

$$\sigma' = 1 - \frac{Kf_{sw}}{\phi_w(f_{sw} + f_{sm})} \tag{57}$$

In view of Eq. (56), we have

$$\sigma = 1 - \frac{\omega v_s}{L_p} - \frac{Kf_{sw}}{\phi_w(f_{sw} + f_{sm})} \tag{58}$$

Using Eq. (57), we get

$$\sigma = 1 - \frac{\omega \bar{v}_s}{L_p} - \frac{\omega \Delta x f_{sw}}{\phi_w} \tag{59}$$

Eqs. (58) and (59) help to understand the physical meaning of the reflection coefficient: (a) Ideal semipermeable membranes prevent the permeation of the solute, hence $\omega = 0$ and $\sigma = 1$. (b) Generally, increase in f_{sm} (due to large molecular weight) decreases ω, and hence increases σ. On the other hand, if K increases simultaneously with f_{sm}, than ω may increase, while σ decreases. In some cases the increase in K is so strong that ω assumes large values, which make σ negative. This causes negative anomalous osmosis since $\Delta P^o (= \sigma\Delta\Pi)$ becomes negative. In coarse nonselective membranes we have

$$\frac{\omega v_s}{L_p} + \frac{\omega \Delta x f_{sw}}{\phi_w} = 1 \tag{60}$$

If we assume that the solute penetrates only through the water filled channels in the membrane (with volume fraction ϕ_w), then the solute-water frictional coefficient is close to that of free diffusion D^o

$$f_{sw} = f_{sw}^o = \frac{RT}{D^o} \tag{61}$$

and the solute flow will be given by

$$J_s = D\phi_w \frac{\Delta\Pi / RT}{\Delta x} \tag{62}$$

where D is the diffusion coefficient in the membrane. The relation between ω and D is given by

$$\omega = \frac{D\phi_w}{RT\Delta x} \tag{63}$$

Using Eqs. (61) and (63) we obtain for σ

$$\sigma = 1 - \frac{\omega \bar{v}_s}{L_p} - \frac{D}{D^o} \tag{64}$$

1.1. Composite membranes

For composite (with compartments) membranes the dissipation function Ψ in terms of flows of volume, salt, and electrical current and corresponding forces is

$$\Psi = \mathbf{j}_v(\Delta P - \Delta\pi) + \mathbf{j}_s \Delta\mu_s^c + I\psi \tag{65}$$

where $\Delta P = P^{(1)} - P^{(2)}$ is the difference in hydrostatic pressure across the membrane, $\Delta\pi$ is the difference in the osmotic pressure, $\Delta\mu_s^c$ is the concentration-dependent part of the chemical potential differences of the salt; $\mathbf{j}_v = \mathbf{j}_w \bar{V}_w + \mathbf{j}_s \bar{V}_s$, where $\bar{V}_w$ and $\bar{V}_s$ are the partial molar volumes of water and salt, respectively. ψ is the electric potential, and I the electrical current. Here $\mathbf{j}_v, \mathbf{j}_w, \mathbf{j}_s$ represent the virtual flows. Experimentally $\mathbf{j}_v$ is determined by measuring the change in volume of one or both compartments at opposite surfaces of the membranes. Eq. (65) yields a set of three-flow linear phenomenological equations of conductance type

$$\mathbf{j}_v = L_{11}(\Delta P - \Delta\pi) + L_{12}\left(\frac{\Delta\pi_s}{c_s}\right) + L_{13}\psi \tag{66}$$

$$\mathbf{j}_s = L_{21}(\Delta P - \Delta\pi) + L_{22}\left(\frac{\Delta\pi_s}{c_s}\right) + L_{23}\psi \tag{67}$$

$$I = L_{31}(\Delta P - \Delta \pi) + L_{32}\left(\frac{\Delta \pi_s}{c_s}\right) + L_{33}\psi \tag{68}$$

where $\Delta \pi_s$ is the difference in the osmotic pressure due to the permeant solute (salt), and c_s is the mean concentration of the salt, and given by

$$c_s = \frac{\Delta \pi_s}{\Delta \mu_s^c} \tag{69}$$

As Eq. (65) is an appropriately derived dissipation function consisting of the conjugate flows and forces, the Onsager reciprocal rules states that $L_{ij}=L_{ji}$.

Eqs. (66)-(68) can be also expressed in terms of the flows using the resistance coefficients K_{ij}, and we have the resistance type of formulation

$$(\Delta P - \Delta \pi) = K_{11}\mathbf{j}_v + K_{12}\mathbf{j}_s + K_{13}I \tag{70}$$

$$\frac{\Delta \pi_s}{c_s} = K_{21}\mathbf{j}_v + K_{22}\mathbf{j}_s + K_{23}I \tag{71}$$

$$= K_{31}\mathbf{j}_v + K_{32}\mathbf{j}_s + K_{33}I \tag{72}$$

where the coefficients K_{ij} are the inverse of the conductance coefficients L_{ij}, and are symmetrical $K_{ij} = K_{ji}$.

Eqs. (66)-(68) are related to various classical studies of electrokinetic phenomena, since they describe the coupled processes and yield naturally a number of symmetry relationships, which have been observed experimentally. Therefore, they provide a practical use of the linear nonequilibrium thermodynamic approach. For example, we may consider studies with identical solutions at each surface of the membrane, so that $\Delta \pi = \Delta \pi_s = 0$. Then the system has only two degrees of freedom, and we have

$$\Psi = \mathbf{j}_v \Delta P + I\psi \tag{73}$$

and the linear phenomenological equations become

$$\mathbf{j}_v = L_{11}\Delta P + L_{12}\psi \tag{74}$$

$$I = L_{12}\Delta P + L_{22}\psi \tag{75}$$

Eqs. (74)-(75) represent the whole classical electrokinetics. For example, the magnitude of the electroosmotic volume flow per unit potential at zero pressure difference, $\Phi = (j_v / \psi)_{\Delta P=0}$, and the streaming current per unit pressure difference at short circuit, $(I / \Delta P)_{\psi=0}$, must be identical.

We also observe the well-established *Saxen relations* between the ratios of force and flow

$$\left(\frac{\mathbf{j}_v}{I}\right)_{\Delta P=0} = -\left(\frac{\psi}{\Delta P}\right)_{I=0} \tag{76}$$

$$\left(\frac{\mathbf{j}_v}{I}\right)_{\psi=0} = -\left(\frac{\psi}{\Delta P}\right)_{\mathbf{j}_v=0} \tag{77}$$

These symmetry relationships observed do not depend on the specific features of any given model but follow quite generally from the linear phenomenological equations of nonequilibrium thermodynamics. Therefore any linear model, which does not predict these relations, is likely to be incorrect.

Sometimes we are interested in processes in which a significant separation of salt from water occurs in the volume flow. Considering solutions that $L_{11}\Delta\pi_i << \Delta\pi$, so that $\mathbf{j}_v\Delta\pi \cong \mathbf{j}_v\Delta\pi_s$, we have

$$\Psi = \mathbf{j}_v\Delta P - \mathbf{j}_w\overline{V}_w\Delta\pi_s + \left(\frac{\mathbf{j}_s}{c_s}\right)(1 - c_s\overline{V}_s)\Delta\pi_s \tag{78}$$

If we have dilute solutions, $c_s\overline{V}_s << c_w\overline{V}_w \cong 1$, we have approximately

$$\Psi = \mathbf{j}_v\Delta P + \left(\frac{\mathbf{j}_s}{c_s} - \frac{\mathbf{j}_w}{c_w}\right)\Delta\pi_s = \mathbf{j}_v\Delta P + \mathbf{j}_D\Delta\pi \tag{79}$$

The expression in the parenthesis that shows the velocity of the salt relative to water is called the diffusional flow $\mathbf{j}_D$. The dissipation function provides a natural basis for the analysis of systems in which mechanical energy derived from the volume flow and the hydrostatic pressure gradient is utilized to produce a separation of salt from water in the face of an adverse concentration gradient.

In Eqs. (66)-(72), a system with three degrees of freedom is characterized by the six independent phenomenological coefficients. The conductance coefficients L_{ij} would be readily evaluated if it proved possible to control the two forces, for example

$$L_{12} = \left(\frac{\mathbf{j}_v}{\Delta \pi_s / c_s} \right)_{\Delta P - \Delta \pi = 0, \psi = 0}$$

or (80)

$$L_{12} = \left(\frac{\partial \mathbf{j}_v}{\partial (\Delta \pi_s / c_s)} \right)_{\Delta P - \Delta \pi, \psi}$$

For the resistance coefficients K_{ij}, it would be desirable to control the two flows.

It is sometimes more useful to consider alternative expressions of phenomenological equations without the need for further transformation of the dissipation function. For example consider Eqs. (70)-(72), expressing the forces as functions of the flows. For practical purposes it is desirable to use relations in which the independent variables are readily controlled experimentally. We may rewrite Eqs. (66)-(68) in such a way that $\Delta \pi_s / c_s$ becomes an independent variable, while $\mathbf{j}_s$ becomes a dependent variable, so that we have

$$(\Delta \pi - \Delta P) = \frac{K_{11} K_{22} - K_{12} K_{21}}{K_{22}} \mathbf{j}_v + \frac{K_{12}}{K_{22}} \frac{\Delta \pi_s}{c_s} + \frac{K_{22} K_{13} - K_{12} K_{23}}{K_{22}} I \tag{81}$$

$$\mathbf{j}_s = -\frac{K_{21}}{K_{22}} \mathbf{j}_v + \frac{1}{K_{22}} \frac{\Delta \pi_s}{c_s} + \frac{K_{23}}{K_{22}} I \tag{82}$$

$$= \frac{K_{22} K_{31} - K_{21} K_{32}}{K_{22}} \mathbf{j}_v + \frac{K_{32}}{K_{22}} \frac{\Delta \pi_s}{c_s} + \frac{K_{22} K_{33} - K_{32} K_{23}}{K_{22}} I \tag{83}$$

It is useful to replace the complex coefficients of Eqs. (81)-(83) by the practical transport coefficients; they may be evaluated experimentally under the conditions in which two of the independent variables $\mathbf{j}_v$, $\Delta \pi_s / c_s$, and I are set equal to zero. Such a set of coefficients may be identified with six of the coefficients of Eqs. (81)-(83). Because of the Onsager reciprocal relations, the remaining three coefficients may be evaluated as follows

$$\left(\frac{\mathbf{j}_s}{\mathbf{j}_v} \right)_{\Delta P_s, I} = -\left(\frac{\Delta P - \Delta \pi}{\Delta \pi_s / c_s} \right)_{\mathbf{j}_v, I} = \mathrm{c_s} (1 - \sigma) \tag{84}$$

$$-\left(\frac{\psi}{\Delta \pi_s / c_s} \right)_{\mathbf{j}_v, I} = \left(\frac{\mathbf{j}_s}{I} \right)_{\Delta \pi_s, \mathbf{j}_v} = \frac{\tau_1}{v_1 z_1 F} \tag{85}$$

$$\left(\frac{\psi}{\mathbf{j}_v}\right)_{\Delta\pi_s,I} = \left(\frac{\Delta P - \Delta\pi}{I}\right)_{\mathbf{j}_v,\Delta\pi_s} = -\frac{\beta}{L_p} \qquad (86)$$

where σ is the *reflection coefficient*, and given by

$$\sigma = \left(\frac{\Delta P - \Delta\pi}{\Delta\pi_s}\right)_{\mathbf{j}_v,I} \qquad (87)$$

τ_I is the *transport number*

$$\tau_1 = -v_1 z_1 F\left(\frac{\psi}{\Delta\pi_s / c_s}\right)_{\mathbf{j}_v,I} \qquad (88)$$

β is the *electoosmotic permeability*

$$\beta = \left(\frac{j_v}{I}\right)_{\Delta P - \Delta\pi,\Delta\pi_s} \qquad (89)$$

and L_p is the *filtration coefficient*

$$L_p = \left(\frac{j_v}{\Delta P - \Delta\pi}\right)_{\Delta\pi_s,I} \qquad (90)$$

With these coefficients, Eqs. (81)-(83) can be expressed in a more useful form

$$(\Delta\pi - \Delta P) = \left(\frac{1}{L_p}\right)\mathbf{j}_v - c_s(1-\sigma)\frac{\Delta\pi_s}{c_s} - \left(\frac{\beta}{L_p}\right)I \qquad (91)$$

$$\mathbf{j}_s = c_s(1-\sigma)\mathbf{j}_v + c_s\omega\frac{\Delta\pi_s}{c_s} + \left(\frac{\tau_1}{v_1 z_1 F}\right)I \qquad (92)$$

$$= -\left(\frac{\beta}{L_p}\right)\mathbf{j}_v - \left(\frac{\tau_1}{v_1 z_1 F}\right)\frac{\Delta\pi_s}{c_s} + \left(\frac{1}{\kappa}\right)I \qquad (93)$$

where ω is the *solute permeability*

$$\omega = \left(\frac{\mathbf{j}_s}{\Delta \pi_s} \right)_{j_v, I} \tag{94}$$

and κ is the *electric conductance*

$$\kappa = \left(\frac{I}{\psi} \right)_{j_v, \Delta \pi_s} \tag{95}$$

Here a molecule of the salt dissociates into ν_1 cations of charge z_1, and ν_2 anions of charge z_2, and F is the Faraday constant. Eqs. (91)-(93) are useful for the treatment of a composite membrane consisting of compartments in series. The practical phenomenological coefficients, seen in Eqs. (91)-(93), were derived from observations long before the studies based on the linear nonequilibrium thermodynamics approach. Combination of them in a self-consisted formulation provides a sound basis for the analysis.

In a simple example, the permeant solute may be nonelectrolyte. In this case there is no current flow to be considered, and Eqs. (91)-(93) can be expressed in terms of the flows as a function of the forces

$$\mathbf{j}_v = L_p(\Delta P - \Delta \pi) - \sigma L_p \Delta \pi_s \tag{96}$$

$$\mathbf{j}_s = c_s(1 - \sigma)\mathbf{j}_v - \omega \Delta \pi_s \tag{97}$$

These equations correspond respectively to Eqs. (70)-(72) for the case of dilute solutions. The value of reflection coefficient σ must depend on the nature of both the solute and the membrane. For the case of volume flow in the absence of concentration gradient of permeant solute ($\Delta \pi_s = 0$), we see that the quantity (1-σ) is a direct measure of the extent of coupling between the solute flow and the volume flow. If the membrane is completely nonselective, then $\sigma = 0$; if the membrane is perfectly selective, permeable only to the solvent, $\sigma = 1$. In most cases σ will lie between 0 and 1.

A diffusion flow against its conjugate gradient driven by the dissipation of another diffusional process would be called the “incongruent” diffusion. For example the flow of the *i*th component across a membrane may be expressed by

$$j_i = L_{ii} \Delta \mu_i + \sum_{k=1}^{n} L_{ik} \Delta \mu_k \tag{98}$$

If $\Delta \mu_i = 0$, but $\Delta \mu_k \neq 0$, a flow of substance *i* may still be possible.

1.2. Electrokinetic effect

Electrokinetic effects are consequence of interaction between the flow of matter and flow of electricity through a porous membrane. The linear phenomenological equations for the simultaneous transport of matter and electricity are [Eqs. (74) and (75)]

$$J_v = L_{11}\Delta P + L_{12}\Delta\psi$$

$$I = L_{21}\Delta P + L_{22}\Delta\psi$$

where I is the electrical current per unit area, J_v is the volume flow of matter, $\Delta\psi$ is the potential difference, ΔP is the pressure difference, and L_{ij} are the phenomenological coefficients defined as

$$L_{11} = \left(\frac{J_v}{\Delta P}\right)_{\Delta\psi=0} \quad \textit{(hydraulic permeability, filtration coefficient)} \tag{99}$$

$$L_{12} = \left(\frac{J_v}{\Delta\psi}\right)_{\Delta P=0} \quad \textit{(electroosmosis)} \tag{100}$$

$$L_{21} = \left(\frac{I}{\Delta P}\right)_{\Delta\psi=0} \quad \textit{(streaming current)} \tag{101}$$

$$L_{22} = \left(\frac{I}{\Delta\psi}\right)_{\Delta P=0} \quad (\textit{conductance of permeant}\text{-filled electroosmotic cell}) \tag{102}$$

Eqs. (74) and (75) indicate that even if no electromotive force acts on the clay $\Delta\psi = 0$, the existence of a pressure difference will produce an electrical flow if the coupling coefficient is nonvanishing; when no pressure is applied $\Delta P = 0$, the action of the electrical force will cause a volume flow of water.

Since $L_{12} = L_{21}$, electrical current per unit pressure force at $\Delta\psi=0$, equals to the volume flow J_v per unit potential difference at $\Delta P = 0$,

$$\left(\frac{J_v}{\Delta\psi}\right)_{\Delta P=0} = \left(\frac{I}{\Delta P}\right)_{\Delta\psi=0} \tag{103}$$

The efficiency of electrokinetic energy conversion for two operations modes, namely electroosmosis η_{eo} and streaming potential η_{sp} are expressed as

$$\eta_{eo} = -\frac{J_v \Delta P}{I \Delta \psi} = -\frac{\text{output}}{\text{input}} = -\frac{J_v \Delta P}{(\Delta \psi)^2 / R} \tag{104}$$

$$\eta_{sp} = -\frac{I \Delta \psi}{J_v \Delta P} = -\frac{\text{output}}{\text{input}} = -\frac{(\Delta \psi)^2 / R}{J_v \Delta P} \tag{105}$$

where R is the resistance. The maximum values of energy conversions occur when the output forces equals to half of their steady state values. For example, η_{eo} is maximum when ΔP equals to half the value of electroosmosis pressure

$$\Delta P = -\frac{1}{2}(\Delta P)_{J_v=0} \tag{106}$$

The maximum energy conversion is related to the *merit* β through the degree of coupling q for the linear phenomenological relations [11]

$$\eta_{\max} = \frac{(1+\beta)^{1/2} - 1}{(1+\beta)^2 + 1} \tag{107}$$

where $\beta = \left(\frac{1}{q^2} - 1\right)^{-1}$ and $q = \frac{L_{ij}}{(L_{ii} L_{jj})^{1/2}}$

with $L_{ij} = L_{ji}$. Since the value of β is much smaller than unity, Eq. (107) approximately becomes

$$\eta_{\max} \cong \beta / 4 \tag{108}$$

As a consequence of Onsager's reciprocal relations, we have

$$\eta_{\max,eo} = \eta_{\max,sp} \tag{109}$$

Tables 1 show the efficiency of electrokinetic energy conversion values.

2. FACILITATED TRANSPORT

Facilitated transport in a membrane involves a chemical agent as a carrier to increase the passive transport with a carrier-mediated transport. A chemical agent

Table 1
Efficiency of electrokinetic energy conversion for different mixed-lipid membranes [19].

Type of membrane	β_{eo} 10^5	β_{sp} 10^5	$\eta_{\max,eo}$ 10^5	$\eta_{\max,sp}$ 10^5
Cephalin-serine	2.108	1.714	0.527	0.428
Cephaline-inositol	2.235	2.255	0.558	0.563
Lecithin-serine	3.687	3.686	0.921	0.920
Lecithin-inosithol	3.946	3.942	0.986	0.985
Lecithin-cephalin	5.855	5.852	1.463	1.463
Inositol-serine	10.044	10.156	2.511	2.539

can react reversibly with a permeant, and yields high selectivity and permeability, which makes the facilitated transport a very attractive separation technique. The chemical agent carries the substance in the form of carrier-bounded substance; carrier releases the substance on the other side of the membrane due to chemical conditions (mainly pH, electric charge) and diffuses back. Mainly a carrier with high association and dissociation rate constants, which are similar in magnitude is desirable. Various substances such as amino acids, organic acids, NaOH, NaCl, carbon dioxide, oxygen, metals, and various ions such as Cd(II), Cu(II), Co(II), Fe(III), can be separated by using suitable carrier agents in liquid or in solid composite membranes.

Membranes for the facilitated transport can be prepared by impregnating the pores of a microporous support with a chemical agent solution. This type of membrane is called the supported liquid membrane. Liquid membranes behave like the double liquid-liquid extraction systems where the usage of organic solvent is minimized. Such devices are generally prepared as bulk liquid, emulsion liquid, and supported liquid membranes. For example, amines can be used as carriers for the facilitated transport of carbon dioxide [27].

Analysis of facilitated transport by the nonequilibrium thermodynamic approach is reported for nonisothermal facilitated transport in ion exchaged membranes in a reasonable range of chemical potential and temperature difference [4-8]. Linear phenomenological coefficients have been determined for the facilitated transport of liquid ortoboric acid by polystyrene-di-vinylbenzene and carbon dioxide by monoprotonated ethylenediamide (EDAH) in a cation exchange membrane [3,4].

A set of example chemical reactions for boric acid that take place in an anion exchange membrane is

$$\mathrm{HB} + \mathrm{OH}^- \Leftrightarrow \mathrm{B}^- + \mathrm{H_2O} \tag{110}$$

$$(a - b)B^- + bHB \Leftrightarrow H_bB_a{}^{(a-b)-} \tag{111}$$

where HB denotes boric acid, and B^- is the borate, which is the carrier ion , a and b are the stoichiometric coefficients. Assuming that the system is in mechanical equilibrium and at stseady state, the dissipation function is expressed by

$$\Psi = J_q \frac{\Delta T}{T_{av}} + J_B \Delta\mu_B + \sum_i A_i J_{r,i} \tag{112}$$

where

$$\Delta\mu_B = \mu_{B,I} - \mu_{B,II}$$

$$\Delta T = T_I - T_{II} \text{ and } T_{av} = (T_I + T_{II})/2$$

J_q is the heat flow which takes into account the enthalpy of mixing and the heat of reactions, $J_{r,i}$ is the flow for reaction i, $\Delta\mu_B$ is the chemical potential potential difference of H_3BO_3, ΔT is the temperature difference acros the membrane, and J_B is the absolute flow of boric acid directed from compartment I to compartment II, assuming that the absolute flow of water through the membrane is negligible.

Eq. (112) shows the three contributions of dissipations due to heat flow, mass flow, and chemical reaction, respectively. As the membrane is assumed as an isotropic medium, there will be no coupling between flows and scalar quantity of chemical reaction according to the Curie-Prigogine principle.

For carbon dioxide transport in a membrane with the chemical agent EDAH the following chemical reaction is used

$$CO_2 + 2EDAH^+ \Leftrightarrow EDACO_2 + HEDAH_2^{2+} \tag{113}$$

Supported liquid membranes and emulsion liquid membranes [15] are widely used in the facilitated transport systems. However they are not suitable for large-scale industrial applications, and are not stable since chemical mediators are easily lost in liquid membranes. One of the methods for overcoming this disadvantage is the use of an ion exchange membrane as the support. Also the solid polymer composite membranes containing a chemical agent are being tested successfully in separation technology [29-37]. Some of the solid composite membranes are Nafion-poly(pyrrole) films with silver or sodium [32], activated composite membrane [29,33], and solid polymer electrode composite membranes [31]. Composite membranes are stable and suitable for industrial applications, and are usually made of a support polymer matrix of porous structure in which a chemical carrier is added. Preparation of polymer layers containing different amounts of carrier agents may need special polymerization techniques, such as interfacial polymerization. In ion separation, composite membranes utilize a chemical agent as carrier dissolved in an organic solvent contained in a polymeric matrix or within the pores of a polymer membrane. For example pseudo-crown ethers can be used as the fixed site carriers in ion separations [37].

The use of composite membranes allow wide scale application in industrial separations of substances from mixtures leading to far less expensive separation. For example, olefins, amino acids, heavy metals, gases, fatty acids, water, inorganic salts can be separated selectively by the facilitated transport.

Transport in membranes is mostly complex and coupled process, and coupling between solute and membrane, and coupling between diffusion and chemical reaction may play important role in the efficiency. It is important to understand and quantify the coupling to describe the transport in membranes. Kinetic studies may also be helpful. However thermodynamics might be a new and rigorous approach towards understanding the coupled transport in composite membranes without the need of detailed mechanism of diffusion through the solid structure.

3. ACTIVE TRANSPORT

Active transport mainly implies a transport of substances against the direction imposed by the thermodynamic force without chemical modification. In group translocations, the transported substrate is chemically modified. The transport systems are either coupled to a scalar system of chemical reaction, such as the hydrolysis of ATP in the living cells, or an ion gradient is used as the energy source for the substrate transport.

In specialized membranes, such as biomembranes, metal ions of Na^+, K^+, and Ca^{2+} can be transported uphill against their concentration gradients. Organic compound such as amino acids and saccharides can also be transported in the same way. Active transport is one of the coupled processes between chemical reaction and diffusion of ions and other substance in the living cells.

Some artificial membranes with the function of biomembranes are prepared and used for active transport. For example active transport of metal ions through the synthetic polymer membranes; some specific examples are the cation exchange membranes from 2,3-epithiopropyl methacrylate (ETMA) and 2-acrylamide-2-methyl propane sulfonic acid (AMPS) copolymers for active transport of alkali and alkali earth metal ions [25,26], and ETMA-AMPS copolymer membranes for active transport of amino acids of glycine, penylalanine, and lycine by using the hydrochloric acid as the receiving solution [21]. One possible mechanism is that glycine and water are transferred by osmotic pressure into the membrane, then glycine is protonated with H^+ released from sulfonic acid groups in the membrane, later the protonated glycine is transported by means of sulfonic acid groups to the other side of membrane regardless of the smaller electric potential difference. The transport of amino acids depends on the composition of the membranes and the structure of the amino acids.

A membrane with the function of active transport can recover uranyl ions UO_2^{2+} in the eluate. Uranyl ions form anion complexes with sulfate ions in sulfuric acid, and can be transported against its concentration gradient through a liquid membrane with tertiary amine by using carbonate solution as receiving solution. Polymeric anion exchange membranes can also transport uranyl ions selectively from the eluate of sulfuric acid containing alkali earth metal ions or cupric ions [22].

REFERENCES

[1] B. Baranowski, J. Memb. Sci., 57 (1991) 119.

[2] S.R. Caplan and A. Essig, Bioenergetics and Linear Nonequilibrium Thermodynamics, The Steady State, Harvard University Press, Cambridge, 1983.

[3] E. Selegny, J.N. Ghogoma, R. Roux, D. Langevin and M. Metayer, J. Memb. Sci., 93 (1994) 217.

[4] E. Selegny, J.N. Ghogoma, D. Langevin, R. Roux and C. Ripoll, J. Memb. Sci., 123 (1997) 147-159.

[5] A. Narebska, W. Kujawski and S. Koter, J. Memb. Sci., 30 (1987) 125.

[6] A. Narebska and S. Koter, J. Memb. Sci., 30 (1987) 141.

[7] A. Narebska, W. Kujawski, S. Koter and T.T. Le, J. Memb. Sci., 106 (1995) 25.

[8] A. Narebska, W. Kujawski, S. Koter and T.T. Le, J. Memb. Sci., 106 (1995) 39.

[9] N. Bachelier C. Chappey, D. Langevin, M. Metayer, and J.-F. Verchere, J. Memb. Sci., 119 (1996) 285.

[10] R.L. Blokhra and S. Kumar, J. Memb. Sci., 43 (1989) 31.

[11] R.L. Blokhra and S.S. Thakur, J. Non-Equlib.Thermodyn., 7 (1982) 145.

[12] R. Chakraborty and S. Datta, J. Memb. Sci., 115 (1996) 129.

[13] E.L. Cussler, R. Aris and A. Brown, J. Memb. Sci., 43 (1989) 149.

[14] A. Katchalsky and P.F. Curran, Nonequilibrium Thermodynamics in Biophysics, Harvard University Press, Cambridge, 1967.

[15] X.Liu and D. Liu, Sep. Sci., Tech., 33 (1998) 2597.

[16] A. Narebska and A. Warszawski, J. Memb. Sci., 88 (1994) 167.

[17] A. Narebska and S. Koter, Polish J. Chem., 71 (1997) 1707.

[18] J.D. Ramshaw, J. Non-Equilib. Thermodyn., 18 (1993) 121.

[19] S. A. Rizvi and S. B. Zaidi, J. Memb. Sci., 29 (1986) 259.

[20] W. Yang, A. Yamamuchi and H. Kimizuka, J. Memb. Sci., 70 (1992) 277.

[21] T. Nonaka, T. takeda and H. Egawa, J. Memb. Sci., 76 (1993) 193.

[22] T. Nonaka and M. Kawamoto, J. Memb. Sci., 101 (1995) 135.
[23] C. Nigon, P. Michalon, B.Perrin and B. Maisterrena, J. Memb. Sci., 144 (1998) 237.
[24] B. Maisterrena ,C. Nigon, P. Michalon and R. Couturier, J. Memb. Sci., 144 (1997) 85.
[25] T. Nonaka, H. Ogawa and H. Egawa, J. Appl. Polym. Sci., 41 (1990) 2869.
[26] T. Nonaka, H. Ogawa, M. Morikawa and H. Egawa, J. Appl. Polym. Sci., 45 (1992) 285.
[27] H. Matsuyama, M. Teramoto, H. Sakakura and K. Iwai, J. Memb. Sci., 117 (1996) 251.
[28] Y. Demirel, AIChE's 2000 Annual Meeting, Nov. 12-17, Los Angeles, CA.
[29] J.A. Calzado, C. Palet and M. Valiente, Anal. Chim. Acta, 431 (2001) 59.
[30] Y.S. Park, J. Won and Y.S. Kang, J. Memb. Sci., 183 (2001) 163.
[31] I. Pinnau and L.G. Toy, J. Memb. Sci., 184 (2001) 39.
[32] A. Sungpet, J.D. Way, C.A. Koval and M.E. Eberhart, J. Memb. Sci., 189 (2001) 271.
[33] M. Oleinikova, M. Munoz, J. Benavente and M. Valiente, Analytica Chim. Acta, 403 (2000) 91.
[34] T. Gumi, M. Oleinikova, C. Palet, M. Valiente and M. Munoz, Analytica Chim. Acta, 408 (2000) 65.
[36] K.M. White, B.D. Smith, P.J. Duggan, S.L. Sheahan and E.M. Tyndall, J. Memb. Sci., 194 (2001) 165.
[37] B.J. Eliott, W.B. Willis and C.N. Bowman, J. Memb. Sci., 168 (2000) 109.

Chapter 12

Thermodynamics and biological systems

INTRODUCTION

Classical thermodynamics analyzes the conversion of energy for systems in equilibrium and provides a set of inequalities describing the direction of change. It can specify impossibilities and prescribe maximum limits of attainable efficiencies or of useful work; however it cannot describe adequately the coupled physical, chemical, and biological processes powered by chemical degradation and oxidations. These chemical reactions are maintained in nonequilibrium state by the influx of reactants and efflux of products. Nonequilibrium and unstable biological systems do not proceed with decaying towards an equilibrium state, but increasing in size, structuring, and complexity.

Systems may exhibit two different types of behavior: (i) the tendency towards maximum disorder, or (ii) the spontaneous appearance of a high degree of organization in space, time, and/or function. The best examples of the latter are dissipative systems at nonequilibrium conditions, such as Benard cell, Krebs cycle, ecosystems, and living systems. As living systems grow and develop, a constant supply of energy is needed for organized structures for the ability of reproduction and surviving in changing conditions. To maintain organized structures requires a number of coupled metabolic reactions and transport processes that control the rate and timing of the life processes. Schrodinger proposed that these processes appear to be at variance with the second law of thermodynamics, which states that a finite amount of organization may be obtained at the expense of a greater amount of disorganization in a series of interrelated (coupled) spontaneous changes.

Many physical and biological processes occur in nonequilibrium, open systems with irreversible changes, such as the transport of matter, energy and electricity, nerve conduction, muscle contractions, and complex coupled phenomena. Many of these have coupled processes implying the driving and driven processes. The kinetic equations and statistical models can describe such processes satisfactorily. However, these procedures often require more detailed information, which may be only sometimes available. The nonequilibrium thermodynamics theory may be another useful approach to describe the physical and biological processes.

An evolved and adapted biological system converts energy in the efficient manner for transport of substances across a cell membrane, the synthesis and assembly of the proteins, muscular contraction, reproduction and survival. In this process the source of energy is adenosine triphosphate (ATP), which has been produced by oxidative phosphorylation in the inner membrane of the mitochondria. This is a coupled-membrane bound process.

There exist a large number of 'phenomenological laws' describing irreversible processes in the form of proportionalities, such as Fick's law between the flow of a substance and its concentration gradient, and the mass action law between the reaction rate and affinities. When two or more of these phenomena occur simultaneously in a system, they may couple; hence a flow can occur without or against its conjugate force. For example, in active transport a substrate can be transported against the direction imposed by its thermodynamic force. If the coupling does not take place, such "uphill" transport would be in violation of the second law of thermodynamics. As explained by the nonequilibrium thermodynamics theory, dissipation due to either diffusion or chemical reaction can be negative, only if these two processes couple in an anisotropic medium and produce a positive total dissipation.

The linear nonequilibrium thermodynamics theory is valid for near equilibrium systems in which the Gibbs free energy change is small that is $\Delta G <<$ 2.5 kJ/mol at room temperature and atmospheric pressure; linear relationships exist between the flows and forces identified by an appropriate dissipation function, which is obtained from the Gibbs relation and the general transport equations of mass, momentum, energy and entropy balances. The matrix of coefficients of the linear phenomenological equations becomes symmetric according to the Onsager reciprocal relations. The cross coefficients naturally relate the coupled flows. This theory does not require the detailed mechanisms of biological process, while a complete analysis requires a quantitative description of the mechanisms of energy conversion.

This chapter includes a short description of the mitochondria and energy transduction in the mitochondrion. The proper pathways and the study of multi-inflection points in bioenergetics are summarized. We also summarize the concept of thermodynamic buffering caused by soluble enzymes and some important processes of bioenergetics using the linear nonequilibrium thermodynamics formulation.

1. MITOCHONDRIA

Mitochondria are organelles typically ranging in size from 0.5 micrometer to 1 micrometer in length, found in the cytoplasm of eukaryotic cells. Mitochondria contain the inner and outer membranes, separated by a space. Both the inner and outer membranes are constructed with tail-to-tail bilayers of phospholipids into which mainly hydrophobic proteins are embedded. One portion of the lipid molecule is hydrophilic (water-attracting) and the other portion is hydrophobic (lipid-attracting). As a result of this unique character, lipids spontaneously form a bimolecular lipid bilayer in aqueous solution. The self-assembled lipid bilayer is in a dynamic and liquid-crystalline state. The outer membrane contains proteins and lipids. The smooth outer membrane holds numerous transport proteins, which shuttle materials in and out of the mitochondrion The outer membrane is 60-70 Å thick and permeable to small molecules including salts, adenine and nicotinamide nucleotides, sugars and coenzyme. The inner membrane contains all the enzymes and less lipid than the outer membrane. These membranes produce two separate compartments creating the intermembrane space (C-side) and the space enclosed by the inner membrane called matrix (M-side) (Fig.1). The intermembrane space is usually 60-80 Å in width and contains some enzymes. The matrix however is very viscous and rich in protein, enzymes and fatty acids. A membrane component exhibits allotropy and changes its property when separated. Experimental evidence shows that the mitochondria exhibit anisotropy.

The inner membrane houses the respiratory chain and ATP synthesis, and is permeable to small neutral molecules such as water, oxygen, and carbon dioxide. Its permeability to charged molecules such as proton and ions is limited. The inner membrane has numerous folds called *cristae*, which have folded structure greatly increasing the surface area where ATP synthesis occurs. Transport proteins, molecules called electron transport chains, and enzymes that synthesize ATP are among the molecules embedded in the cristae (Figs. 1 and 2). The cristae have the major coupling factors F_1 (a hydrophilic protein) and F_o (a hydrophobic lipoprotein complex), which together comprise the ATPase complex activated by Mg^{+2}. ATPase catalyses hydrolysis of ATP to adenosine diphosphate (ADP) and phosphate, while ATPsynthase produces ATP using the energy released by the redox reactions of the respiratory chain. Both reactions are inhibited by the antibiotics such as oligomycin.

Mitochondria contain deoxyribonucleic acid (DNA) and ribosomes, protein-producing organelles in the cytoplasm. Within the mitochondria, the DNA directs the ribosomes to produce proteins as enzymes, or biological catalysts, in ATP production. Mitochondria are responsible for converting nutrients into the energy-yielding ATP to power the cell's activities. The number of mitochondria in a cell depends on the cell's function. Cells with particularly heavy energy demands, such as muscle cells, have more mitochondria than other cells.

The main function of the mitochondria is to provide energy for cellular activity by the process of aerobic respiration. In this process, glucose is broken down in the cell's cytoplasm to form pyruvic acid, which is transported into the mitochondrion. In a series of reactions, part of which is called the citric acid cycle or Krebs cycle, the pyruvic acid reacts with water to produce carbon dioxide and hydrogen atoms. These hydrogen atoms are transported with special carrier molecules called coenzymes to the cristae, and some eventually combine with oxygen to form water.

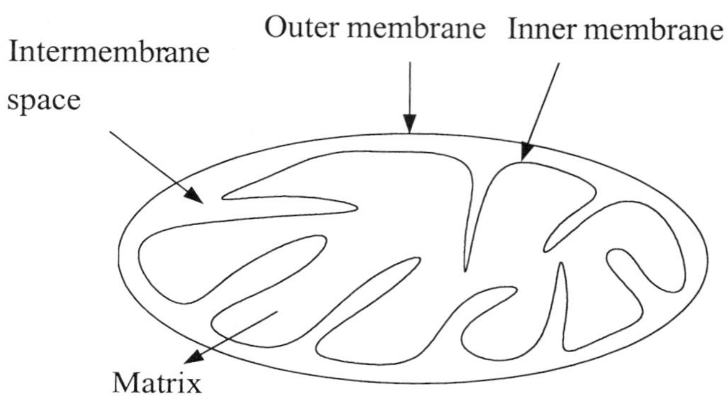

Fig. 1. Structure of the mitochondria.

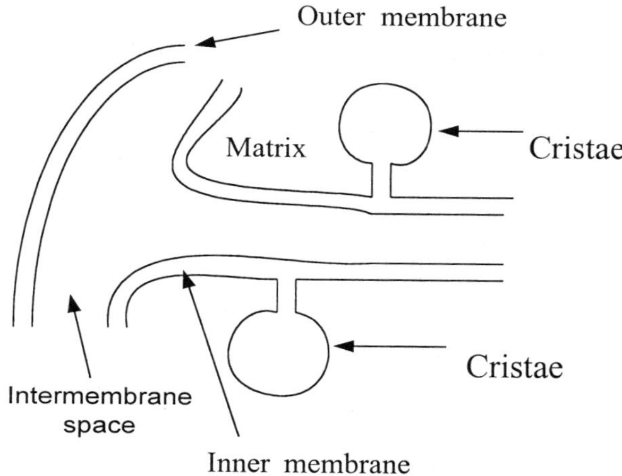

Fig.2. Inner membrane structure of the mitochondria.

The electrons flow from the coenzymes down to the oxygen atoms, and protons are pumped from the matrix to the intermembrane space. When the protons flow back into the matrix, a phosphate group is added to ADP to form ATP, which is transported to the cytoplasm of the cell and hydrolyzed into ADP for virtually every energy-requiring reaction and process. ADP is returned to the mitochondrion to be reutilized.

Food containing sugars or carbohydrates is converted to basic chemicals that the cell can use. Sugars are broken down by enzymes into the glucose, which is broken down further to make ATP in two pathways. The first pathway is called glycolysis, which occurs in the cytoplasm outside the mitochondria, and requires no oxygen. During glycolysis, glucose is broken down into pyruvate, which is a 3-carbon molecule. After it enters the mitochondria, it is broken down to a 2-carbon molecule by a special enzyme, and carbon dioxide is released. The 2-carbon molecule is called Acetyl CoA and it enters the Kreb's cycle by joining to a 4-carbon molecule called oxaloacetate. Once the two molecules are joined, they produce citric acid. This is the first reaction that makes citric acid, and the citric acid cycle gets its name from that. Only 4-ATP molecules can be produced by one molecule of glucose.

There are two types of carrier molecules for the electrons: one is called the nicotinamide adenine dinucleotide (NAD^+) and the other is called the flavin adenine dinucleotide (FAD^+). The third molecule, of course, is oxygen. Eventually, the process produces the 4 carbon oxaloacetate again, and is called a cycle, because it ends up always where it started with oxaloacetate available to combine with more acetyl CoA. Oxidation of NADH and the flow of electrons through the electron transport system leads the transfer of protons from the matrix into the intermembrane space. This creates the vital proton electrochemical gradient to power the synthesis of ATP.

Membrane proteins transfer material and information between the cells and their environment and between the compartments housing the organelles. Some of these proteins selectively transport specific molecules and ions, and some others are receptors for chemical signals from outside the cell. They can support the transport of ion and electron, and the energy conversion and conservation. They act as transducers capable of gathering information, processing it, and delivering a response. This indicates the electronic and molecular character of their functions. Their electrical activities are measurable as an electric potential difference across the membrane. Changes in the membrane permeability would yield a change in the potential difference. In the cotransport system the movement of one permeant is dependent on the simultaneous movement of a different permeant either in the same direction called the *symport* or in the opposite direction called the *antiport*. The best-known antiport system is the Na^+/K^+-ATPase pump that is present in the plasma membrane of all animal cells. This pump transports sodium ions out of the cell and potassium ions into the cell through the lipid bilayer against their electrochemical potential gradients, and operates as an antiport. This antiport, like every active transport, needs to couple to a dissipative process in the metabolic activity, like ATP hydrolysis. Transmembrane activities are thermodynamically driven by the gradients of chemical and electrochemical potentials. The metabolic processes are generally able to maintain steady nonequilibrium conditions across the cell membranes by generating the flows of ions or electrons. Therefore transport and rate equations occur in the formulation of metabolic activities.

Cytoplasm houses many metabolic cycles and synthetic pathways, as well as the protein synthesis. Beside the matter and information transfer across the cell membrane, there is the essential interaction of living bodies with the external surroundings. A coordinated body action requires integration of respiratory, nerve, sensing, muscle, etc. At the cellular level, communications via the membrane are called the *signal transduction*, and facilitated with the ligands or messengers, such as proteins, peptide hormones. These ligands facilitate the communication by directly entering the cell, or interacting with a specific receptor situated on/in the lipid bilayer of the membrane.

Electron transfer and associated reactions leading to the ATP synthesis are completely membrane-bound. Photosynthetic energy conservation occurs in the thylakoid membrane of plant chloroplasts; oxidative phosphorylation takes place in the mitochondrial inner membrane. These membranes facilitate the interactions between the redox system and the synthesis of ATP, and are referred to as coupling membranes. The coupling mechanisms of oxidative phosphorylation may change during development. Firstly, the membrane is an efficient and regulated energy-transducing unit as it organizes the redox systems and associated enzymes. Secondly, the membrane is a permeability barrier of the cell, controls the transport of certain solutes and the effects of osmotic imbalance.

2. BIOENERGETICS IN MITOCHONDRIA

Bioenergetics is concerned with energy metabolism in biological systems, which compromise the mechanism of energy coupling and control within the cells. The clusters of orthologous genes database has identified 210 protein families involved in energy production and conversion; the protein families show complex phylogenetic patterns and cause diverse strategies of energy conservation [37]. The optimum conditions for biological energy conversion do not necessarily imply the maximum thermodynamic efficiency; the rate of ATP production or maintenance of the cellular phosphate potential may be more important than the efficiency of the energy conversion process. Photosynthetic energy conservation occurs in the thylakoid membrane of plant chloroplasts.

Mitochondria are an efficient and regulated energy-transducing unit as it organizes the electron transfer and the associated reactions leading to the ATP synthesis. Cytchrome c oxidase, terminal enzyme of the chain, (i) reduces dioxygen to water with four electrons from cytochrome c and four protons taken up from the matrix of mitochondria, and (ii) pumps protons from the matrix into the intermembrane space, causing an electrochemical proton gradient across the inner membrane, which is used by ATPase for the synthesis of ATP [38]. There is also another cycle (the Q cycle) around cytochrome bc_1 complex, which causes a substantial proton pumping. Synthesis of ATP is an endergonic reaction, and hence conserves the energy released during biological oxidation-reduction reactions. Photosynthesis, driven by light energy, leads to the production of ATP, through the electron transfer and the photosynthetic phosphorylation. Hydrolysis of one mole of ATP is pH and $[Mg^{+2}]$ dependent, and is an exergonic reaction releasing 31 kJ/mol at pH = 7. This energy drives various metabolic reactions and the transport of various ions such as H^+, K^+, and Na^+ [37,38].

Mitochondria have different complexes for converting substrate (nutrient) to energy. Complex I performs the cellular NADH production from fatty acid oxidation, acetic acid cycle, and glycolsis. Complex II receives $FADH_2$ from succinate dehydrogenase and is dependent on the acetic acid cycle. Cristae have the major coupling factors F_1 and F_o protein parts, which together comprise the large ATPase protein complex. ATPase can catalyze the synthesis and the hydrolysis of ATP, depending on the change of electrochemical potential of proton $\Delta\bar{\mu}_H$. The measured ratio of ATP production to oxygen consumption, P/O ratio, can vary according to the various physiological processes: (i) maximizing the ATP production, (ii) maximizing the cellular phosphate potential, (iii) minimizing the cost of production, and (iv) combination of these three processes.

Living systems adjust the energy demands by coupling the respiration to the rate of ATP production and utilization [38]. The values of P/O change in the range of 1 to 3, and is characteristic of the substrate undergoing oxidation and the physiological organ role [66]. In the case of excess oxygen and inorganic phosphate, the respiratory activity of the mitochondria is controlled by the amount of ADP available. In the controlled state called the state 4, the amount of ADP is low. With the addition of ADP, the respiratory rate increases sharply; this is the active state called the state 3. The ratio of the respiratory rates of state 3 to state 4 is called the respiratory control index.

Mitochondria involve in the functioning of the cell, such as transport and regulation of Ca^{+2}, protein import, cell death and aging, and obesity. Mitochondria from different organ systems, such as liver, heart, and brain, display morphological and functional differences, and are capable of economizing energy production against energy demands of oxidative phosphorylation [66].

Three-dimensional images show that inner membrane involutions (cristae) have narrow and long tubular connections to the intermembrane. These openings lead to the possibility that lateral gradients of ions, molecules, and macromolecules may occur between the compartments of mitochondria. This type of structure may influence the magnitude of local pH gradients produced by chemiosmosis, and internal diffusion of adenine nucleotides. The information on the spatial organization of mitochondria is important to understand and describe the bioenergetics. The mitochondria have elongated tubes aligned approximately in parallel and are embedded in a multilamellar stack of endoplasmic reticulum, which could be related to specific function of the mitochondria [65].

Mitochondria are the major source of reactive oxygen species through the respiratory chain. These oxygen radicals may affect the function of the enzyme complexes involved in energy conservation, electron transfer and oxidative phosphorylation [45].

3. OXIDATIVE PHOSPHORYLATION

Oxidative phosphorylation occurs in the mitochondria of all animal and plant tissues, and is a coupled process between the oxidation of substrates and production of ATP. As the Kreb's cycle runs, hydrogen ions (or electrons) are carried by the two carrier molecules NAD or FAD to the electron transport pumps. The protons are pumped to the intermembrane region where they accumulate in a high enough concentration to phosphorylate the ADP to ATP.

According to the chemiosmotic coupling hypothesis, ATP synthesis decreases proton electrochemical gradient hence stimulates the respiratory chain to pump more protons across the mitochondrial inner membrane and maintaine the gradient. However, electron supply to the respiratory chain also affects respiration and ATP synthesis. For example, calsium stimulates mitochondrial matrix dehydrogenase, and increases the electron supply to the respiratory chain and hence the rate of respiration and ATP synthesis [36].

Control of respiration and ATP synthesis shift as the metabolic state of the mitochondria changes. In state 4 of mitochondria, respiration is low, proton electrochemical gradient is high, and there is no ATP synthesis. However, in state 3 respiration is high, proton electrochmical gradient is lowered, and ATP synthesis is high. In an isolated mitochondria, the control over state 4 respiration is mainly due to the proton leak through the mitochondrial inner membrane. This type of control decreases as mitochondria change from the state 4 to state 3, and the control by the adenine nucleotide and the dicarboxylate carriers, cytochrome oxidase, the ATP utilizing reactions, and transport activities increase. Therefore in state 3, most of the control is due to respiratory chain and substrate transport [36].

It is generally assumed that 2,4-dinitrophenol acts as protonophores, and carries the proton by increasing the proton conductance of the inner membrane. This regulates coupling efficiency between the ATP synthesis and the respiratory rate. According to the chemiosmotic theory, the proton pump is connected by a proton electrochemical potential difference between intermembrane space and matrix. The electrons are carried from complex to complex by ubiquinone and cycochrome c. The ATP synthase uses the proton gradient to form ATP from ADP and phosphate. The cristae house and organize the electron transport chain and the ATP pumps. A terminal enzyme in the respiratory chain, cytochrome c oxidase, reduces oxygen to water. Thus, each compartment in the mitochondrion is specialized for one phase of these reactions (Fig. 3).

Without oxygen only 4 molecules of ATP energy packets are produced for each glucose molecule (in glycolysis). Oxidative phosphorylation produces 24-28 ATP molecules from the Kreb's cycle from one molecule of glucose converted into pyruvate.

Two theoretical approaches applied to the oxidative phosphorylation are metabolic control analysis and nonequilibrium thermodynamics. These approaches are helpful for quantitative description and understanding of control and regulation of the oxidative phosphorylation [36,46]. For example metabolic control theory can provide a quantitative decription for the microbial growth [79].

The application of nonequilibrium thermodynamics is one of the early attempts to fornulate oxidative phosphorylation in a quantitative manner. This application assumes a linear flow-force relationships for the oxidation and phosphorylation flows. Such a linear dependence between the flows and thermodynamic forces has been established by measurements during state 4 to state 3 transistion as a linear part of a more general sigmodial relationship. Nonequilibrium thermodynamics has proved to be useful especialy in describing the energetics aspects of oxidative ATP production and the transport of substrates coupled to ATP hydrolysis.

The thermodynamics approach is an effort of a unified description of a system. However, living systems are complex, and it is not realistic to assume that a simple mathematical formalism can yield a complete description of such complex and coupled systems. Beside thermodynamics tool, mathematical tools such as metabolic control analysis can also be useful to interpret the measured properties of oxidative phosphorylation. A kinetic model of oxidative phosphorylation may not be fully and definetely completed, because of the several assumptions and simplifications associated with it. A proper kinetic approach can relate macroscopic behavior to microscopic properties, and hence allows a deeper insight into the mechanisms related to the control and regulation of oxidative phosphorylation; it may provide a modeling procedure and methodological approach to describe dynamic and stationary properties of energy coupling in membranes [104]. The application of metabolic control analysis to mitochondria may describe how control is distributed throughout the mitochondria, but does not predict how the systems are regulated, which may be improved by developing of a regulation analysis [36].

The linear nonequilibrium thermodynamic theory and quasi-linear flow-force relations may be useful for decribing a simplified example of oxidative phosphorylation [69], and the dissipation function is given by

$$\Psi = J_P A_P + J_H \Delta\mu_H + J_O A_O \tag{1}$$

Here the subscripts *P*, *H*, and *O* refer to the phosphorylation, H^+ flow, and substrate oxidation respectively, and $\Delta\mu_H = \Delta\mu_H^i - \Delta\mu_H^o$. We consider only steady states. The dissipation function can be transformed as

$$\Psi = J_P A_P^{\text{ex}} + J_H \Delta\mu_H + J_O A_O^{\text{ex}} \tag{2}$$

Where A^{ex} is the external affinity. When the interior of the mitochondrion is in a stationary state, it suffices to measure the changes in the external solution only.

From Eq. (2) the appropriate phenomenological equations in terms of the resistance formulations are expressed by

$$A_P^{\text{ex}} = K_P J_P + K_{PH} J_H + K_{PO} J_O \tag{3}$$

$$\Delta\mu_H = K_{PH} J_P + K_H J_H + K_{OH} J_O \tag{4}$$

$$A_O^{\text{ex}} = K_{PO} J_P + K_{OH} J_H + K_O J_O \tag{5}$$

For the oxidative phosphorylation we have the three degrees of couplings, q_{PH}, q_{OH}, and q_{PO},. If A_o^{ex} is kept constant and $\Delta\mu_H$ is not controlled, $J_H = 0$ in the stationary state, which is also called the *static head*, and Eqs. (3) and (5) become

$$A_P^{\text{ex}} = K_P J_P + K_{PO} J_O \tag{6}$$

$$A_O^{\text{ex}} = K_{PO} J_P + K_O J_O \tag{7}$$

Consequently, the degree of coupling is given by.

$$q = -\frac{K_{PO}}{(K_P K_O)^{1/2}} \tag{8}$$

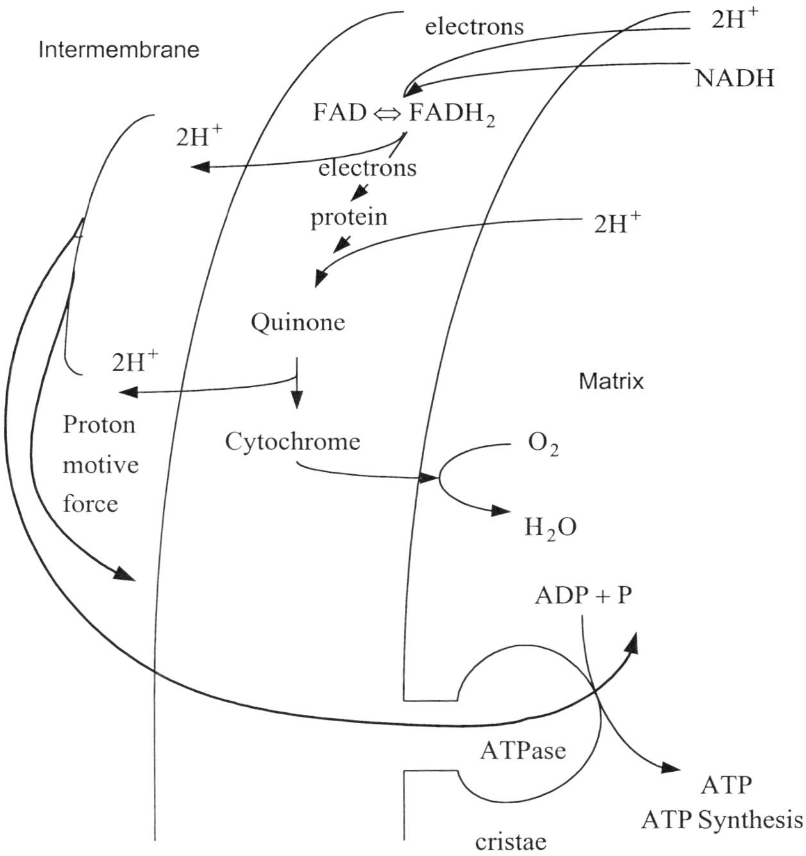

Fig.3. Electron transport in phosphorylation. [43]

When we have the *level flow*, the force vanishes, $\Delta\mu_H = 0$, and Eqs. (3) to (5) give

$$A_P^{ex} = K_P(1-q_{PH}^2)J_P - (K_P K_O)^{1/2}(q_{PO} + q_{PH}q_{OH})J_O \tag{9}$$

$$A_O^{ex} = -(K_P K_O)^{1/2}(q_{PO} + q_{PH}q_{OH})J_P + K_O(1-q_{OH}^2)J_O \tag{10}$$

where the term q is given by

$$q = \frac{q_{PO} + q_{PH}q_{OH}}{(1-q_{PH}^2)(1-q_{OH}^2)} \tag{11}$$

When the rate of performance of electroosmotic work is appreciable, we can define the effectiveness of energy conversion, which is expressed by

$$\eta = -\frac{J_P X_P}{J_r A} \tag{12}$$

where X_p is the force for proton transportation.

It is also useful to consider the force developed per given rate of expendeture of metabolic energy, which is caled the efficiacy of force

$$\varepsilon_{Xp} = -\frac{X_P}{J_r A} \tag{13}$$

4. PROPER PATHWAYS

Linear phenomenological equations require that the flows are zero at equilibrium, and each flow depends on all forces, or each force depends on all flows. For a steady state far from equilibrium, there may be pathways in the vicinity of this state along which the forces are varied linearly to lead a linear flow-force relation. Caplan and Essig [2] elaborated the matter both for uncoupled and coupled processes. The linearity of flows observed in the experimental studies of active transport in epithelia suggests the existence of proper pathways where the phenomenological coefficients become nearly constant.

Formulation of the relationships between forces and flows is an important step in the theoretical and experimental analysis of biological reactions and transport processes. This step will lead to understanding the change of affinity of a reaction driving the transepithelial active transport, tissue anisotropy (compartmentalization), free energy, and activity. Experiments show that biological processes take place in many steps, each of which is thought to be nearly reversible, and exhibit linear relationship between the steady-state flows and the conjugate thermodynamics forces, such as the transepithelial active Na^+ and H^+ transports and the oxidative phosphorylation in mitochondria [2,8,53,69]. The linear relations between the rate of respiration and specific growth rate are observed for many microbial systems [2].

Conventional phenomenological equations may constitute an incomplete description of the processes, because the forces can be controlled in various ways; for example, it may be possible to find a proper pathway leading to quasi-linear force-flow relationships so that the theory of linear nonequilibrium thermodynamics can be applied. For a first order reaction S $\rightarrow$ P, doubling the concentrations of S and P will double the reaction rate for an ideal system, although the affinity remains the same. In this case a distinction must be made between thermodynamic and kinetic linearity. Proper pathways are associated with the thermodynamic linearity. The rate of a process depends not only on the force also on the reference state; the flow of a solute across a membrane depends on its chemical potential and also on its thermodynamic state on both sides of the membrane.

The constancy of phenomenological coefficients L may be accomplished by the appropriate constraints to vary the force X in the relationship $J = LX$. The values of L reflect the nature of the membrane, and can control the force X. If a homogeneous thin membrane is exposed to the same concentrations at each surface, flow is induced solely by the electric potential difference, and L is constant with the variation of X. However, if X is the chemical potential difference, dependent upon the bath solute concentrations, then the L becomes

$$L = \frac{\alpha u c_m}{\Delta z} \tag{14}$$

where u is the mobility, α is the solvent-membrane partition coefficient, and c_m is the logarithmic mean bath concentration, $\Delta c / \Delta \ln c$. If a value of c_m is chosen and the concentrations are then constrained to the locus $\Delta c = (c_m)\Delta \ln c$, then L becomes constant. The logarithmic mean concentration can be used in the linear formulation of membrane transport [2,44]. If the force is influenced by both the concentrations and the electrical potential difference, then L becomes more complex, yet it is still possible to obtain a constant L by measuring J and X in a suitable experiment [4].

For a first order chemical reaction of S $\rightarrow$ P, the reaction rate is given by

$$J_r = k_f c_S - k_b c_P = k_b c_P (e^{A/RT} - 1) \tag{15}$$

where k_f and k_b are the rate constants for forward and backward reactions respectively. At steady state, far away from equilibrium, the reaction rate may be

$$J_r = LA = \frac{L^* A}{RT} \tag{16}$$

where $L^* = TRL$, and can be evaluated my measuring J_r and A

$$L^* = k_b c_P \left(\frac{A/RT}{e^{A/RT} - 1} \right)^{-1} \tag{17}$$

Eq. (16) shows that for different values of A at various stationary states, the same values of L^* will describe the chemical reaction by choosing the concentrations appropriately. For a specified value of A, Eqs. (15) and (16) determine c_P and the ratio of c_P/c_S respectively, and a constant L can be found by limiting the c_P and c_S to an appropriate locus. As the system approaches equilibrium, A tends to vanish and $k_b c_P$ approaches the value L^*. This procedure can also be used in more complex reaction systems.

Proper pathways can be identified in the neighborhood of a reference steady state far from equilibrium by varying the forces X_1 and X_2 in such a manner to lead to linearity of flows and forces. Highly coupled systems show similarity to a single uncoupled flow, and the linear dependencies on conjugate and nonconjugate forces exist. In the vicinity of static head, where the transmembrane flows are zero, linearity would be expected when the degree of coupling is close to unity. Sometimes kinetic nonlinearity may occur because of a feedback and not by the large affinities, and the sustained oscillations may occur near equilibrium [18].

Beside the forces controlling, alternatively flows controlling for a certain pathway is also important. The *metabolic control analysis* can be used to evaluate the flow control within the biochemical pathways, and provide information on the regulation of pathway fluxes. From the measurements of fluxes exchanged through the cell membrane, it is possible to quantify the pathway fluxes and alternative pathways to the same metabolite with the metabolic flux analysis [58]. The methodology is limited only to the analysis of simple two-step pathways, although the larger pathways can be lumped into two overall changes. The influence of the individual reaction rates (enzyme activities) on the overall flux through the pathway is called the flux control coefficients C_r, which are expressed by

$$C_{r,i} = \frac{\partial \ln(J)}{\partial \ln(J_{r,i})} \qquad (i=1,\ldots,m) \tag{18}$$

where J is the steady-state flux through the pathway, $J_{r,i}$ is the rate of ith reaction, and m is the enzymic steps. The flux control coefficients are related to the *elasticity coefficients* ε as follows

$$\sum_{i=1}^{m} \varepsilon_{i,j} C_{r,i} = 0 \qquad (j=2,..,m) \tag{19}$$

where the *elasticity coefficients* are defined by

$$\varepsilon_{i,j} = \frac{\partial \ln(J_{r,i})}{\partial \ln(c_j)} \qquad (j = 2,..,m)\ (i = 1,..,m) \tag{20}$$

Determination of flux control coefficients is difficult, and requires the independent variation of the activity of all the enzymes within the pathway. On the basis of linear nonequilibrium thermodynamics, the kinetics of enzyme reactions can be described by the linear functions of the change in Gibbs free energy. This yields a direct relation between the elasticity coefficients and the change in Gibbs free energy for the reactions in a simple two-step pathway.

5. MULTIPLE INFLECTION POINTS

A common intermediate of two enzymes may catalyze consecutive reactions of a pathway, and diffuse from one active center to the other without dissociation. This is called *metabolic channeling*, which could lead to a decrease in the steady-state concentration of the intermediate metabolite even at constant flux [82].

Rothschild et al. [19] found the existence of a multidimensional inflection point well outside of the equilibrium in the force-flow space of enzyme-catalyzed reactions indicating linear behavior between the logarithm of reactant concentrations and enzyme-catalyzed flows. Thus enzymes, operating near this multidimensional point and leading to a particular choice of the state variables, may produce some linear coupled biological systems. This range of kinetic linearity may be far from equilibrium. Neglecting the electrical effects, the conditions for the existence of a multidimensional inflection point are: (i) each reactant with varying activity influences the transition rates for leaving one state only; (ii) the kinetics of the reaction involving the given reactant are of fixed order with respect to that reactant; and (iii) for various concentrations of reactants, at least a certain cycle is present containing only that combination and no others. The first condition excludes the autocatalytic systems, however for many biological energy transducers it may well be satisfied.

Since the local asymptotic stability is supported by the local symmetry, and experimental evidence of the linear behavior of some highly coupled biological energy-transducing systems suggest that the kinetic linearity may lead to the thermodynamic linearity and cause a proper pathway.

Consider an ensemble of enzyme molecules or membrane proteins in the coupled processes of reactions and vectorial flows. Such systems consist of a set of cycles and subcycles due to cyclic flows that relate to the coupled processes. For the cycle flux in cycle k J_k ($k = a,b,..,h$), the first two steady-state fluxes are given by the following relations

$$J_1 = J_a + J_b + J_f \tag{21}$$

$$J_2 = J_a + J_b + J_c + J_g \tag{22}$$

Expanding these two flows as functions of their conjugate forces in Taylor series about some reference steady state, and assuming all other forces as constant, yields the finite differences from the first order terms

$$\delta J_i = \left(\frac{\partial n \sum\limits_{k(i)} J_k}{\partial X_i} \right) \delta X_i + \left(\frac{\partial n \sum\limits_{k(i)} J_k}{\partial X_j} \right) \delta X_j + .. \tag{23}$$

$$\delta J_j = \left(\frac{\partial n \sum\limits_{k(j)} J_k}{\partial X_i} \right) \delta X_i + \left(\frac{\partial n \sum\limits_{k(j)} J_k}{\partial X_j} \right) \delta X_j + .. \tag{24}$$

where $k(i)$ and $k(j)$ show the sets of cycles associated with J_i and J_j, respectively. The expansion of flows in a Taylor series about a multidimensional inflection point yields expressions linear in $\ln(c_i)$ and $\ln(c_j)$ up to third order in either one if changing concentrations of components i and j are the reactant concentrations. In Eqs. (23) and (24) the proper conjugate forces, X_i and X_j, appear explicitly. The reference state may not be an inflection point except with respect to c_i and c_j

Caplan and Essig [2] provided a simple model of active ion transport, having properties consistent with a multidimensional inflection point when one of the variables was the electrical potential difference across the membrane.

A multiple inflection point may not be unique; other conditions may exist where flows J_1 and J_2 simultaneously pass through an inflection point on variation of X_1 with constant X_2, and vice versa. It is frequently not possible to vary both forces independently in biological systems. However, if X_1 can be controlled experimentally along a proper pathway while X_2 is kept constant, the response of the flows to change in X_1 will permit a thermodynamic characterization of the system. For highly coupled biological systems approximately satisfy the Onsager relations [2,8].

Stucki [8] demonstrated that in mitochondria, variation of the phosphate potential, while maintaining the oxidation potential constant, yields linear flow-force relationships. Extensive ranges of linearity are found for the reaction driven active sodium transport in epithelial membranes, where the sodium pump operates close to a stationary state with zero net flow [8]. In the vicinity of such stationary state, kinetic linearity to a limited extent simulates thermodynamic linearity at the multidimensional inflection point. There may be a physiological advantage in the near linearity for a highly coupled energy transducer at the multiple inflection point, since the local asymptotic stability is guaranteed by these conditions. This could be achieved by the thermodynamic regulation (buffering) of enzymes, and could be interpreted that the intrinsic linearity would have an energetic advantage and may have emerged as a consequence of evolution. The thermodynamic approach may provide a theoretical framework for understanding the effects of changes in the environment on the stability of cellular systems

6. COUPLING IN MITOCHONDRIA

A two-flow coupling implies an interrelation between flow i and flow j, so that a flow can take place without a force or against its conjugate driving force. In a well-known biological activity of active transport, an ion flows against its electrochemical potential gradients due to interactions with metabolic chemical reactions. Active transport is a universal property of the cells and tissues; almost all cells maintain a high concentration of K^+ and a low concentration of Na^+ inside. The Na^+/K^+ pump actively transports $3Na^+$ out and $2K^+$ into the cell against their respective electrochemical potential gradients. The Na^+/K^+ pump is powered by the energy resulting from the hydrolysis of ATP. This phenomenon has been long observed in especially the nerve and muscle cells [2].

In the early experimental work, inside and outside cell concentrations of K^+, Na^+ and Cl^- ions were measured and compared against with those obtained from the Nernst potentials, given by

$$ {}_j = \left(\frac{RT}{z_j F}\right) \ln\left(\frac{a_{j,i}}{a_{j,o}}\right) \tag{25} $$

where ψ_j is the equilibrium potential, z_j the valence, and a_j the activity of the species j in both sides of the membrane (i and o). The measured potentials were different from the Nernst potentials indicating that a cell maintained the concentration difference at a steady state diffusional flow. The basis of an active transport is the coupled metabolic reaction to external diffusion in the animals, while most of the chloride flow in plant cells depends on the photosynthesis.

One of the conventional methods for establishing the existence of active transport is to analyze the effects of metabolic inhibitors. The second is to correlate the level or rate of metabolism with the extent of ion flow or the concentration ratio between the inside and outside of cells. The third is to measure the current needed in a short-circuited system having similar solutions on each side of the membrane; the measured flows contribute to the short-circuited current. Any net flows detected should be due to the active transport, since the electrochemical gradients of all ions are zero ($\Delta\psi = 0$, $c_o = c_i$). Experimental analysis indicates that the level of sodium ion within the cells is low in comparison with potassium ions. Kedem [44] presented the earlier analysis of the interactions between the flow and the chemical reactions by the nonequilibrium thermodynamics approach.

The generalized force of chemical affinity shows the distance from equilibrium of the ith reaction

$$X_i = RT \ln \frac{K_{i,\text{eq}}}{\prod_{j=1}^{m} c_j^{\nu ji}} \tag{26}$$

where R is the gas constant, K is the equilibrium constant, c_j is the concentration of the jth chemical species and ν_{ji} are the stoichiometric coefficients, negative for reactants and positive for products, for the ith reaction.

The phosphate potential in mitochondria is expressed as

$$X_i = -\Delta G_p^o - RT \ln[ATP/(ADP * P)] \tag{27}$$

Stucki [8] applied the linear nonequilibrium thermodynamics theory to the process of oxidative phosphorylation without the need of complex enzyme mechanisms. This analysis is based on the linear relations between the flows and forces, and the phenomenological coefficients, which obey the Onsager relations. The cross-phenomenological coefficients $L_{ij} (i \neq j)$ reflect the coupling effect. This approach enables one to assess the oxidative phosphorylation with H^+ pumps as a process driven by respiration by assuming that the steady state transport of ions. A set of representative linear phenomenological relations are given by

$$J_1 = L_{11}X_1 + L_{12}X_2 \tag{28}$$

$$J_2 = L_{12}X_1 + L_{22}X_2 \tag{29}$$

where J_1 is the net flow of ATP, J_2 is the net flow of oxygen, X_1 is the phosphate potential as given by Eq. (27), and X_2 is the redox potential, which is the difference in redox potentials between electron-accepting and electron-donating redox couples. Stucki [8] showed the linearity of oxidative phosphorylation experimentally within the practical range of phosphate potentials.

The degree of coupling is defined as

$$q = \frac{L_{12}}{(L_{11}L_{22})^{1/2}} \qquad 0 < |q| < 1 \tag{30}$$

and it indicates the extent of overall coupling for the various individual degrees of coupling of the different reactions driven by the respiration in the mitochondria.

With an angle α whose *sine* is q, we have

$$\alpha = \arcsin(q) \qquad \pi < \alpha < \pi/2 \tag{31}$$

By defining the phenomenological stoichiometry Z

$$Z = \left(\frac{L_{11}}{L_{22}}\right)^{1/2} \tag{32}$$

and by dividing Eq. (28) by Eq. (29), we can obtain the flow ratio $j = J_1/(J_2 Z)$ in terms of the force ratio $x = X_1 Z / X_2$ as follows

$$j = \frac{x+q}{qx+1} \tag{33}$$

Assuming that the oxidation drives the phosphorylation then $X_1 < 0$ and $X_2 > 0$, and J_1/J_2 is the conventional P/O ratio, while X_1/X_2 is the ratio of phosphate potential to the applied redox potential. The following relations are from Stucki [8]. At *static head* (sh), analogous to an open circuited cell, the net rate of ATP vanishes, and the rate of oxygen consumption and the force (the phosphate potential) are expressed in terms of α as follows

$$(J_2)_{\mathrm{sh}} = L_{22} X_2 \cos^2 \alpha \tag{34}$$

$$(X_1)_{\mathrm{sh}} = -\frac{X_2 \sin \alpha}{Z} \tag{35}$$

where L_{22} is the phenomenological conductance coefficient of the respiratory chain. Therefore energy is still converted and consumed by the mitochondria. Nonequilibrium phosphate value is a distant from the equilibrium value ($X_1 = -X_2 Z$) by a factor of q, and given by

$$(X_1)_{\mathrm{sh}} - (X_1)_{\mathrm{eq}} = \frac{X_2}{Z}(1-q) \tag{36}$$

The dissipation at the static head can be obtained from

$$\frac{\Psi_{\mathrm{sh}}}{T} = L_{11}\left(\tan^2 \alpha\right)^{-1} (X_1)_{\mathrm{sh}}^2 \tag{37}$$

Eq. (37) shows that the energy needed at the static head is a quadratic function of the phosphate potential.

At *level flow* (lf), analogous to a short-circuited cell, the phosphate potential vanishes. Hence no net work is performed by the mitochondria, and we have

$$\left(\frac{J_1}{J_2}\right)_{\mathrm{lf}} = qZ \tag{38}$$

which shows the maximal P/O ratio measurable in mitochondria at a zero phosphate potential. Eq. (38) also indicates that at level flow, flow ratio does not yield the phenomenological stoichiometry Z but approaches this value within a factor of q. Therefore, if the degree of coupling q is known, it is then possible to calculate Z from the P/O measurements at a closed-circuited cell.

Obviously in state 3 the phosphate potential is not zero, however for the values of q approaching unity, the dependence of flow ratio on the force ratio is weak according to Eq. (33). Therefore the state 3 is only an approximation to the level flow at values of q close to unity, and the dissipation function to maintain a level flow is given by

$$\frac{\Psi_{\mathrm{lf}}}{T} = \frac{(J_1)^2_{\mathrm{lf}}}{q^2 L_{11}} \tag{39}$$

The efficiency of substrate to energy conversion is related to the degree of coupling, and given by

$$\eta = -\frac{J_1 X_1}{J_2 X_2} = -\frac{x+q}{q+(1/x)} \tag{40}$$

The efficiency reaches a maximum value between the static head and the level flow, which is the function of degrees of coupling only, and expressed by

$$\eta_{\mathrm{opt}} = \frac{q^2}{\left[1+\left(\sqrt{1-q^2}\right)\right]^2} = \tan^2\left(\frac{\alpha}{2}\right) \tag{41}$$

The value of x at $\eta_{\max}$ is given by

$$x_{\mathrm{opt}} = -\frac{q}{1+\sqrt{1-q^2}} = -\tan\left(\frac{\alpha}{2}\right) \tag{42}$$

The dissipation function Ψ can be expressed in terms of force ratio x and degree of coupling q

$$\frac{\Psi}{T} = (x^2 + 2qx + 1)L_{22}X_2^2 \tag{43}$$

If we assume X_2 as constant, the dissipation function has a minimum at the static head force ratio

$$x_{sh} = -q \tag{44}$$

For linear phenomenological equations, the theorem of minimal entropy generation or the dissipation at steady state is a general evolution and stability criteria. Static head is the natural steady state where the net ATP flow vanishes, and minima of Ψ occur along the loci of the static head states

$$\frac{\Psi_{sh}}{T} = \cos^2 \alpha L_{22}X_2^2 \tag{45}$$

The dissipation at the state of optimal efficiency is obtained using x_{opt} in Eq. (43), and we have

$$\frac{\Psi_{opt}}{T} = 2\frac{\cos^2 \alpha}{1+\cos\alpha} L_{22}X_2^2 \tag{46}$$

The dissipation for the level flow is given by

$$\frac{\Psi_{lf}}{T} = L_{22}X_2^2 \tag{47}$$

Without a load Stucki [8] suggested the following order

$$\Psi_{sh} < \Psi_{opt} < \Psi_{lf} \tag{48}$$

This inequality means that the minimum dissipation and the natural steady state do not imply the optimal efficiency of oxidative phosphorylation.

At the level flow there is a load hence a load conductance corresponding to the state of optimal efficiency between the static head and the level flow. The dissipation of oxidative phosphorylation with a coupled process (load) utilizing ATP is given as follows

$$\frac{\Psi_c}{T} = J_1X_1 + J_2X_2 + J_3X_3 \tag{49}$$

Assuming that the ATP utilizing processes are driven by the phosphate potential, $X_3 = X_1$, and a linear relation between the net rate of ATP utilization and X_1, we have

$$J_3 = L_{33}X_1 \tag{50}$$

Here L_{33} is the phenomenological conductance of load, and the dissipation function in terms of the force ratio x becomes

$$\frac{\Psi_c}{T} = \left[x^2\left(1 + \frac{L_{33}}{L_{11}}\right) + 2qx + 1)\right]L_{22}X_2^2 \tag{51}$$

It can be found that only if the following equation is satisfied

$$\frac{L_{33}}{L_{11}} = \sqrt{1-q^2} = \cos\alpha \tag{52}$$

then Eq. (51) is minimal at x_{opt}.

For the mitochondria, L_{33} is an overall phenomenological coefficient lumping all the conductances of ATP utilizing processes, while L_{11} shows the conductance of phosphorylation. If these two coefficients match according to Eq. (52) then the natural steady of oxidative phosphorylation is at the optimal efficiency. Stucki [8] called Eq. (52) the condition of *conductance matching* of oxidative phosphorylation, and presented an experimental verification for perfused livers.

The dissipation with conductance matching at static head is given by

$$\frac{(\Psi_c)_{\text{sh}}}{T} = (\cos\alpha + \cos^2\alpha - \cos^3\alpha)L_{22}X_2^2 \tag{53}$$

The dissipation function at the state of optimal efficiency of oxidative phosphorylation is given by

$$\frac{(\Psi_c)_{\text{opt}}}{T} = \cos\alpha L_{22}X_2^2 \tag{54}$$

At the conductance matching state, the dissipation Eq. (51) is minimum at the loci of the optimal efficiency states. At the optimal efficiency the P/O ratio is given by

$$\left(\frac{J_1}{J_2}\right)_{\text{opt}} = \left(\frac{J_1}{J_2}\right)_{\text{lf}} \frac{1}{1+\cos\alpha} \tag{55}$$

Eq. (55) shows that unless $q = 1$, a maximal P/O ratio is incompatible with the optimal efficiency, and we have the inequality

$$\frac{1}{2}\left(\frac{J_1}{J_2}\right)_{\text{lf}} < \left(\frac{J_1}{J_2}\right)_{\text{opt}} < \left(\frac{J_1}{J_2}\right)_{\text{lf}} \tag{56}$$

Therefore, the low P/O ratios do not necessarily mean a poor performance of the oxidative phosphorylation. Similarly the net rate of ATP synthesis at optimal efficiency is given by

$$(J_1)_{\text{opt}} = (J_1)_{\text{lf}} \frac{\cos\alpha}{1+\cos\alpha} \tag{57}$$

with the boundaries

$$0 < (J_1)_{\text{opt}} < \frac{1}{2}(J_1)_{\text{lf}} \tag{58}$$

This inequality means that a maximal net rate of ATP production is incompatible with the optimal efficiency. Cellular pathways balance the rate and efficiency of ATP production with respect to the energy needs of the cell. For example, heart and brain mitochondrial systems utilize more oxygen and produce ATP at a faster rate compared with the systems in liver. However, liver mitochondria can produce ATP more efficiently based on a higher P/O ratio and a higher degree of coupling in oxidative phosphorylation [66].

Cairns et al. [66] reported an experimental mean degree of coupling for isolated liver mitochondria as 0.955, which is close to the value of 0.952 found by Sobol and Stucki [15] using the isolated perfused whole liver from fasted rats. Under similar cytoplasmic ATP levels, the ATP utilization for muscular contraction and the ion transport would be much higher in a beating heart than an arrested heart. ADP supply to the mitochondria of the beating heart would lead to the higher rates of oxidative phosphorylation and ATP production.

As Eq. (57) shows that when the degrees of coupling is close to unity, $(J_1)_{\text{opt}}$ vanishes. For a favorable ATP production at optimal efficiency of oxidative phosphorylation, we should have $q < 1$. Therefore it is important to explore the relationships between the optimized production functions and the degree of coupling of oxidative phosphorylation. This can be based on the thermodynamic tradeoff between reduced efficiency and increased output or between increased efficiency and optimized output. The specific degrees of coupling correspond the following set values: q_f is the maximum net output flow of ATP at optimal efficiency, q_p is the maximal net output power, q_f^{ec} is the economic net output flow, and q_p^{ec} is the economic net output power at optimal efficiency.

With the consideration of conductance matching Stucki [7,8] determined four such production functions, which are given in Table 1.

Table 1

Production functions with the consideration of conductance matching [4,7,8]. Reprinted with permission from Elsevier, Biophysical Chemistry 97 (2002) 87.

Production Function	Loci of the Optimal Efficiency States	q	With Energy Cost
1. Optimum rate of ATP production: $J_1 = (q+x)ZL_{22}X_2$ (59)	From the plot of J_1 vs x: $(J_1)_{opt} = \tan(\alpha/2)\cos\alpha ZL_{22}X_2$ (60)	$q_f = 0.786$ α = 51.83°	No η= cons.
2. Optimum output power of oxidative phosphorylation: $J_1X_1 = x(x+q)L_{22}X_2^2$ (61)	From the plot of J_1X_1 vs x: $(J_1X_1)_{opt} = \tan^2(\alpha/2)\cos\alpha L_{22}X_2^2$ (62)	$q_p = 0.910$ α = 65.53 °	No η= cons.
3. Optimum rate of ATP production at minimal energy cost: $J_1\eta = -\dfrac{x(x+q)^2}{xq+1}ZL_{22}X_2$ (63)	From the plot of $J_1\,\eta$ vs. x: $(J_1\eta)_{opt} = \tan^3(\alpha/2)\cos\alpha ZL_{22}X_2$ (64)	$q_f^{ec} = 0.953$ α = 72.38 °	Yes
4. Optimum output power of oxidative phosphorylation at minimal energy cost: $J_1X_1\eta = -\dfrac{x^2(x+q)^2}{qx+1}L_{22}X_2^2$ (65)	From the plot of $J_1\,X_1\,\eta$ vs x: $(J_1X_1\eta)_{opt} = \tan^4(\alpha/2)\cos\alpha L_{22}X_2^2$ (66)	$q_p^{ec} = 0.972$ α = 76.34 °	Yes

6.1. Variation of coupling

The degrees of coupling depend on the nature of the output required from the energy conversion system in the mitochondria. For an optimal efficiency state, the condition of conductance matching must also be satisfied. Experiments with liver perfused at a metabolic resting state suggest that the conductance matching is satisfied on a time average, and the degree of coupling q^{ec} yields an economic process of oxidative phosphorylation. Optimization may be based on other constraints different from the efficiency, such as the production of thermal energy in the mitochondria, which requires low degrees of coupling.

Mitochondria in the liver are optimized for the maximum production of ATP at minimum cost regardless the cellular phosphate potential. In the heart and brain, the experimental value of q for cellular respiration pathway is close to the value of $q_f^{ec} = 0.953$, which suggests that the pathway is optimized to ATP economically for the cellular processes. In the brain, the coupling of the acetic acid cycle approaches to $q_p = 0.91$ suggesting a maximized cellular energy state. However, in the heart, the acetic acid cycle coupling is 0.786, which is between q_p and q_f, and consistent with the maximum ATP production necessary for preserving the cellular energy state [66].

Optimal flow ratios are also a characteristic of oxidative phosphorylation, and may provide additional information into the relationships between respiratory response and energy demand simulation by ADP. Most metabolic processes in living cells are dynamic systems, and the behavior of flows may reflect complex system mechanisms than do the models dependent on end-point measurements. For example, the ratio of ADP/O describes the state of the end-point capacity of oxidative phosphorylation based on the input flow of ADP [66].

Figure 4 shows the effect of degree of couplings on the characteristics of four different output functions f given by

$$f = \tan^m(\alpha/2)\cos(\alpha) \qquad (m = 1,2,3,4) \qquad (67)$$

The optimum production functions and the associated constants are described previously in Table 1. If the system has to maximize the ATP production at optimal efficiency then $f = (J_1)_{opt}$ and $q_f = 0.786$. Instead, if the system has to maximize the power output at optimal efficiency, we have the output function $f = (J_1X_1)_{opt}$ occurring at $q_p = 0.91$. If an efficient ATP synthesis at minimal energy cost is imposed, then the functions for economic ATP flow and economic power output occur at $q_f^{ec} = 0.953$ and $q_p^{ec} = 0.972$, respectively. The optimum output power $(J_1X_1)_{opt}$ and the efficiency $(J_1X_1\,\eta)_{opt}$ are calculated from the plots of J_1X_1 vs. x and from the plot of $J_1X_1\eta$ vs. x, respectively. A transition from q_p and q_p^{ec} causes a 12% drop in output power (J_1X_1) and 51% increase in efficiency [8].

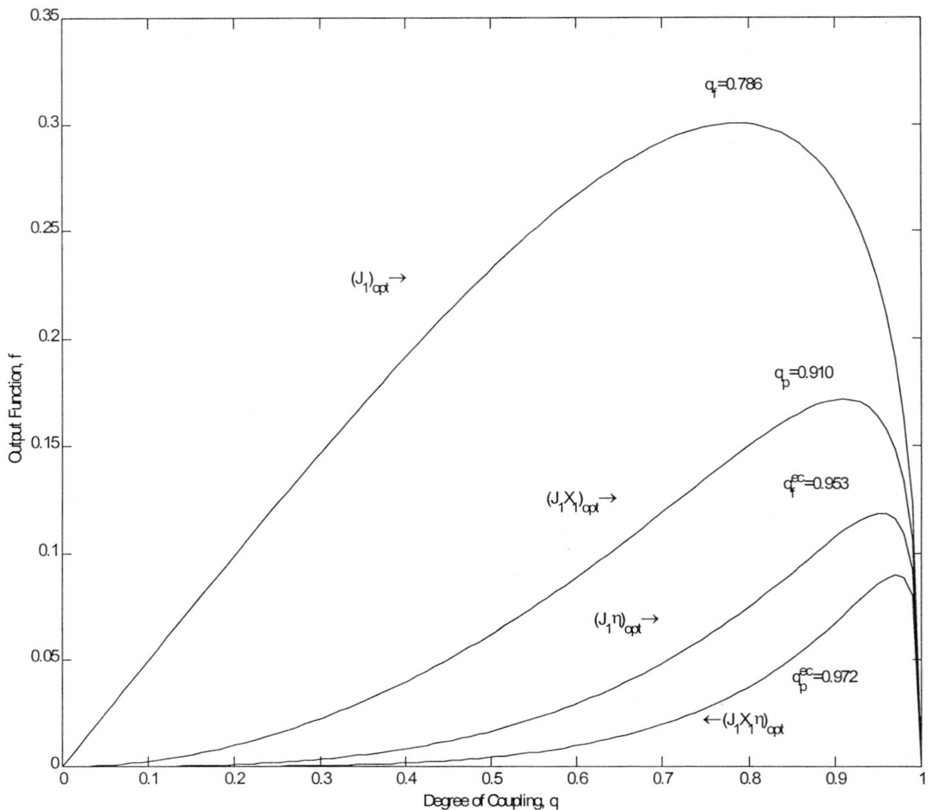

Fig.4. Effect of the degree coupling q on the output functions f, (Eq.67). Reprinted with permission from Elsevier, Biophysical Chemistry 97 (2002) 87.

Stucki [22] analyzed the sensitivity of the force (the phosphate potential) to the fluctuating cellular ATP utilization, and found that the sensitivity is minimal at $q = 0.95$. This is based on an eigenvalue sensitivity analysis of the experimentally supported linear nonequilibrium thermodynamics model of oxidative phosphorylation, and indicates that the phosphate potential is highly buffered with respect to the changing energy demand. The value of q is adopted at which the net ATP production occurs at an optimal efficiency. This leads to simultaneous maximization of kinetic stability and thermodynamic efficiency at the same degree of coupling. For H^+-translocating ATPase, the H^+/ATP coupling ratio is important for mechanistic, energetic and kinetic consequences, and a value of 4 has been adopted for the ratio, and the standard reaction Gibbs energy of ATP production is obtained as 31.3 kJ/mol at $T = 20$ °C, pH = 8.0, pMg = 2.5 and 0.08 M ionic strength; standard enthalpy of the reaction is 28.1 kJ/mol [103]. Differences in the rates of proton pumping, ATPase activities, and degrees of coupling are adjusted by each biologic system in order to survive and compete in its environment [25].

Experiments with liver perfused at a metabolic resting state suggest that the conductance matching is satisfied over a time average, and the degree of coupling q^{ec} yields an efficient oxidative phosphorylation [8]. Optimization may be based on other constraints different from the efficiency. For example, the production of heat in the mitochondria requires rather low degrees of coupling. ATP production with a low rate and high yield results from a cooperative resource use and may evolve in the spatially structured environments [53]. The degrees of coupling are also measured in Na^+ transport in epithelial cells, and growing bacteria where the maximization of net flows are the most important for the system. On the other hand, for a fed rat liver in a metabolic resting state, economy of power output has the priority, while a starved rat liver has to produce glucose, and the priority of energy conversion is replaced by the maximum ATP production. Theoretical and experimental studies indicate that the degree of coupling is directly related to the metabolic regulation and stability in the living organisms [8,23,66]. If the system cannot cope with instabilities then fluctuations ,such as pH and pressure of the blood, could irreversibly harm the organism.

7. THERMODYNAMIC REGULATION IN BIOENERGETICS

Regulation implies a physiological outcome as a result of manipulating mitochondrial function, and hence it is different from control. However, many physiological signals to mitochondria cause large changes in activity associated with control, such as in the shift from state 4 to state 3. Mitochondrial function is regulated by a number of factors over a time scale. These include calcium stimulating NADH supply to the respiratory chain, or oxidative phosphorylation complex activities, nitric oxide inhibiting cytochrome oxidase, and thyroid hormones binding to cytochrome oxidase. The physiological reasoning to these regulations of mitochondria are mainly to match ATP supply efficiently to changes in workload, modulate the thermogenesis, biogenesis or cell death.

The bioenergetics of warm-blooded animals have to coop with variable energy needs and environmental conditions, such as increasing thermogenegis at low temperatures, responding under stress conditions, efficient usage of nutrients under starvation, and degredation of excess food. According to the chemiosmotic theory the electrochemical proton gradient across the membrane is one of the important mechanism to regulate the rate of respiration and ATP synthesis. Respiratory control mainly means simulation of mitochondrial respiration by ADP and its decrease because of conversion of ADP to ATP.

Various substrates regulate the energy metabolism; fatty acids may regulate and tune the degree of coupling [87-89] by inducing uncoupling to set the optimum efficiency of oxidative phosphorylation [88,90]. Experiments with incubated rat-liver mitochondria show that the adenylate kinase reaction can buffer the phosphate potential to a value suitable for the optimal efficiency of oxidative phosphorylation in the presence of very high rate of ATP hydrolysis. Stucki [4] called this class of enzymes, such as adenylate kinase and creatine-kinase, the thermodynamic buffer enzymes. A fluctuating ATP/ADP ratio and deviations from the optimal efficiency of oxidative phosphorylation are largely overcome by the thermodynamic buffering [22.24].

As a terminal component of the respiratory chain cytochrome oxidase catalyzes transfer of electrons from cytohcrome c to oxygen, which is coupled with proton pumping. Although the reaction catalyzed by cytochrome oxidase is far from equilibrium, it has commonly assumed that the flow-force relationship is unique and proportional. The reaction becomes more nonlinear in the thermodynamic branch as the oxygen concentration decreases. This is interpreted as a thermodynamic cost of kinetic regulation [30]. When the oxygen concentration decreases, the cell prefers to optimize the flow through cytochrome oxidase rather than the thermodynamic force of the reaction in order to maintain the oxygen flow and ATP production constant [102]. Since the ATP synthase is assumed to work in the vicinity of equilibrium, the value of the internal phosphorylation potential should follow changes in the proton-motive force. The external phosphorylation potential is related to the internal phosphorylation potential mainly through the ATP/ADP carrier, which acts as a thermodynamic buffer, keeping the external phosphorylation potential as constant as possible, when oxygen concentration and the internal phosphorylation potential decrease. Therefore, electrochemical potential of proton may be considered as a universal regulatory factor of cytochrome oxidase [30].

The enzymes are capable of causing certain reaction pathways by catalyzing a conversion of a substance or a coupled reaction [13,32,38]. For example, on adding nigericin to a membrane, the system reaches a steady state in which the gradients of H^+ and K^+ are balanced. On the other hand, if we add valinomycin and protonophore, both the gradients rapidly dissipate. The mitochondrial creatine kinase is a key enzyme of aerobic energy metabolism [30,85], and involves in the buffering, transport, and reducing the transient nature of the system [68]. This can be achieved: (i) by increasing the enzymatic activities in a pathway, (ii) by metabolic channeling of substrates, and (iii) by damping oscillations of ATP and ADP flows upon sudden changes in the workload.

7.1. Uncoupling

Uncoupling proteins are a subgroup of the mitochondrial anion transporter family, and are identified in prokaryotes, plants and animal cells. Three mammalian uncoupling proteins are called UCP1, UCP2 and UCP3. The proton electrochemical gradient developed across the innermembrane during electron transport of respiratory chain is used to phosphorylate ADP to ATP by F_0-F_1-ATP synthase, and hence the respiration is coupled to phosphorylation. However ATP synthesis is matched to cellular ATP utilization for osmotic work of (downhill and uphill) transport, or mechanical work such as muscle contraction and rotation of bacterial flagellum. Uncoupling of mitochondrial electron transport chain from phosphorylation of ADP is physiological and optimizes the efficiency and tune the degree of coupling of oxidative phosphorylation, and prevents reactive oxygen species generation by the respiratory chain in the resting state [88,91]. Uncontrolled production of reactive oxygen molecules can cause the collapse of mitochondrial energy conservation, loss of membrane integrity, and cell death by necrosis [91].

The respiratory chain is a powerful source of reactive-oxygen molecules, which include oxygen free radicals (O_2^-) and hydroxyl radical hydrogen peroxide, and nitric oxide; they are very reactive and able to damage cellular components and macromolecules, and influence programmed cell death or apoptosis. Cells have developed various strategies to dissipate reactive oxygen molecules and remove its oxidation products. Mitochondria are protected from oxidative stress by both low oxygen levels and a defense system against reactive oxygen species [102]. Uncoupling proteins are capable of modulating reactive oxygen molecules.

Fatty acids facilitate the net transfer of protons from intermembrane space into the mitochondrial matrix, hence lowering the gradient of proton electrochemical potential and mediating weak uncoupling. Possible mechanisms of fatty acid uncoupling have been reviewed in various studies [23,88,101].

Uncoupling proteins generally facilitate the dissipation of the transmembrane electrochemical potentials of H^+ or Na^+ produced by the respiratory chain, and results in an increase in the H^+ and Na^+ permeability of the coupling membranes. They provide adaptive advantages, both to the organism and to individual cell, and also increase vulnerability to necrosis by compromising the mitochondrial membrane potential [24,69,87]. Some uncoupling is favorable for the energy-conserving function of cellular respiration [23]. The linear nonequilibrium thermodynamics formulation is used for a macroscopic description of some inhibitors of oxidative phosphorylation, by considering protonophores, some ATPase inactivators, and some electron-chain inhibitors.

In oxidative phosphorylation leaks cause a certain uncoupling of two consecutive pumps, such as electron transport and ATP synthase, and may be described as membrane potential-driven backflow of protons across the bilayer.

7.2 Slippage

A slip means a decreased H^+/e- stoichiometry of proton pumps [32]. The nonequilibrium thermodynamic approach may be useful in describing the variation for the coupling and hence the notion of slippage. Mainly the slippage results when one of two coupled reactions in a cyclic process proceeds alone without its counterpart that causes intrinsic uncoupling. On the microscopic level, individual enzymes cause slippage by either passing a proton without contributing to ATP synthesis, or hydrolyze ATP without contributing to proton pumping. On the macroscopic level, the measured degree of coupling may be different from the expected coupling as these microscopic slips are averaged across a population of enzymes.

Slippage is not a membrane leak, rather an intrinsic property of the enzyme, and hence to be related to enzyme mechanism and structure. In terms of the thermodynamic energy conversion, a slip may decrease efficiency, it may however allow dynamic control and regulation of the enzyme over the varying ranges of electrochemical gradient of protons and the chemical potential of ATP in equilibrium with ADP and Pi. It is possible that the slips have evolved to enhance the function of particular coupling enzymes in particular conditions [25].

Mitochondrial energy metabolism may be regulated by the ATP/ADP ratio and the slip of proton pumping in cytochrome c oxidase at high proton motive force [38]. The proton pumping in cytochrome c oxidase, but not complex III of the respiratory chain (cytochrome bc_1 complex) does not occur at a constant stoichiometry; it varies depending on the level of proton motive force. In transportation leaks can be found in the proton-sugar symport in bacteria where a protein mediates the transport of protons and sugar across the membrane, and adding a protonophore, a parallel pathway occurs causing a leak in the transport.

A slip can occur when one of two coupled processes (i.e., reactions) in a cyclic process proceeds without its counterpart, which is also called intrinsic uncoupling [93]. Leaks and slips may affect the metabolic rate [27,75]. Schuster and Westerhoff [27] developed a theory for the metabolic control by enzymes that catalyze two or more incompletely coupled reactions. Control by the coupled reactions is distinguished quantitatively from the control by the extent of slippage using the linear nonequilibrium thermodynamics formulations; here the limits of coupling, or the tightness of coupling [63,64] may be an important parameter, and be obtained as the ratio of coupled-to-uncoupled rates, which is a function of the binding energy of the substrate and the carrier protein. Evolution favors the coupling in organisms as tight as possible, since the loss of available metabolic energy is a disadvantage. For example, the rather loose coupling of calcium and sugar transport systems may represent a tradeoff between efficiency and the rate that can be achieved in primary and secondary active transport [74]. One other concern in an interconnected biological network is the behavior of a subsystem (e.g. glycolysis), which may become unsteady and chaotic, so that the output of this subsystem (e.g. ATP production) is adversely affected, and becomes an external noise for other subsystems causing inhibition and desynchronization.

7.3. Potassium channels

A class of cardiac potassium channels operates in smooth and skeletal muscle, brain, and pancreatic cells. Potassium channels are activated when intracellular ATP levels decrease, and are important link between the cellular excitability and the metabolic status of the cell. The ratio of ATP/ADP, pH, lactate and divalent cations determine and modulate the channel activity. Opening of the potassium channels leads to membrane hyperpolarization and potential decrease as the potassium ions flow out of the cell. Since phosphorylation changes the activity of potassium channels, it modulates cellular excitability.

Potassium channels play an important role in the control of insulin secretion in β-pancreatic cells. In a resting β-pancreatic cell, the membrane potential is maintained below the threshold for insulin secretion by an efflux of potassium ions through the open potassium channels. As glucose levels rise within the cell, ATP production increases, and the change in APT/ADP ratio leads to the closure of potassium channels. This causes the calcium channels to open, and the entering calcium ions signals insulin secretion. Factors, which are able to modulate the β-pancreatic potassium channels, can facilitate a fine-tuning for the insulin secretion. Regulation by the potassium channels plays an important role in the coupling of cellular excitability to the metabolic state of the cell [105].

7.4. Metabolic control analysis

Metabolic control analysis determines quantitatively the effects of various metabolic pathways reactions on the flows and on the metabolic concentrations. The analysis defines two coefficients: (i) the control coefficients, which characterize the response of the system flows, concentrations, and other variables after parameter perturbations; and (ii) the elasticity coefficients, which quantify the changes of reaction rates after perturbations of substrate concentrations or kinetic parameters under specified conditions.

Metabolic control analysis can be used to study diseases caused by the enzyme functions or disfunctions, and helps to understand the certain pathways. It may be critical to determine the enzymes with highest flow control coefficient, in order to inhibit or control the enzyme functions. This may lead to the quantification of rate limitations in the complex enzyme systems.

Steady state metabolic flows J_j depend on the total concentrations of the enzymes E_k, and flow (flux) control coefficients $C_{J,E}$ are defined by

$$C_{J,E} = \left(\frac{E_k \Delta J_j}{J_j \Delta E_k} \right)_{\Delta E_k \to 0} = \frac{E_k}{J_j} \frac{\partial J_j}{\partial E_k} \quad (68)$$

The flow control coefficients relate the fractional changes in the steady-state flows to the changes in the total enzyme concentrations. The partial derivatives of reaction rates $J_{r,i}$ with respect to the substrate concentrations S_j are called the elasticity coefficients ε_{ij}, and given by

$$\varepsilon_{ij} = \frac{S_j}{J_{r,i}} \frac{\partial J_{r,i}}{\partial S_j} \quad (69)$$

The flow or concentration control coefficients are related to elasticity coefficients through the conservation relations and connectivity theorems [71].

Metabolic systems usually consist of a number of functional units and metabolic pathways. Modular control analyses are developed to streamline the analysis of control and regulation of the metabolic systems. Enzyme-enzyme interactions and metabolic channeling are parts of a metabolic network as a module. The slipping enzyme may be considered as a module catalyzing two reactions of exergonic and endergonic processes providing a biological energy transudation. Control coefficients related to slipping enzymes can be calculated by the linear nonequilibrium thermodynamics approach. The overall control coefficients in the modular approach express the control exerted by the particular degrees of freedom of a module on the measurable variables at steady-state [27].

8. FACILITATED TRANSPORT

In the facilitated transport in a membranes, a carrier agent can specifically interact with one of the substances in the feed mixture, the substance-carrier complex diffuses across the membrane, the carrier dissociates at the end of the membrane, and finally returns to its original position (Figure 5). Red blood cell membrane transports oxygen with the hemoglobin as the carrier. Carrier-facilitated transport is used successfully to extract various organic and inorganic substances from a feed mixture in the liquid membranes. Liquid membranes are generally employed as bulk liquid membranes, emulsion liquid membranes and supported liquid membranes [93].

Many biological mass transfer processes occur as a result of the combination of a substance with a membrane constituent to form a complex.. For example, myoglobin carries oxygen

$O_2(s)$+Myoglobin (c) $\rightarrow$ Oxymyoglobin (bc) (70)

8.1. Kinetic formulation

Since the biological membranes act as barriers for hydrophilic and large molecules, a mobile carrier molecule due to increased mobility of the substrate-carrier complex may increase the transport. Facilitated transport may be described by the jumping mechanism [93] for a fast reaction between the carrier and substrate. If the transport of substance-carrier across the membrane is not fast enough, then conventionally diffusion-reaction system of Eq. (70) is described by

$$\frac{\partial c_s}{\partial t} = D_s \frac{\partial^2 c_s}{\partial x^2} + k_2 c_{bc} - k_1 c_s c_c \tag{71a}$$

$$\frac{\partial c_c}{\partial t} = D_c \frac{\partial^2 c_c}{\partial x^2} + k_2 c_{bc} - k_1 c_s c_c \tag{71b}$$

$$\frac{\partial c_{bc}}{\partial t} = D_{bc} \frac{\partial^2 c_{bc}}{\partial x^2} - k_2 c_{bc} + k_1 c_s c_c \tag{71c}$$

where c_s, c_c and c_{bc} denote the concentrations of substrate, free carrier and bound carrier respectively, D_s, D_c and D_{bc} are corresponding diffusion coefficients, and k_1 and k_2 are the chemical reaction constants. Initially there is no oxymyoglobin, and the concentration of myoglobin c_{co} is uniform across the membrane. With a membrane thickness of l, the boundary conditions are

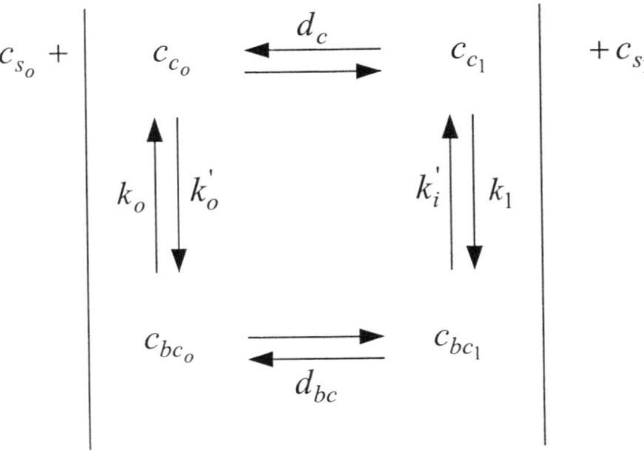

Fig. 5. Facilitated transport of a substrate.

$$\frac{\partial c_c(0,t)}{\partial t}=\frac{\partial c_c(l,t)}{\partial x}=\frac{\partial c_{bc}(0,t)}{\partial x}=\frac{\partial c_{bc}(l,t)}{\partial x}=0$$
$$c_s(0,t)=\alpha_o c_{so},\quad c_s(l,t)=\alpha_1 c_{s1},\quad c_c(x,0)=c_{co} \tag{72}$$

where c_{so}, c_{s1} are the concentrations of substrate on the two sides of membrane, and α_o and α_l are the partition coefficients. The carrier cannot leave the membrane, and there is no considerable resistance to the diffusion of substrate at the interface. Sugar transport in the red blood cells, and oxygen transport by hemoglobin and myoglobin are modeled by the set of Eq. (71). When the carrier molecule such as hemoglobin is much larger than the substrate, it is usually assumed that $D_c = D_{bc}$.

Membrane thickness is an important parameter in the unsteady state facilitated transport of oxygen across membranes, and the transport, association and dissociation processes from a red cell can be solved using the finite difference methods based on a time-liberalization approach [106].

In membrane transport usually one-dimensional, steady-state models are used. If the permeates move independently of one another and with the ideal interface permeability, the simple diffusion across the membrane is given by the boundary value problem

$$\frac{\partial c_s}{\partial t}=D\frac{\partial^2 c_s}{\partial x^2}=0 \tag{73}$$

$$c_s(0,t)=\alpha c_{so},\quad c_s(l,t)=\alpha c_{s1} \tag{74}$$

At steady state the flow J is constant, then

$$J=-\alpha\left(\int_0^l \frac{1}{D(x)}dx\right)^{-1}(c_{s1}-c_{so}) \tag{75}$$

Permeability is $\alpha\left(\int_0^l \frac{1}{D(x)}dx\right)^{-1}$. If D is constant then the permeability is $\alpha D/l$.

The interaction of particles with each other and with the membrane is the common situation, and there are several models proposed for such interactions. One model assumes a membrane as an energy profile for the moving molecules across the barriers of binding sites; the transport takes place in a series of reactions, and the energy profile consists of a series of conformation alterations of the particles. If a membrane has a homogeneous interior, the diffusion coefficient is constant and the energy profile consists of the energy barriers of equal size.

In the facilitated diffusion, the carrier molecules are in a limited number, therefore the transport rate is not controlled by the concentration gradient, and shows saturation. The flow is expressed by

$$J = -P(c_{s1} - c_{so}) \tag{76}$$

where the permeability P is dependent on the concentration of substrate and carrier molecules c_{s1}, c_{so}. In a simple model, it is usually assumed that the membrane is thin, and a stationary state is established; all flows are determined by the substrate concentrations on both sides, and the reaction takes place on the surface.

For a facilitated transport seen in Fig. 5, if $k_o'/k_o = k_1'/k_1$, the transport is passive. If $k_o'/k_o \neq k_1'/k_1$ the cycle is preferring one direction, and the substrate will accumulate on one side. Since the carrier and bound carrier molecules do not leave the membrane, we have

$$(c_{c1} + c_{co} + c_{bco} + c_{bc1})/2 = C \tag{77}$$

The steady state facilitated transport [73] (Fig. 5) may be described by

$$d_c(c_{c1} - c_{co}) = -k_o' c_{bco} + k_o c_{so} c_{co} \tag{78}$$

$$d_c(c_{c1} - c_{co}) = k_1' c_{bc1} - k_1 c_{s1} c_{c1} \tag{79}$$

$$d_{bc}(c_{bc1} - c_{bco}) = k_o' c_{bco} - k_o c_{so} c_{co} \tag{80}$$

$$d_{bc}(c_{bc1} - c_{bco}) = -k_1' c_{bc1} + k_1 c_{s1} c_{c1} \tag{81}$$

The flow of oxygen J is expressed as

$$J = -d_{bc}(c_{bc1} - c_{bco}) = d_c(c_{c1} - c_{co}) \tag{82}$$

By solving Eqs. (77)-(82), we have

$$J = \frac{-CD(\sigma_1 - \sigma_o)}{1 + D(\sigma_o / k_1' + \sigma_1 / k_o') + (D/d)(\sigma_o + \sigma_1) + (D/d_c)\sigma_o \sigma_1} \tag{83}$$

where

$d = 2(1/d_{bc} + 1/d_c)^{-1}$, $D = (1/d_{bc} + 1/k_o' + 1/k_1')^{-1}$, $\sigma_1 = c_{s1}/K_1$, $\sigma_o = c_{so}/K_o$,

$K_1 = k_1^{'} / k_1, \; K_o = k_o^{'} / k_o$

Since the concentrations inside the membrane are generally not known, we can express the equilibrium oxygen concentration at steady state using the diffusion-reaction relations of Eqs. (71a)-(71c); with the assumption of $D_c = D_{bc}$,

$$D_s \frac{d^2 c_s}{dx^2} = -k_2 c_{bc} + k_1 c_s c_c \tag{84}$$

$$D_c \frac{d^2 c_c}{dx^2} = -k_2 c_{bc} + k_1 c_s c_c \tag{85}$$

$$D_c \frac{d^2 c_{bc}}{dx^2} = k_2 c_{bc} - k_1 c_s c_c \tag{86}$$

By adding Eqs. (85) and (86), we have

$$\frac{d^2 c_c}{dx^2} + \frac{d^2 c_{bc}}{dx^2} = 0 \tag{87}$$

After two successive integrations, we have

$$c_c + c_{bc} = a_o x + a_1 \tag{88}$$

where a_o and a_1 are the constants, and after applying the boundary and initial conditions a_o vanishes, and we obtain

$$c_c + c_{bc} = c_{co} \tag{89}$$

By subtracting Eq. (84) from Eq. (85), we have

$$D_c \frac{d^2 c_c}{dx^2} - D_s \frac{d^2 c_s}{dx^2} = 0 \tag{90}$$

First integration of Eq. (90) yields

$$D_c \frac{dc_c}{dx} - D_s \frac{dc_s}{dx} = b_o = J \tag{91}$$

Myoglobin does not leave the membrane, and the myoglobin flow is zero $dc_c/dx=$, therefore b_o or J represents the flow of oxygen at steady state. The second integration of Eq. (90) yields

$$D_c c_c - D_s c_s = Jx + b_1 \tag{92}$$

where the constant b_1 is determined from the boundary conditions. Using Eqs. (89) and (92), we express c_c in terms of c_{bc}

$$c_{bc} = c_{co} - \frac{1}{D_c}(Jx + b_1 + D_s c_s) \tag{93}$$

By using Eqs. (91) and (92), we eliminate the concentrations c_c and c_{bc} from Eq. (84), and we have

$$D_s \frac{d^2 c_s}{dx^2} = \frac{1}{D_c}(D_s c_s + Jx + b_1)(k_2 + k_1 c_s) - k_2 c_{co} \tag{94}$$

This single nonlinear equation is called the Wyman equation, which can represent experimental data of the facilitated oxygen transport by hemoglobin. Carbon monoxide possesses an approximately 250 times higher affinity than that of oxygen for hemoglobin, and facilitated transport of carbon monoxide is very small. The last term on the right hand side of Eq. (94) $k_2 c_o$ is defined as the "driving force" of the facilitated diffusion. Whenever the concentration of carrier and the dissociation constant of the substrate-carrier reaction system are finite, we expect to have facilitated transport. We can determine the oxygen flow at steady state by solving Eq. (94).

8.2. Nonequilibrium thermodynamic approach

In experimental studies [34,44], a membrane composed of a filter soaked in a solution of hemoglobin is used, as shown in Fig. 6. Oxygen gas, at different pressures $P_1 > P_2$, was placed in the two compartments and the steady state flow of oxygen across the membrane was measured. The presence of hemoglobin enhanced the flow of oxygen at low oxygen pressure, however the facilitation of oxygen transfer disappeared at higher pressure of oxygen. Katchalsky and Curran [5] used the linear nonequilibrium thermodynamics theory for the facilitated oxygen transport by the hemoglobin based on the linear flow-force relations. The following reaction is used for the facilitated oxygen transport by the carrier hemoglobin

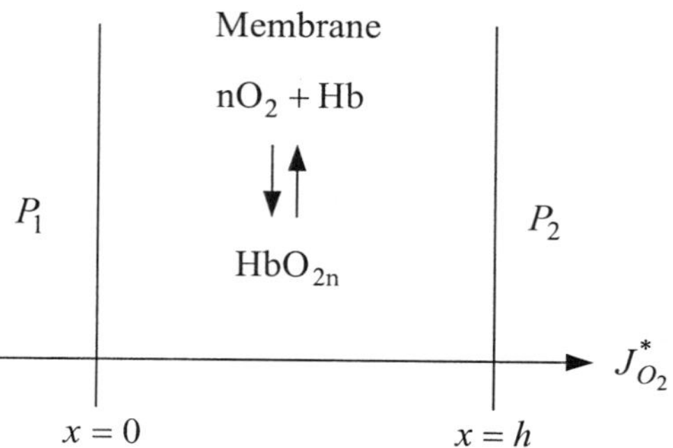

Fig.6. Facilitated transport of oxygen by hemoglobin.

$$nO_2 + Hb \rightarrow HbO_{2n} \tag{95}$$

The affinity A for Eq. (95) is given by

$$A = n\mu_1 + \mu_2 - \mu_3 \tag{96}$$

where the subscripts 1, 2 and 3 refer to oxygen, hemoglobin (Hb), and oxyhemoglobin (HbO_{2n}) respectively. We assume that the rate of reactions more rapid than that of diffusion so that the reaction is at equilibrium, and hence $A = 0$.

Applying the gradient operator to Eq. (96), we obtain

$$n\nabla\mu_1 + \nabla\mu_2 = \nabla\mu_3 \tag{97}$$

Eq. (97) expresses a relationship among the forces that lead to a coupling of flows. The flows passing at any point in the membrane are those of free oxygen J_1, of hemoglobin J_2, and of oxyhemoglobin J_3. The externally measured flow of oxygen $J_1{}^*$ equals to the flows of free oxygen and oxygen carried by hemoglobin.

$$J_1^* = J_1 + nJ_3 \tag{98}$$

Since no external flows of hemoglobin $J_2{}^*$ takes place, we have

$$J_2^* = J_2 + J_3 = 0 \tag{99}$$

The dissipation function takes the form

$$\Psi = J_1\nabla(-\mu_1) + J_2\nabla(-\mu_2) + J_3\nabla(-\mu_3) \tag{100}$$

Using Eq. (97), we transform the dissipation equation as follows

$$\Psi = J_1^* \nabla(-\mu_1) + J_2^*(-\mu_2) \tag{101}$$

From Eq. (101) the following linear phenomenological equations are obtained

$$J_1^* = -L_{11}\nabla\mu_1 - L_{12}\nabla\mu_2 \tag{102}$$

$$J_2^* = 0 = -L_{21}\nabla\mu_1 - L_{22}\nabla\mu_2 \tag{103}$$

Eq. (102) can also be expressed in terms of the diffusivity coefficients

$$J_1^* = -D_1\frac{dc_1}{dx} + nD_2\frac{dc_2}{dx} \tag{104}$$

By integrating Eq. (104) between $x = 0$ and $x = h$, and assuming that J_1^*, D_1 and D_2 are constant, we get

$$J_1^* = D_1\frac{c_1^o - c_1^h}{h} + nD_2\frac{c_2^h - c_2^o}{h} \tag{105}$$

The external oxygen pressure determines the first term on the right-hand side of Eq. (105). If $P_1 > P_2$, $c_1^o > c_1^h$, contribution of the carrier is considerable. The transport of oxygen increases with increasing hemoglobin concentration. When $P_2 = 0$, ($c_1^h = 0$) the facilitated oxygen transfer decreases with increasing P_1.

9. ACTIVE TRANSPORT

Normally, diffusion occurs spontaneously from a region of higher chemical potential μ_1 to lower chemical potential region μ_2, and the direction of flow is the same with the direction of decreasing chemical potential. The total Gibbs energy change for such a system is expressed by

$$-dG = -dN\mu_1 + dN\mu_2 = dN(\mu_2 - \mu_1) \tag{106}$$

However, if a compensating process with a coupling mechanism is added into the isolated system causing the total free energy of the complete system to decrease, then it is possible to envisage diffusion against a potential gradient. This type of mass diffusion is called the *active transport* for which the product of the flow J_i by the generalized force X_i acting upon is negative

$$J_i X_i < 0 \tag{107}$$

This inequality implies that the flow is occurring in the direction opposite to the direction of the force. If the compensating process is a chemical reaction, then the coupling will only be allowed in an anisotropic medium according to the Currie-Prigogine principle. Since chemical potential gradient is a vector quantity

$$\text{grad}\,\mu_i = \mathbf{i}\frac{\partial \mu_i}{\partial x} + \mathbf{j}\frac{\partial \mu_i}{\partial y} + \mathbf{k}\frac{\partial \mu_i}{\partial z} \tag{108}$$

and chemical reaction is a scalar quantity, no chemical reaction can impose directional properties to the flow of substances unless the spatial gradient of a thermodynamic potential is altered.

Free energy for the active transport is supplied by a driving system from a high potential state to low potential state. In *primary active transport* the driving system is a chemical reaction away from equilibrium, and in *secondary active transport* it is a concentration gradient [64].

The factors affecting the flow of a substance across a membrane are the potential gradient and the permeability coefficients of a biological membrane, which are not necessarily physical constants. If the membrane is simply a static physical barrier, an enzyme might change its structure, and hence its permeability is affected. If the membrane is maintained in a particular structure by metabolic processes, then metabolic activity might affect its permeability. Information on permeability coefficients may be obtained from the thermodynamic considerations and structural properties.

Active transport is a universal property of the living systems; it is closely connected with the cybernetic system of the cells; it plays important role in osmoregularity and in the adaptation of organisms to their environment. Therefore, the development of suitable mechanisms for active transport is the prerequisite for the evolution of living organisms.

Conventional methods for establishing the existence of active transport are to analyze the effects of metabolic inhibitors, to correlate the rate of metabolism with the extent of ion flow or the concentration ratio between inside and outside of the cells, and to measure the current needed in a short-circuited system having identical solutions on each side. Measurements indicate the flow contributing to the short-circuited current, and any net flow detected is due to the active transport, since the electrochemical gradients of all ions are zero ($\Delta\psi = 0$, $c_o = c_i$).

There are several mechanisms proposed to explain how biological membranes can transport charged or uncharged substrates against their thermodynamic forces [33]. It is widely accepted that cross transports by a protein are discreet events. Biomembranes contain enzymes, pores, charges or membrane potentials, and catalytic activities associated with the transport of substrates. It is well established that the electrostatic interactions between the membrane and a charged solute may play important role in the transport. Therefore, we have to establish the reliable links among the membrane's charge effect, the pore size and lengths distribution, and the pore density to describe the interactions.

From the local conservation of mass, we obtain

$$\frac{\partial c_i}{\partial t} = -\frac{dJ_i}{dx} + \nu_i J_i \tag{109}$$

At steady state the local concentrations do not vary with time, and we have

$$\frac{dJ_i}{dx} = \nu_i J_r \tag{110}$$

Eq. (110) shows that a stationary state imposes a relation between the diffusion and chemical reaction, and is of special interest in isotropic membranes where the coupling coefficients vanish.

For a homogeneous and isotropic media the phenomenological equations are expressed as the uncoupled system for diffusion and chemical reaction [31]

$$-\frac{d\mu_i}{dx} = \sum_k K_{ik} J_k \tag{111}$$

$$A = K_r J_i \tag{112}$$

where A is the affinity. If the coefficients K are independent of position, we can differentiate Eq. (111) with respect to x, and insert Eq. (110) to obtain

$$-\frac{d^2 \mu_i}{dx^2} = \sum_k K_{ik} \frac{dJ_k}{dx} = \left(\sum_k \nu_k K_{ik} \right) J_r \tag{113}$$

After multiplying both sides by ν_i, and summation, Eq. (113) becomes

$$-\sum_i \nu_i \frac{d^2 \mu_i}{dx^2} = \left(\sum_{i,k} \nu_i \nu_k R_{ik} \right) J_r \tag{114}$$

From the definition of affinity A, we find

$$\frac{d^2A}{dx^2} = -\left(\sum_i \frac{\nu_i d^2\mu_k}{dx^2}\right) \tag{115}$$

After substituting J_r from Eq. (112), we have

$$\frac{d^2A}{dx^2} = \left(\sum_{i,k} \frac{\nu_i \nu_k K_{ik}}{K_r}\right)A \tag{116}$$

The expression

$$\left(\sum_{i,k} \frac{\nu_i \nu_k K_{ik}}{K_r}\right) \tag{117}$$

has the dimension of cm^{-2} and will be denoted by λ^{-2} to obtain

$$\frac{d^2A}{dx^2} = \lambda^{-2}A \tag{118}$$

The characteristic parameter λ is called the relaxation length of the coupled reaction-diffusion processes within the membrane [107].

For a membrane thickness of Δx, dimensionless number $\lambda/\Delta x$ is closely related to the Thiele modulu used for the characterization of heterogeneous reaction columns. This dimensionless quantity is also related to the relaxation time of chemical reaction τ_r and the average relaxation time of diffusional processes τ_d as follows

$$\left(\frac{\lambda}{\Delta x}\right)^2 = \frac{\tau_r}{2\langle \tau_d \rangle} \tag{119}$$

If a reaction relaxes faster than the time necessary for diffusion across the membrane $\tau_r < \langle \tau_d \rangle$, then λ will be smaller than Δx, and the reaction will reach equilibrium on the surface after the reactants have diffused only a short distance within the membrane. On the other hand, when Δx is very small, which is the case in biological systems, then τ_r and $\langle \tau_d \rangle$ may be of the same order of magnitude, hence the equilibrium treatment is not possible.

Active transport is found in biological desalination and ion separation. Salt excretion by the glands of desert plants, and accumulation of potassium in certain bacteria against a very large concentration gradient are some of the examples of active transport. Active transport is a highly selective process; it can be prevented by specific metabolic inhibitors, and is closely related to facilitated transport.

The total turnover per unit membrane per unit time is given by

$$J_{r,\text{tot}} = \int_{x}^{\Delta x} J_r dx \neq 0 \tag{120}$$

where J_r is the local reaction coupled to the transport of substrates in moles/m^3 s, x is a coordinate normal to the membrane surface, and Δx is the membrane thickness. In facilitated transport $J_{r,\text{tot}} = 0$

Biological membranes show anisotropy as their molecules are preferentially ordered in a definite direction in the plane of membrane, and the coupling between chemical reactions (scalar) and diffusion flow (vectorial) can take place. Almost all outer and inner membranes of the cell have the ability of active transport. The sodium and potassium pumps take place almost in all cells, especially nerve cells, while the active transport of calcium takes place in muscle cells (Fig. 6). Active transport of protons is important in mitochondrial membranes, chloroplasts, and retina.

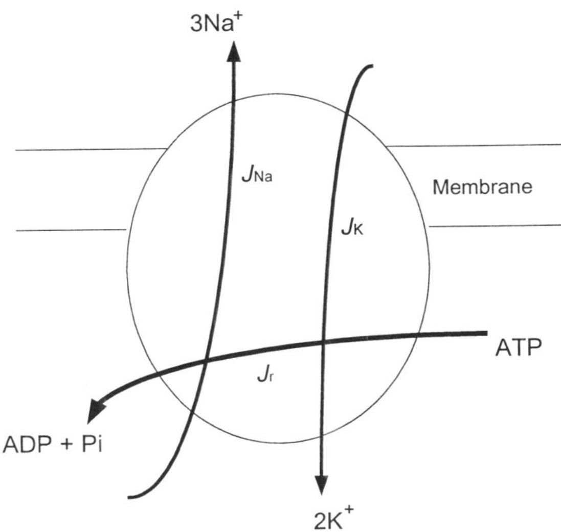

Fig. 6. Schematic coupling of sodium and potassium pumps with ATPase operating in membrane.

For a two-flow system of diffusion and chemical reaction a representative dissipation function is expressed by

$$\Psi = T\Phi_s = J_i X_i + J_r A \geq 0 \tag{121}$$

where J_r is the rate of chemical reaction, and J_i is the flow of a substance. Eq. (121) represents the simplest thermodynamic description of the coupled flow-chemical reaction systems. The total dissipation function does not require that both the terms J_iX_i and J_rA must be positive; if one is positive and sufficiently large, the other term can be negative, provided that they are coupled, and hence the second law of thermodynamics is not violated. If J_iX_i is negative, then the signs of J_i and X_i are opposite, causing the flow of substrate in a direction opposite to that imposed by its conjugate force X_i, leading to the active transport. This is possible only if the flow is coupled to the chemical reaction with a large positive dissipation J_rA, and hence producing a positive overall dissipation.

For a simple derivation of the dissipation function, consider an isothermal composite system with three compartments consisting of two external chambers (I and II) and a membrane compartment (m) in between. The volumes of the compartments are constant ($dV_{\mathrm{I}} = dV_{\mathrm{II}} = dV_{\mathrm{m}} = 0$). The Gibbs relations for the compartments are

$$dU_{\mathrm{I}} = TdS_{\mathrm{I}} + \sum_i \mu_{i,\mathrm{I}} dN_{i,\mathrm{I}} \tag{122}$$

$$dU_{\mathrm{II}} = TdS_{\mathrm{II}} + \sum_i \mu_{i,\mathrm{II}} dN_{i,\mathrm{II}} \tag{123}$$

$$dU_{\mathrm{m}} = TdS_{\mathrm{m}} + \sum_i \mu_{i,\mathrm{m}} dN_{i,\mathrm{m}} \tag{124}$$

It is assumed that reactions take place only in the membrane, and the net change of number of moles is expressed by

$$dN_{i,\mathrm{m}} = dN_{i,\mathrm{exch}} + dN_{i,\mathrm{react}} \tag{125}$$

where $dN_{i,\mathrm{exch}}$ is the number of moles of the component i exchanged with the surrounding compartments, and $dN_{i,\mathrm{react}}$ denotes the number of moles of component i produced by the chemical reaction, which is expressed for the kth chemical reaction as follows

$$dN_{i,\mathrm{react}} = \sum_k \nu_{ik} d\xi_k \tag{126}$$

where ν_{ik}'s are the stoichiometric coefficients of the *i*th component in the *k*th reaction, and ξ is the extent of advancement of the *k*th chemical reaction.

Combining Eqs. (125) and (126) with Eq. (124), we obtain

$$dU_{\mathrm{m}} = TdS_{\mathrm{m}} + \sum_i \mu_i dN_{i,\mathrm{exch}} - \sum_k A_{k,\mathrm{m}} d\xi_k \tag{127}$$

where $A_{k,\mathrm{m}}$ is the affinity of the *k*th reaction within the membrane.

Since the composite system is a closed system, we have

$$dN_{i,\mathrm{I}} + dN_{i,\mathrm{II}} + dN_{i,\mathrm{m}} = 0 \tag{128}$$

$$dU_{\mathrm{I}} + dU_{\mathrm{II}} + dU_{\mathrm{m}} = 0 \tag{129}$$

where $-dN_{i,\mathrm{I}}$ is the number of moles of the *i*th component gained by the membrane, and $dN_{i,\mathrm{II}}$ is the number of moles of component *i* lost by the membrane through diffusional processes, and the input and output flows are expressed by

$$J_{i,\mathrm{in}} = -\frac{dN_{i,\mathrm{I}}}{dt} \quad \text{and} \quad J_{i,\mathrm{out}} = \frac{dN_{i,\mathrm{II}}}{dt} \tag{130}$$

The macroscopic driving forces for the flows are given by

$$\Delta\mu_{i,\mathrm{in}} = \mu_{i,\mathrm{I}} - \mu_{i,\mathrm{m}} \quad \text{and} \quad \Delta\mu_{i,\mathrm{out}} = \mu_{i,\mathrm{m}} - \mu_{i,\mathrm{II}} \tag{131}$$

The total rate of the *k*th chemical reaction within the membrane is

$$J_{r,k} = \frac{d\xi_k}{dt} \tag{132}$$

The entropy contributions of the three compartments are expressed by

$$dS_{\mathrm{I}} + dS_{\mathrm{II}} + dS_{\mathrm{m}} = d_i S_{\mathrm{I}} + d_i S_{\mathrm{II}} + d_i S_{\mathrm{m}} = d_i S \tag{133}$$

The contributions from the exchangeable entropy terms $d_e S$ cancel in Eq. (133). By adding Eqs. (122) to (124) and Eqs. (127) to (129), we obtain

$$Td_i S + \sum_i (\mu_{i,\mathrm{I}} - \mu_{i,\mathrm{m}}) dN_{i,\mathrm{I}} - \sum_i (\mu_{i,\mathrm{m}} - \mu_{i,\mathrm{II}}) dN_{i,\mathrm{II}} - \sum_k A_{k,\mathrm{m}} d\xi_k = 0 \tag{134}$$

Dividing Eq. (134) by *dt* and using Eqs. (130) to (132), we obtain an expression for the dissipation function

$$\Psi = T\frac{d_iS}{dt} = \sum_i J_{i,\text{in}}\Delta\mu_{i,\text{in}} + \sum_i J_{i,\text{out}}\Delta\mu_{i,\text{out}} + \sum_k A_{k,\text{m}}J_{r,k} \tag{135}$$

Assuming that under steady state operations the difference between the output and input flows of the ith component is equal to the amount of ith component reacted in all chemical processes, and we have

$$J_{i,\text{out}} - J_{i,\text{in}} = \sum_k \nu_{ik}J_{r,k} \tag{136}$$

The last term in Eq. (135) can be expressed by the difference of the output and input flows

$$\begin{aligned}\sum_k A_{k,\text{m}}J_{r,k} &= -\sum_k\left(\sum_i \nu_{i,k}\mu_{i,\text{m}}\right)J_{r,k} - \sum_i \mu_{i,\text{m}}\sum_k \nu_{i,k}J_{r,k} \\ &\quad -\sum_i \mu_{i,\text{m}}(J_{i,\text{out}} - J_{i,\text{in}})\end{aligned} \tag{137}$$

After introducing Eq. (137) into Eq. (135), we obtain

$$\Psi = \sum_i J_{i,\text{in}}\mu_{i,\text{I}} - \sum_i J_{i,\text{out}}\mu_{i,\text{II}} \tag{138}$$

However, only the difference in chemical potentials is measurable, and Eq. (138) can be further transformed for practical use by introducing the affinity of the kth reaction in compartment II, $A_{k,c}$

$$A_{k,c} = -\sum_i \nu_{ik}\mu_{i,\text{II}} \tag{139}$$

By introducing $J_{i,\text{out}}$ from Eq. (130) we obtain

$$\Psi = \sum_i J_{i,\text{in}}(\mu_{i,\text{I}} - \mu_{i,\text{II}}) + \sum_k J_{r,k}A_{k,c} = \sum_i J_{i,\text{in}}\Delta\mu_i + \sum_k J_{r,k}A_{k,c} \tag{140}$$

where $\Delta\mu_i = \mu_{i,\text{I}} - \mu_{i,\text{II}}$ is the cross-membrane difference of chemical potentials, which is measurable.

The dissipation function in Eq. (140) can be used in the thermodynamic formulation of active transport. Kedem and Caplan [34] presented an early analysis of the interactions between the ion transport and chemical reactions using the linear nonequilibrium thermodynamics approach with the dissipation function

$$\Psi = J_j\Delta\tilde{\mu}_i + J_rA_r \tag{141}$$

where $\Delta\tilde{\mu}_i$ is the difference in electrochemical potentials, J_j and J_r are the diffusional flow and the rate of reaction respectively, and A the conjugate affinity within the cell. The phenomenological relations in terms of the forces are expressed by

$$\Delta\tilde{\mu}_i = \sum_{j=1}^{n} K_{ij} J_j + K_{ir} J_r \tag{142}$$

$$A_r = \sum_{j=1}^{n} K_{rj} J_j + K_{rr} J_r \tag{143}$$

The cross-coefficient K_{ir} represents the coupling between the diffusion flow and the rate of reaction. Since diffusion flow is a vector and the rate of reaction is scalar, the coefficient K_{ir} must be a vector. For nonvanishing values of K_{ir}, biological membranes must have a structural anisotropy in the direction of diffusional flow. Biomembranes have a topological organization, which determines the pattern of flow across the membrane, hence Eqs. (142) and (143) are applicable with the local flows and forces in biomembrane system. We can integrate the expressions over the membrane thickness at steady state so that parameters do not change with time. The rate of metabolic reaction J_r may be taken as the rate of oxygen intake by a tissue, and J_i may be obtained from Eq. (142)

$$J_i = \frac{\Delta\tilde{\mu}_i}{K_{ii}} - \sum_{j=1}^{n} \frac{K_{ij}}{K_{ii}} J_j - \frac{K_{ir}}{K_{ii}} J_r \tag{144}$$

The last term on the right-hand side represents the active transport of substance *i*.

Eq. (142) and (143) can be applied to analyze sodium flow in the frog skin [5]. The flow of sodium chloride across the skin comprises the flow of sodium ions J_{Na}, which is coupled to the metabolic process $J_{r,\mathrm{tot}}$, while the flow of the chloride ions J_{Cl} may be assumed to be passive transport. The driving forces for the ionic flows are the electrochemical potential differences, and is given for a component i in a simple system as follows

$$\Delta\tilde{\mu}_i = RT(\ln c_{i,\mathrm{I}} - \ln c_{i,\mathrm{II}}) + z_i F(\psi_{\mathrm{I}} - \psi_{\mathrm{II}}) \tag{145}$$

where I and II denote the surrounding compartments adjacent to the membrane, z_i is the valence of the ion, ψ is the electrical potential, and F is the Faraday constant.

Using the dissipation function, given in Eq. (141), the phenomenological equations for the sodium pump can be expressed by

$$\Delta\tilde{\mu}_{\text{Na}} = K_{\text{Na}} J_{\text{Na}} + K_{\text{Na}r} J_{r,\text{tot}} \tag{146}$$

$$\Delta\tilde{\mu}_{\text{Cl}} = K_{\text{Cl}} J_{\text{Cl}} \tag{147}$$

$$A = K_{\text{Na}r} J_{\text{Na}} + K_r J_{r,\text{tot}} \tag{148}$$

Eq. (147) implies that no coupling exist between sodium and chloride flows, while the sodium flow J_{Na} is coupled to the metabolic reaction J_r. Equations (146)-(148) can be applied to these experimental cases:

(i) Short circuit: Consider two electrodes inserted in each compartment. If these electrodes are short circuited, the potential difference ($\psi_{\text{I}} - \psi_{\text{II}}$) is made zero, and an electrical current I is allowed to flow across the membrane. If the experiment is carried out at equal salt concentrations in I and II, so that ($lnc_{\text{I}} - lnc_{\text{II}}$) = 0, and hence we have $\Delta\tilde{\mu}_{\text{Na}} = 0$, $\Delta\tilde{\mu}_{\text{Cl}} = 0$, the only remaining driving force is the affinity A of the metabolic reaction in Eq. (146), and we have

$$J_{\text{Na}} = -\frac{K_{\text{Na}r}}{K_{\text{Na}}} J_{r,\text{tot}} \tag{149}$$

Since the flow of electricity is determined by the ionic flows

$$I = (J_{\text{Na}} - J_{\text{Cl}})F \tag{150}$$

from Eqs. (149) and (150) we have

$$I = -\frac{K_{\text{Na}r}}{K_{\text{Na}}} J_{r,\text{tot}} F \tag{151}$$

Under the short-circuited measurements, electrical current is linearly related to the overall rate of reaction, and the coupling coefficient $K_{\text{Na}r} \neq 0$.

(ii) Open circuit: In an open-circuited potentiometric experiment where $I = 0$, both J_{Na} and J_{Cl} vanish, and Eq. (146) becomes

$$\Delta\tilde{\mu}_{\text{Na}} = K_{\text{Na}r} J_{r,\text{tot}}$$

At steady state the concentrations and electrical potentials in compartments I and II have to be different, and we have

$$RT \ln \frac{c_{\text{Na,I}}}{c_{\text{Na,II}}} + F(\psi_{\text{I}} - \psi_{\text{II}}) = K_{\text{Na}r} J_{r,\text{tot}} \tag{152}$$

$$RT\ln\frac{c_{\mathrm{Cl,I}}}{c_{\mathrm{Cl,II}}} - F(\psi_{\mathrm{I}} - \psi_{\mathrm{II}}) = 0 \tag{153}$$

Addition of Eqs. (152) and (153) yields

$$RT\ln\frac{c_{Na,\mathrm{I}}c_{\mathrm{Cl,I}}}{c_{\mathrm{Na,II}}c_{\mathrm{Cl,II}}} = K_{\mathrm{Nar}}J_{r,\mathrm{tot}} \tag{154}$$

Since $c_{\mathrm{Na}} = c_{\mathrm{Cl}}$ in the compartments, and using the salt concentration denoted by c_s, we have

$$2RT\ln\frac{c_{s,\mathrm{I}}}{c_{s,\mathrm{II}}} = K_{\mathrm{Nar}}J_{r,\mathrm{tot}} \tag{155}$$

Eq. (155) shows the salt distribution due to nonvanishing coupling coefficient K_{Nar}. If the total rate of chemical reaction is known, short-circuit and open-circuit experiments allow us to determine the straight and cross coefficients.

The chemical and electrochemical composition of the extracellular medium is essentially different from the intracellular medium. Equilibrium is achieved when the electrochemical potential of ions is the same on inside and outside, and expressed by

$$RT(\ln c_i - \ln c_o) = Fz(\psi_o - \psi_i) \tag{156}$$

This equation is called *Nernst's law* and relates the diffusion to electrochemical flow. We can reexpress Nerst's law by dividing the numerator and denominator of the right hand side by Avogadro's number, and we get

$${}_o - \psi_i = \frac{kT}{Q}\ln\frac{c_i}{c_a} \tag{157}$$

where k is the Boltzmann constant (k = 8.62 10^{-5} eV/K), and Q is the charge of each ion of the substance (Q = +e for potassium).

Potassium leaves the cell, while the net flow of sodium is inward. A nonequilibrium stationary state for the cell at rest is maintained by the sodium and potassium pumps, which pump out the entering sodium ions and pump the leaking potassium ions back into the cell interior using the certain metabolic output.

The sodium transfer is coupled with the chemical reaction. The electrochemical potential difference for sodium ions is expressed as

$$X_{\mathrm{Na}} = \tilde{\mu}_{\mathrm{Na,ext}} - \tilde{\mu}_{\mathrm{Na,int}} \tag{158}$$

The related phenomenological equations are

$$J_{Na} = L_{Na} X_{Na} + L_{Nar} A \tag{159}$$

$$J_r = L_{Nar} X_{Na} + L_r A \tag{160}$$

Due to the Onsager relations, three coefficients are to be determined. They are the passive permeability to sodium L_{Na}, the metabolic reaction coefficient if there is no sodium transport L_r, and the cross coefficent between the chemical reaction and the sodium flow L_{Nar}. The linear nonequilibrium thermodynamics formulation for the active transport of sodium and the associated oxygen consumption in the frog skin and toad urinary bladders are studied experimentally [2]. Sodium flow J_{Na} is taken as positive in the direction from outer to the inner surface of the tissue. The term J_r is the rate of suprabasal oxygen consumption assumed to be independent of the oxygen consumption associated with the metabolic functions.

It is essential that the phenomenological coefficients evaluated remain near-constant in the course of perturbations of the external variables. In general, these coefficients and A will be the functions of state, and may be influenced by perturbation, which alters the tissue configuration and composition. If X_{Na} is changed by the perturbation of $\Delta\psi$ alone, then Eqs. (159) and (160) become

$$J_{Na} = L_{Na}(-F\Delta\psi) + L_{Nar} A \tag{161}$$

$$J_r = L_{Nar}(-F\Delta\psi) + L_r A \tag{162}$$

where $\Delta\psi = \psi_i - \psi_o$. For constant affinity A, the phenomenological coefficients are

$$L_{Na} = -\left(\frac{1}{F}\right)\left(\frac{dJ_{Na}}{d(\Delta\psi)}\right) \tag{163}$$

$$L_{Nar} = -\left(\frac{1}{F}\right)\left(\frac{dJ_r}{d(\Delta\psi)}\right) \tag{164}$$

$$L_r = \frac{J_{ro}}{A} \tag{165}$$

where J_{ro} is the rate of oxygen uptake. The quantitative values of the phenomenological coefficients indicate the character of the coupled transport.

We can express the phenomenological Eqs. (159) and (160) in terms of the resistance formulation

$$X_{Na} = K_{Na} J_{Na} + K_{Nar} J_r \tag{166}$$

$$A = K_{rNa} J_{Na} + K_r J_r \tag{167}$$

For analysis we can define the phenomenological stoichiometry Z, and the degree of coupling q as follows

$$Z = \left(\frac{K_r}{K_{Na}} \right)^{1/2} \tag{168}$$

$$q = -\frac{K_{Nar}}{(K_{Na} K_r)^{1/2}} = \left(1 - \frac{(J_r)_{J_{Na}=0}}{J_{ro}} \right) \tag{169}$$

As for the general case we have for the degree of coupling $-1 \le q \le 1$.

In active transport, the coupling of the driving reactions to the uphil transport of substrates cannot be perfectly tight. There will be a certain amount of slippage. Evolution in organisms implies a tight coupling to avoid wasting metabolic energy. However the rather loose coupling for the calcium and sugar transport systems could indicate the most adventageous trade off between the efficiency and the rate that can be accomplished in primary and secondary active transports [64].

Two processes will be coupled if they are combined in a single overall reaction system. When the coupling protein guides the reaction along a path involving both processes, its specificity and transport properties have to be altered through the substrate binding energy level. The ratio of coupled to uncoupled rates, called the thightness of coupling [64], is a function of the binding energy; a strong binding may limit the tightness of coupling by reducing the rate of substrate dissociation after it has been transferred through the membrane.

From Eqs. (166) and (167) we can express the flows as functions of the forces and resistance coefficients

$$J_{Na} = \frac{X_{Na} + (q/Z)A}{K_{Na}(1-q^2)} \tag{170}$$

$$J_r = \frac{(q/Z)X_{Na} + (1/Z)^2 A}{K_{Na}(1-q^2)} \tag{171}$$

The above model is useful, however biological membranes, which transport various substances, are complex systems. Such membranes are close to composite membranes with series and parallel elements. A value of q < 1 shows an incomplete coupling, where a metabolic energy must be expended to maintain an electrochemical potential difference of sodium even in the absence of active transport, that is $(J_r)_{JNa=0} \neq 0$. The efficiency of energy conversion is a sensitive function of the degree of coupling. In principle, incomplete coupling could come about in several ways. They are the possibility of leakage and recirculation of transported sodium ions. Also we may expect incomplete coupling of oxidative phosphorylation or incomplete coupling of the mechanism linking ATP utilization and translocation of Na^+.

Another approach to the problem of energy transformation in biomembranes assumes that the coupling device operates as an enthalpic molecular machine, which directly utilizes the energy released during the energy-donating process. For tightly coupled energy-donating and energy-accepting systems, the energy released has not enough time for dissipation over all degrees of freedom of the macroscopic compartment, being at first localized predominantly on certain specific degrees of freedom. A fast local chemical change is accompanied by the fast vibrational relaxation of the active center and its immediate surroundings, while the whole macromolecular structure remains practically unchanged. The subsequent changes are of a relaxation nature and might require the formation of weak secondary bonds, and the overcoming of a large entropy barrier. Conformational relaxation can last for microseconds, and even seconds. After a fast chemical transformation of the enzyme, active center is already changed, while the structure of the main volume of protein globule remains the same, i.e., initially the globule exists in an out-of-equilibrium state relative to the changed active center. There appears a structural strain between the relaxed and unchanged parts of the protein globule. This strain is slowly released in the course of protein conformation relaxation.

For a specific membrane, the phenomenological equations relating the flows and forces of either vectorial or scalar character may be written. Such flows and forces must be derived from an appropriate dissipation function. Kedem [44] reported a membrane transport based on the following dissipation function

$$\Psi = \mathbf{j}_w \nabla \mu_w + \mathbf{j}_1 \nabla \bar{\mu}_1 + \mathbf{j}_2 \nabla \bar{\mu}_2 + J_r A \tag{172}$$

Where the subscripts w, 1, and 2 refer to water, cation, and anion, respectively, and A and J_r refer to chemical affinity and chemical rate, respectively. Therefore we have the corresponding linear relations in the resistance type formulations where the forces are expressed as the function of flows

$$\Delta \mu_w = K_{ww} \mathbf{j}_w + K_{w1} \mathbf{j}_1 + K_{w2} \mathbf{j}_2 \tag{173}$$

$$\Delta\bar{\mu}_1 = K_{1w}\mathbf{j}_w + K_{11}\mathbf{j}_1 + K_{12}\mathbf{j}_2 + K_{1r}J_\mathrm{r} \tag{174}$$

$$\Delta\bar{\mu}_2 = K_{2w}\mathbf{j}_w + K_{21}\mathbf{j}_1 + K_{22}\mathbf{j}_2 + K_{2r}J_\mathrm{r} \tag{175}$$

$$A = K_{r1}\mathbf{j}_1 + K_{r2}\mathbf{j}_2 + K_{rr}J_\mathrm{r} \tag{176}$$

The reason for choosing the resistance formulation is that in systems in which several flows interact, the resistance coefficients can reflect the extent of the interactions directly. Also, the resistance formulation utilizes the flows as independent variables, and it is easy to measure and control the flows rather than the forces. The nonzero values of K_{1r} and K_{2r} indicate coupling between the ionic flows and reaction. The transport of an ion i has to be active if the coefficient K_{ir} is not zero. For example, in certain species the electrical current across the short-circuited frog skin mounted between identical solutions consists of the flow of sodium ions. There is a difference between the coefficients K_{ir} and the other coefficients appearing in Eqs. (173)-(176). Since both J_r and A are scalars, while $\Delta\bar{\mu}_i$ and $\boldsymbol{j}_i$ are vectors, the coefficients K_{ir} must have the vectorial character.

For a model of the living cell, Prigogine showed that a stationary nonequilibrium distribution of matter determined by the rate of metabolic reaction can arise within the cell. Consider that initially uncharged species M and O move into the cell, where M is reacted and transformed to N. Some of the N flows out the cell. The transformation is mediated by the action of an enzyme confined to the interior of the cell. The component O does not take part in the transformation. However the flow of O is coupled with the flow of M. For this system the phenomenological equations in the linear range are

$$\Delta\mu_M = K_M\mathbf{j}_M + K_{MO}\mathbf{j}_O \tag{177}$$

$$\Delta\mu_N = K_N\mathbf{j}_N \tag{178}$$

$$\Delta\mu_O = K_{MO}\mathbf{j}_M + K_O\mathbf{j}_O \tag{179}$$

Here $\mathbf{j}$ refers to inward fluxes, while $\Delta\mu_M = \mu_i^{\mathrm{ex}} - \mu_i^{\mathrm{in}}$ and A^{in} refer to the affinity of the reaction in the cell. Eqs. (177)-(179) indicate the coupling between certain flows and lack of coupling between others. The metabolic reaction occurring in the cell is not coupled to any of the flows. After a certain time the system reaches a state such that the concentrations of M and N, but not that of O become constant, so that we have in the stationary state

$$J_r = \mathbf{j}_M = -\mathbf{j}_N \tag{180}$$

If we denote by A^{ex} the affinity of the reaction measured externally, where the requisite enzyme is absent, we have

$$A^{ex} - A^{in} = (\mu_M^{ex} - \mu_N^{ex}) - (\mu_M^{in} - \mu_N^{in}) = \Delta\mu_M - \Delta\mu_N \quad (181)$$

By substituting Eq. (180) into Eqs. (177) to (179), we get

$$\Delta\mu_M = K_M J_r + K_{MO}\mathbf{j}_O \quad (182)$$

$$\Delta\mu_N = -K_N J_r \quad (183)$$

$$\Delta\mu_O = K_{MO} J_r + K_O \mathbf{j}_O \quad (184)$$

$$A^{in} = K_{ri} J_r \quad (185)$$

Set of Eqs. (182)-(185) is combined with Eq. (181), and we have

$$\Delta\mu_O = K_O \mathbf{j}_O + K_{MO} J_r \quad (186)$$

$$A^{ex} = K_{MO}\mathbf{j}_O + K_{re} J_r \quad (187)$$

where

$$K_{re} = K_{ri} + K_M + K_N$$

Only constrained is the fixing of A^{ex}, and the system eventually reaches a state, in which $\mathbf{j}_O = 0$. In this state an accumulation of O occurs, which is given by

$$\Delta\mu_O = K_{MO} J_r = \frac{K_{MO}}{K_{re}} A^{ex} \quad (188)$$

The stationary state of the whole cell, represented by Eq. (180), yields to a new dissipation function corresponding to Eqs. (186) and (187)

$$\Psi = \mathbf{j}_O \Delta\mu_O + J_r A^{ex} \quad (189)$$

In this example, a stationary state coupling occurs between the flow of component O and the reaction. The system is a two-compartment system. One compartment is accessible only through the membrane and contains an enzyme. The system therefore is basically unsymmetrical, and the anisotropy is not necessary. Active transport in the cells such as muscle cells or red cells is not associated with the mechanism considered above. The coupling is generally a property of the membrane, and associated with enzymes that are integral part of the membrane.

10. MOLECULAR EVOLUTION

The living systems utilize a set of genetic instructions to develop a physical characteristics. The quantitative theories for describing the information transfer have focused on phenomenological models, and generally assume that the organization and transfer of information, while constrained by the laws of chemistry and physics, may not be necessarily a consequence of these laws. Proteins are synthesized as linear polymers with the covalent attachments of successive amino acids, and many of them fold into a three-dimensional structure defined by the information contained within the characteristic sequence [95]. This folding results largely from an entropic balance between hydrophobic interactions and configurational constraints. The information content of a protein structure is essentially equivalent to the configurational thermodynamic entropy of the protein relating the shared information between sequence and structure [94,95].

Studies on the evolution of protein sequences suggest that the shared information between sequence and structure can be related to the configurational entropy of the protein. There can be a thermodynamic arrow for the evolution of sequence information, because it will be according to the thermodynamics of the system. From the perspective of the fluctuation-dissipation approach [11], the time evolution of a protein is related to the evolution of the shared information entropy between sequence and structure.

Dewey [94] proposed that the time evolution of a protein depends on the shared information entropy S between sequence and structure, which can be described with a nonequilibrium thermodynamics theory of sequence-structure evolution. The sequence complexity follows the minimal entropy production resulting in a steady nonequilibrium state [11]

$$\frac{\partial}{\partial X_j}\left(\frac{dS}{dt}\right) = 0 \tag{190}$$

A statistical mechanical model of thermodynamic entropy production in a sequence-structure system suggests that the shared thermodynamic entropy is the probability function that weights any sequence average. The sequence information is defined as the length of the shortest string that encodes the sequence; and the connection between sequence evolution and nonequilibrium thermodynamics is that the minimal length encoding of specific amino acids will have the same dependence on sequence as the shared thermodynamic entropy.

Dewey and Donne [94] considered the entropy production of the protein sequence-structure system based on linear nonequilibrium thermodynamics. The change of composition with time is taken as the flow, while the sequence information change with the composition is treated as the thermodynamic affinity, which can be interpreted as the chemical potential of the sequence composition. Since the change of entropy with time (dissipation) is a positive quadratic expression in forces, Eq. (190) shows the regions of the sequence that are conserved; the rest of the sequence is driven to a minimum entropy production, hence toward the lower complexity seen in the protein sequence, creating a stable state away from equilibrium with a specific arrow of time.

At steady state, a system decreases its entropy production and loses minimal amounts of free energy led to the concept of least dissipation, which is the physical principle underlying the evolution of biological systems [5,29]. A restoring and regulating force acts in any fluctuation from the stationary state.

11. MOLECULAR MACHINES

Either the hydrolysis of ATP or the draining of an ion gradient is converted into an osmotic work by a pump or a mechanical work by a motor protein. Substrate binding sites in pumps or motor proteins may occur in two sections: (i) one of them binds the substrate, and (ii) the other converges on the bound substrate molecule, and the protein changes into an altered conformation. Some important biological processes resemble macroscopic machines governed by the action of molecular complexes [96-98]. Pumps are commonly used for the transport of ions and molecules across the biological membranes, while the word motor is used for transducing chemical energy into mechanical work in the form of rotatory or translationary by proteins or protein complexes [98]. Some identified motor proteins such as kinesins and dyneins move along tubulin flaments, and myosin move along filaments These motor protein families have a major role in muscle contraction, cell division, and transport of substances in and out of the cell. Molecular motors are isothermal and hence the Carnot efficiency concept is not applicable; the internal states are in local equilibrium; they operate with a generalized force for the motor/filament system; this may be the external mechanical force f_{ext} applied to the motor and the affinity A, which measures the free-energy change per ATP molecule consumed

$$\text{ATP} \Leftrightarrow \text{ADP} + \text{P} \tag{191}$$

$$A = \Delta\mu = \mu_{\text{ATP}} - \mu_{\text{ADP}} - \mu_{\text{P}} \tag{192}$$

The external forces may be optical tweezers, microneedles, or the viscous load of a substance that is carried. These generalized forces create motion, characterized by an average velocity v, and average rate of ATP consumption J_r. Molecular motors mostly operate far from equilibrium, and the velocity and rate of ATP consumptions are not linear functions of the forces. However in the vicinity of linear region, where $A \ll k_B T$, linear relations hold [96]

$$\Psi = vf_{\text{ext}} + J_r A > 0 \tag{193}$$

$$v = L_{11} f_{\text{ext}} + L_{12} A \tag{194}$$

$$J_r = L_{21} f_{\text{ext}} + L_{22} A \tag{195}$$

Here L_{11} is the mobility coefficient, while L_{22} is a generalized mobility relating ATP consumption and the chemical potential difference, and L_{12} and L_{21} are the mechano-chemical coupling coefficients. A given motor/filament system can work in different regimes, and in a regime where the work is performed by the motor, efficiency is defined by

$$\eta = -\frac{vf_{\text{ext}}}{J_r A} \tag{196}$$

For nonlinear motors operating at far from equilibrium velocity reversal allows the direction reversal without a change in microscopic mechanism [96].

Molecular motors are classified in two groups depending on operating in groups or individually. Collection of motors is relevant to muscle contraction. In principle, muscle fibers could oscillate in appropriate conditions; skeletal muscle myofibrils can oscillate spontaneously; for example, spontaneous oscillations of asynchronous muscles are common in the wings of many insects.

The relationship between Gibbs energy of ATP hydrolysis and performance in muscle is sigmoidal, and the normal operating domaine is in the quasi-linear region of the curve; this is inline with the magnetic resonance spectroscopy results obtained from finger-flexor muscle [74].

Proton translocating ATP synthase (F_oF_1) synthesizes ATP from ADP and phosphate, coupled with an electrochemical proton gradient across the biological membrane. Chemiosmotic potentials are also coupled to rotation in multiprotein subunit systems of bacterial flagellar motors and F_oF_1 –ATPases. Rotation of a subunit assembly of the ATP synthase is considered as an essential feature of the

ATPase enzyme mechanism and that of F_oF_1 as a molecular motor generating a torque [99,100].

The bacterial flagellar motor and F_oF_1 are macromolecular assemblies and utilize 6 to 8 distinct protein components to affect chemiosmotic energy transduction. Both assemblies have the integral membrane modules and extensive cytoplasmic modules. With these systems the work is accomplished outside the membrane, whereas the chemiosmotic pumps and transports take place within the membrane. The flagellar motor rotates an external filament, and according to the intracellular signals, it modulates the direction of rotation. F_oF_1 system activity is modified by the intracellular phosphate potential. The energy transduction reactions are reversible for F_oF_1 and partly reversible for flagellar motor. The tightly-coupled and efficient operation of these machines may be achieved by alternating access of protonable residues to the adjacent bulk phases linked to the conformational motions, which generate force [100].

ATP is synthesized by mitochondrial oxidative phosphorylation $J_{p,\mathrm{op}}$ and glycolysis. The latter contribution may be neglected for little or moderate exercize. Exerted power by a muscle P is directly related to the rate of ATP hydrolyzed by the myosin ATPases J_p, and we have

$$P = \alpha J_p \tag{197}$$

where α is a constant and is independent of muscle operation. Some ATP hydrolysis is not associated with the muscle operation $J_{p,\mathrm{l}}$, and may be considered as leak. For the ATP balance for steady state muscle operation Eq. (197) yields

$$P = \alpha(-J_{p,\mathrm{op}} - J_{p,\mathrm{l}}) \tag{198}$$

The quasilinear variation of power with ATP hydrolysis is observed experimentally as contraction is being activated at the level of actinomyocin activity. Kinetic approach suggests that muscle power output varies hyperbolically with ADP concentration. Both the ADP control and the Gibbs energy of ATP hydrolysis control are similar, and when muscle power is varied voluntarily, muscle energetics may be represented by the linear flow-force relationships [74].

12. EVOLUTIONARY CRITERIUM

Tellegen's theorem can be used in an evolving network, where the forces are allowed to change with time, and after a time interval *dt*, the forces become $X_i + dX_i / dt$. Since, according to the Tellegen's theory, the flows and forces lie in the orthogonal spaces, we have

$$dt\Sigma J_i \frac{dX_i}{dt} = 0 \tag{200}$$

Since *dt* is an arbitrary time interval, we get

$$\Sigma J_i \frac{dX_i}{dt} = 0 \tag{201}$$

Eq. (201) comprises both the reversible and irreversible contributions

$$\Sigma\left(J_i \frac{dX_i}{dt}\right)_{\text{irrev}} + \Sigma\left(J_i \frac{dX_i}{dt}\right)_{\text{rev}} = 0 \tag{202}$$

or Eq. (202) rewritten as

$$\Sigma\left(J_i \frac{dX_i}{dt}\right)_{\text{irrev}} = -\Sigma\left(J_i \frac{dX_i}{dt}\right)_{\text{rev}}$$

Reversible part of Eq. (202) obeys the constitutive relations

$$C_i = \frac{dN_i}{d\mu_i} \text{ and } \mathrm{J_i} = C_i \frac{d\mu_i}{dt} \geq 0 \tag{203}$$

Since ideal capacitor capacity $C_i >> 0$, we have

$$C_i \frac{dX_i}{dt} = \frac{J_i^2}{C_i} \geq 0 \tag{204}$$

Therefore the first term of Eq. (202) is negative definite

$$\Sigma\left(J_i \frac{dX_i}{dt}\right)_{\text{irrev}} \leq 0 \tag{205}$$

Eq. (205) is the evolutionary criterium of Prigogine and Glansdorff obtained from the general dissipative function [21]

$$\Psi = \Sigma\left(J_i X_i\right)_{\text{irrev}} \geq 0 \tag{206}$$

The change of the dissipation function with time yields

$$\frac{d\Psi}{dt} = \Sigma\left(J_i \frac{dX_i}{dt}\right)_{\text{irrev}} + \Sigma\left(X_i \frac{dJ_i}{dt}\right)_{\text{irrev}} \tag{207}$$

In the linear region of the thermodynamic branch and with constant phenomenological coefficients, we have

$$\Sigma\left(J_i \frac{dX_i}{dt}\right)_{\text{irrev}} = \Sigma\left(X_i \frac{dJ_i}{dt}\right)_{\text{irrev}} \tag{208}$$

Combining Eqs. (207) and (208), we have

$$\frac{d\Psi}{dt} = 2\Sigma\left(J_i \frac{dX_i}{dt}\right)_{\text{irrev}} \tag{209}$$

From Eqs. (205) and (210), we obtain

$$\frac{d\Psi}{dt} \leq 0 \tag{210}$$

Eq. (210) is valid only in the linear region, which may be rare in biology. Eq. (205) can be used for the evolution of all biological networks, which can be characterized by thermodynamic considerations. Eq. (205) is valid for both linear and nonlinear constitutive relations, and can be used for the quasi-equilibrium and far-from-equilibrium regions of the thermodynamic branch [108].

Mikulecky [22] discussed the nonequilibrium thermodynamics approach to energy conversion in coupled, linear and nonlinear systems along with the impact of Tellegen's theorem.

REFERENCES

[1] D. Jou and J.E. Llebot., Introduction to the Thermodynamics of Biological Processes, Prentice Hall, Englewood Cliffs, NJ, 1990.

[2] S.R. Caplan and A. Essig, Bioenergetics and Linear Nonequilibrium Thermodynamics, The Steady State, Harvard University Press, Cambridge, 1983.

[3] J. T. Edsall and H. Gutfreund, Biothermodynamics: To study of Biochemical Processes at Equilibrium, Wiley, Chichester, 1983.

[4] J.W. Stucki, Euro. J. Biochem., 109 (1980) 257.

[5] A. Katchalsky and P.F. Curran, Nonequilibrium Thermodynamics in Biophysics, Harvard University Press, Cambridge, 1967.

[6] B. Korzeniewski, Biochem. Mol. Biol. Int., 39 (1996) 415.

[7] J.W. Stucki, Adv. Chem. Phys., 55 (1984) 141.

[8] J.W. Stucki, Euro. J. Biochem., 109 (1980) 269.
[9] H.V. Westerhoff, Trends Biochem. Sci., 7 (1982) 275.
[10] Y. Hatefi, The mitochondrial electron-transport and oxidative phosphorylation system, Ann. Rev. Biochem., 54 (1985) 1015.
[11] D. Kondepudi and I, Prigogine, Modern Thermodynamics, Wiley, New York, 1999.
[12] H.V. Westerhoff, Pure and Appl. Chem., 65 (1993), 1899.
[13] G.C. Brown and M.D. Brand, Biochem. J., 225 (1985) 399.
[14] P. Glansdorff and I. Prigogine, Thermodynamic Theory of Structure, Stability and Fluctuations, Wiley, New York 1971.
[15] S. Soboll and J.W. Stucki, Biochim. Biophys. Acta, 807 (1985) 245.
[16] J.W. Stucki, M. Compiani and S.R. Caplan, Biophys. Chem., 18 (1983) 101.
[17] H.V. Westerhoff and K. Van Dam, Thermodynamics and Control of Biological free Energy Transduction, Elsevier, Amsterdam, 1987.
[18] D. Walz and S.R. Caplan, Biophys. J., 69 (1995) 1698.
[19] K.J. Rothschild, S.A. Ellias, A. Essig and H.E. Stanley, Biophys. J., 30 (1980) 209.
[20] B. Korzeniewski, Mol. Cell. Biochem., 174 (1997) 137.
[21] G. Nicolis and I. Prigogine, Self-Organization in Nonequilibrium Systems, Wiley, New York, 1977.
[22] J.W. Stucki, Proc. Royal Soc. London Series-Biol. Sci., 244 (1991) 197.
[23] V.P. Skulachev, Biochim. Biophys. Acta, 1363 (1998) 100.
[24] J.W. Stucki, L.H. Lehmann and P. Mani, Biophys. Chem., 19 (1984) 131.
[25] J.J. Tomashek and W.S.A. Brusilow, J. Bioenergetics Biomemb., 32 (2000) 493.
[26] D. Jou and F. Ferrer, J. Theor. Biol., 117 (1985) 471.
[27] S. Schuster and H.V. Westerhoff, Biosystems, 49 (1999) 1.
[28] H.V. Westerhoff, J.S. Lolkema, R. Otto and K.J. Hellingwerf, Biochim. Biophys. Acta, 683 (1982) 181.
[29] B. Alberts, D. Bray, J. Lewis, M. Raff, K. Roberts and J.D. Watson, The Molecular Biology of the Cell, Garland, New York, 1994.
[30] B. Korzeniewski, Biophys. Chem., 59 (1996) 75.
[31] A. Katchalsky, In Permeability and Function of Biological Membranes, L. Bolis, A. Katchalsky, R.D. Keynes, W.R. Loewenstein and B.A. Pethica, Eds., Elsevier, Amsterdam, 1970.
[32] H.V. Westerhoff, P.R. Jensen, J.L. Snoep and B.N. Kholodenko, Thermochimica Acta, 309 (1998) 111.
[33] C. Nigon, P. Michalon, B. Perrin and B. Maisterrena, J. Memb. Sci., 144 (1998) 237.
[34] O. Kedem and S.R. Caplan, Trans. Faraday Soc., 61 (1965) 1897.
[35] M. Wyss, J. Smeitink, R.A. Wevers and T. Wallimann, Biochim. Biopys. Acta, 1102 (1992) 119.
[36] M.P. Murphy, Biochim. Biophys. Acta-Bioenergetics, 1504 (2001) 1.
[37] J. Castresana, Biochim. Biophys. Acta, 45098 (2001) 1.
[38] B. Kadenbach, M. Huttemann, S. Arnold, I. Lee and E. Bender, Biology & Medicine, 29 (2000) 211.
[39] O.A. Candia and P.S. Reinach, Am. J. Physiol., 242 (1982) F690.
[40] S. Nath, A, Pure Appl. Chem., 70 (1998) 639.
[41] O.Toussaint and E.D. Schenider, Comparative Biochem. Physiol. Part A, 120 (1998) 3.
[42] U. von Stockar and J.-S. Liu, Biochim. Biophys. Acta, 1412 (1999) 191-211.
[43] T.D. Brock, K.M. Brock and D.M. Ward, Basic Microbiology with applications, 3rd ed., Prentice Hall, Englewood Cliffs, NJ, 1986.

[44] O. Kedem, In Membrane Transport and Metabolism, Ed. by A. Kleinzeller and A. Kotyk, Academic Press, New York, 1961.
[45] S. Cortassa, M.A. Aon and H.V Westerhoff, Biophys. J., 60 (1991) 794.
[46] B. Korzeniewski, Mol. Cell. Biochem., 184 (1998) 345.
[47] J.J. Lemasters, J. Biol. Chem., 259 (1984) 3123.
[48] D. Walz, Biochim. Biophys. Acta, 505 (1979) 279.
[49] D. Walz, Biochim. Biophys. Acta,, 1019 (1990) 171.
[50] F. Baumann, Acta Biotechnol., 11 (1991) 3.
[51] H.V. Westerhoff, B.A. Melandri, G. Venturoli, G.F. Azzone and D.B. Kell, Biochim. Biophys. Acta, 768 (1984) 257.
[52] B.H. Groen, H.G.J. Vanmil, J.A Berden and K. Vandam, Biochim. Biophys. Acta, 1140 (1992) 37.
[53] T. Pfeiffer, S. Schuster and S. Bonhoeffer, Science, 292 (2001) 504.
[54] P.R. Territo, V.K. Mootha, S.A. French and R.S. Balaban, Am. J. Physiol.-Cell Physiol., 278 (2000) 423.
[55] M.V. Mesquita, A.R. Vasconcellos and R. Luzzi, Contemporary Phys., 40 (1999) 247.
[56] G. Lenaz, Biochim. Biophys. Acta, 1366 (1998) 53.
[57] V.A. Etkin, Biofizika, 40 (1995) 668.
[58] J. Nielsen, Biochem. J., 321 (1997) 133.
[59] F. Wagner, R. Falkner and G. Falkner, Planta, 197 (1995) 147.
[60] H.V. Westerhoff, C.J.A. Vanechteld and J.A.L. Jeneson, Biophys. Chem., 54 (1995) 137.
[61] A. Stephani and R. Heinrich, Bull. Math. Biol., 60 (1998) 505.
[62] A. Goel, J. Lee, M.M. Domach and M.M. Ataai, Biotech. Bioeng., 64 (1999) 129.
[63] R.M. Krupka, Biochim. Biophys. Acta, 1183 (1993) 105.
[64] R.M. Krupka, J. Memb. Biol., 165 (1999) 35.
[65] C.A. Mannella, J. Bioenergetic Biomemb., 32 (2000) 1.
[66] C.B. Cairns, J. Walther, A.L. Harken and A. Banerjee, American J. Physiology-Regulatory Integrative and Comparative Physiol., 433 (1998) R1376.
[67] H.T. Tien and A. Ottova-Leitmannova, Membrane Biophysics-As viewed from experimental bilayer lipid membranes, Elsevier, Amsterdam, 2000.
[68] P. M. Bisch, Brazilian J. Med. Biol. Res., 26 (1993) 417.
[69] M. Rigoulet, A. Devin, P. Espie, B. Guerin, E. Fontaine, M.-A. Piqiet, V. Nogueira and X. Lverve, Biochim. Biophys. Acta, 1365 (1998) 117.
[70] C.V. Jones, Biological Energy Conservation, Wiley, New York, 1976.
[71] R. Heinrich and S. Schuster, Biosystems 47 (1998) 61.
[72] G. Falkner, F. Wagner and R. Falkner, Acta Biotheoretica, 44 (1996) 283.
[73] W. Ebel, J. Math. Biology, 21 (1985) 243.
[74] H.V. Westerhoff, C.J.A. van Echteld and J.A.L. Jeneson, Biophysical Chem., 54 (1995) 137.
[75] M.D. Brand, J. Theoret. Biol., 245 (1990) 267.
[76] E.M. Popov, Biochem. Mol. Biol. Int., 47 (1999) 443.
[77] T.L. Hill, Free Energy Transduction in Biology, Academic, New York, 1977.
[78] G. Meszena and H.V. Westerhoff, J. Phys. A: Math. Gen., 32 (1999) 301.
[79] M. Rutgers, K. Vandam and H.V. Westerhoff, Critical Rev. Biotech., 11 (1991) 367.
[80] T.G. Dewey and M.D. Donne, J. Theor. Biol., 193 (1998) 593.
[81] H.H. Dung and C-H Chen, J. Memb. Sci., 56 (1991) 327.
[82] P. Mendes, D.B. Kell and H.V. Westerhoff, Biochim. Biophysica Acta, 1289 (1996) 175.
[83] E.J. Bormann, H.H. Grosse and J. Menz, Biotech. Adv., 8 (1990) 277.

[84] H.V. Westerhoff, K.J. Hellingwerf and K. Vandam, Proc. Nat. Acad. Sci. of USA-Biol. Sci., 80 (1983) 305.
[85] J.J. Lemasters, R Grunwald and R.K. Emaus, J. Biol. Chem., 259 (1984) 3058.
[86] P. Jezek, A.D.T. Costa and A.E. Vercesi, J. Biol. Chem., 271 (1996) 32743.
[87] S. Soboll, J. Bioenerg. Biomemb., 27 (1995) 571.
[88] P. Jezek, J. Bioenerg. Biomemb., 31 (1999) 457.
[89] L. Woitczak and P. Schonfeld, Biochim. Biophys. Acta 1183 (1993) 41.
[90] P. Schonfeld, P. Jezek, E.A. Belyaeva, J. Borecky, V.S. Silshenkov, M.R. Wieckowski and L. Wojtczak, Euro. J. Biochem., 240 (1996) 387.
[91] A.M. Diehl and J.B. Hoek, J. Bioenerg. Biomemb., 31 (1999) 493.
[92] T. Yagi, A. Matsunoyagi, S.B. Vik and Y. Hatefi, Biochem., 23 (1984) 1029.
[93] A.A. Kalachev, L.M. Kardivarenko, N.A. Plate and V.V.C. Bagreev, J. Memb. Sci., 47 (1992) 1.
[94] T.G. Dewey, Phys. Rev. E, 54 (1996) R39.
[95] T.G. Dewey, Phys. Rev. E, 56 (1997) 4545.
[96] F. Julicher and A. Ajdari, J. Prost, Rev. Mod. Phys., 69 (1997) 1269.
[97] T. Kreis and R. Vale, Cytoskeletal and Motor Proteins Oxford, Univ. Press, New York, 1993.
[98] P. Dimroth, G. Kaim and U. Matthey, Biochim. Biophys. Acta-Bioenergetics, 1365 (1998) 87.
[99] Y. Sambongi, I. Ueda, Y. Wada and M. Futai, J. Bioenerg. Biomemb., 32 (2000) 441.
[100] S. Khan, Biochim. Biophys. Acta, 1322 (1997) 86.
[101] K.D. Garlid, M. Jaburek, P.Jezek and M. Varecha, Biochim. Biophys. Acta, 1459 (2000) 383.
[102] G. Gnaier, Resp. Physiology, 128 (2001) 277.
[103] O Panke and B. Rumberg, Biochim. Biophys. Acta, 1322 (1997) 183.
[104] E. Cristina and J.A. Hernandez, Biochim. Biophys. Acta, 1460 (2000) 276.
[105] P. Light, Biochim. Biophys. Acta, 1286 (1996) 65.
[106] S. Davis, J. Memb Sci., 56 (1991) 341.
[107] S.K. Friedlander and K.H.Keller, Chem. Eng. Sci., 20 (1965) 121.
[108] G. Oster, A. Perelson and A. Katchalsky, Quart. Rev. Biophys., 1 (1973) 6.

Chapter 13

Nonequilibrium thermodynamics approaches

INTRODUCTION

Applications of nonequilibrium thermodynamics in biological systems are limited because of the linear phenomenological equations between flows and forces. Most of the biological processes take place in the nonlinear region of thermodynamic branch, which is far from equilibrium.

Nonequilibrium phenomena needs to be described by the simultaneous account of mass, temperature and time of the local states with considering the given time and energy dissipation due to temperature changes. The time scale of observation of microscopic changes would be much smaller than that of macroscopic changes. Temperature fluctuations in a microstate will be different from those in a macrostate in which the properties are the averages of many microstate values.

Linear nonequilibrium thermodynamics has some fundamental limitations: (i) it does not incorporate mechanisms into the formulation, or provide values of the phenomenological coefficients, and (ii) it is based on the local-equilibrium hypothesis, and therefore confined to systems in the vicinity of equilibrium. Also, properties not needed or defined in equilibrium may influence the thermodynamic relations in nonequilibrium situations. For example, the density may depend on the shearing rate in addition to temperature and pressure [20]. The local equilibrium hypothesis holds only for linear phenomenological relations, low frequencies and long wavelengths, which make the application of the linear nonequilibrium thermodynamic theory limited for chemical reactions. In the following sections, some attempts that have been made to overcome these limitations are summarized.

1. NETWORK THERMODYNAMICS WITH BOND GRAPH

Highly structured, organized, and coupled systems cannot be reduced to simple systems without losing their unique behavior. For such complex systems network thermodynamics combines classical and nonequilibrium thermodynamics along with the network theory and kinetics to provide a practical formulation. The formulations contain the information, which is vital for describing the organization of the system. Therefore, network thermodynamics method can provide insight into the system topology and permits a systematic analysis of dynamic of the systems. It acts as a bridge between the classical thermodynamics and the general dynamics theory of modern physics, and allows the introduction of thermodynamic concepts into the system approach. Combined with the bond graph methodology, network thermodynamics provides a graphic representation of the processes that control the system behavior. The governing equations can be formulated from a bond graph description of the system to evaluate the perturbations in system configurations and compositions.

In order to define the governing equations, bond graph method identifies the cause (force) and effect (flow) relation for the energy exchange. The model can be modified easily to account for changes in the system or its environmental perturbations. Initial and boundary conditions can be associated within the formulations.

Network thermodynamics is a combination of the electrical network theory with the formalism of nonequilibrium thermodynamics [5,6,11,13,15]. It can be used in the linear and nonlinear regions of nonequilibrium thermodynamics, and has the flexibility to deal with complex systems in which the transport and reactions occurring simultaneously. The results of nonequilibrium thermodynamics due to Onsager can be interpreted and extended to describe coupled, nonlinear systems in biology and chemistry [25].

The bond graph method defines the structure and constitutive equations of the system prior to appropriate mathematical representation. Standard bond graph elements are used to build a model of the structure of the system. Suitable computer programs are available to generate the governing equations, and alternative methods are also developed for deriving equivalent block diagrams, which can represent nonlinear systems.

1.1. Transport processes

Fig. 1 shows a typical membrane system through which a nonelectrolyte substance flows. The elements of the membrane system comprise two reservoirs of 1 and 2 containing the same substances with different levels of the chemical potentials μ_1 and μ_2. In this system the inner (1) and outer (2) compartments communicate through the membrane by the exchange of substances. The system consists of energetic flow between two regions with different chemical potentials, and results an actual flow of power.

For such a system the dissipation function and accompanying linear phenomenological equations are given by

$$\Psi = J_A \Delta\mu_A + J_B \Delta\mu_B \tag{1}$$

$$J_A = L_{AA} \Delta\mu_A + L_{AB} \Delta\mu_B \tag{2}$$

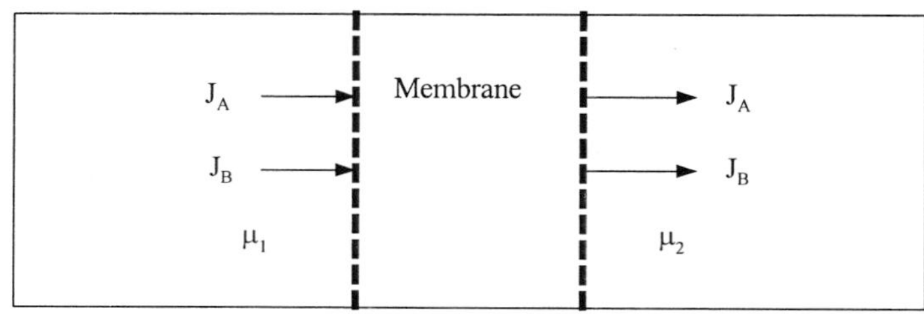

Fig. 1. Membrane-flow system.

$$J_B = L_{BA}\Delta\mu_A + L_{BB}\Delta\mu_B \tag{3}$$

The bond graph of the transport across the membrane is shown in Fig. 2 by a two-port resistance R element. The basic element of the bond graph is the ideal energy bond transmitting power without loss. A bond graph illustrates the system components and their interconnections with arrows, which indicate the positive direction of power flow, associated with the transport processes. All time-dependent processes and all dissipative transformations are localized conceptually in capacity and resistance elements. Two ideal junctions are used in the method; the 0-junction is defined in a way that all forces connected into the junction are equal, so that the sum of all the flows over a 0-junction is zero. At the 1-junction all flows entering or exiting from the junction are equal, and no power accumulates at a 1-junction, and the sum of all the forces is zero.

The flow into the membrane is supplied by reservoir i with a chemical capacity C_i

$$C_i = \frac{dN_i}{d\mu_i} \tag{4}$$

The flow can be defined for the capacitative elements as follows

$$C_i \frac{d\mu_i}{dt} = C_i \frac{dX_i}{dt} = J_i = \frac{dN_i}{dt} \tag{5}$$

The membrane is a resistor, and transmits the flow in a dissipative process. A steady-state membrane relates the thermodynamic force X to the conjugate flow J through a resistance function R, and we have

$$\frac{\partial X_i}{\partial J_i} = \frac{\partial \Delta\mu}{\partial J_i} = R_i \tag{6}$$

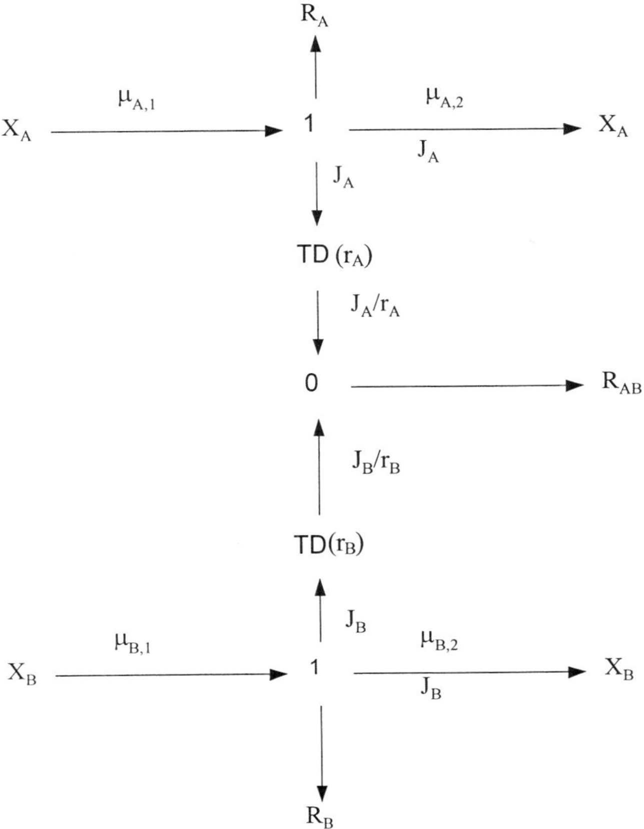

Fig. 2. Bond graph of steady state membrane transport of two substances A and B.

Resistive modules represent the irreversible dissipative processes in the system. This simple model illustrates the general dissipative nature of the flows between the chambers 1 and 2 without any specific indication of the mechanisms involved. The flows J_1 and J_2 are due to the differences in chemical potentials $\Delta\mu_A = \mu_{A,1} - \mu_{A,2}$ and $\Delta\mu_B = \mu_{B,1} - \mu_{B,2}$.

The bond graph in Fig. 2 contains a dissipative coupling between flows A and B, in which only an interacting fractions is involved in the process. Therefore the linear transducer TD, which converts energy from one from to another conserving power, is introduced in the bond graph. Operation of transducer is characterized by a modulus r, which may be a function of the parameter of state, such as temperature or concentration, and is independent of flows and forces.

The scaling of flows and forces by the transducer gives

$$X_2 = rX_1 \text{ and } J_1 = rJ_2 \tag{7}$$

In Fig. 2 there are two transducers, which converts the flows of A and B, and we have J_A/r_A and J_B/r_B, respectively.

At 0-junction the coupled flow J_c is given by

$$J_c = \frac{J_A}{r_A} + \frac{J_B}{r_B} \tag{8}$$

The relation between the force X_c and the flow J_c may be expressed by

$$X_c = R_{AB}(J_c) \tag{9}$$

Similarly, nonlinear relations are assumed for the dissipative elements R_A and R_B

$$X'_A = R_A(J_A) \text{ and } X'_B = R_B(J_B) \tag{10}$$

Summation of the forces around 1-junction yields

$$-\mu_{A,1} + X'_A + \mu_{A,2} + \frac{X_c}{r_A} = 0 \tag{11}$$

$$-\mu_{B,1} + X'_B + \mu_{B,2} + \frac{X_c}{r_B} = 0 \tag{12}$$

Eqs. (11) and (12) may be rearranged using Eqs. (7),. (9), and (10)

$$\Delta\mu_A = R_A J_A + \frac{R_{AB} J_c}{r_A} \tag{13}$$

$$\Delta\mu_B = R_B J_B + \frac{R_{AB} J_c}{r_B} \tag{14}$$

Eqs. (13) and (14) represent the nonlinear phenomenological relations between the external driving forces of permeation flow $\Delta\mu_A$ and $\Delta\mu_B$ and the conjugate flows J_A and J_B.

Linear phenomenological equations obey the Onsager reciprocal relations. For nonlinear region from the symmetry of the Jacobian of forces versus flows, we have

$$\left(\frac{\partial \Delta \mu_A}{\partial J_B}\right)_{J_A} = \left(\frac{\partial \Delta \mu_B}{\partial J_A}\right)_{J_B} \tag{15}$$

As R_AJ_A and r_A are independent of flow J_B, from Eq. (15) we obtain

$$\left(\frac{\partial \Delta \mu_A}{\partial J_B}\right)_{J_A} = \frac{1}{r_A}\frac{\partial R_{AB}}{\partial J_B} \tag{16}$$

Combining Eq. (7) and Eq. (16), we obtain [5]

$$\left(\frac{\partial \Delta \mu_A}{\partial J_B}\right)_{J_A} = \frac{1}{r_A r_B}\frac{dR_{AB}}{dJ_c} \tag{17}$$

$$\left(\frac{\partial \Delta \mu_B}{\partial J_A}\right)_{J_B} = \frac{1}{r_B r_A}\frac{dR_{AB}}{dJ_c} \tag{18}$$

From Eqs. (17) and (18), we have

$$\left(\frac{\partial \Delta \mu_A}{\partial J_B}\right)_{J_A} = \left(\frac{\partial \Delta \mu_B}{\partial J_A}\right)_{J_B} \tag{19}$$

The symmetry of coupled membrane transport holds in a wide range of network structure applications and general behavior of biological networks [5].

A thermodynamic flow system may be fully described in a n-dimensions of flow and n-dimensions of conjugate force. According to Tellegen's theorem, we have

$$\mathbf{J}^{\mathrm{T}} \cdot \mathbf{X} = \sum \mathbf{J}_i \mathbf{X} \geq 0 \tag{20}$$

Eq. (20) is a fundamental outcome of network theory, and show that a thermodynamic flow can be described fully in a 2n-dimensional space in terms of n-dimensions of flow and n-conjugate dimensions of force. Eq. (20) also shows that the general space consists of two orthogonal subspaces: subspace of flow (vector of flow) and subspace of forces orthogonal to that of the flows. One of the consequences of this consideration is the network equivalent of the evolutionary principles of Prigogine and Glansdorff [21]. Tellegen's theorem can be used to demonstrate the existence of Onsager reciprocity and topological justification in the network formulation [22].

The network thermodynamics modeling has been applied for understanding the effects of diffusion coupling in membrane transport of binary flows [7]. In the formalism of network thermodynamics a membrane is treated as a sequence of discrete elements called the lumps, where both dissipation and storage of energy may occur. These lumps are joined in the bond graphs, and have a resistance R_i and capacitance (volume) C, which are defined by

$$R_i = \frac{L}{D_i A n} \tag{21}$$

$$C = \frac{aL}{n} \tag{22}$$

Where D_i is the diffusion coefficient, which may change from lump to lump, n is the number of lumps, a and L are the membrane area and thickness, respectively.

In analogy to the electrical circuits the energy flow in the bond graph can be determined. Diffusion flow J and force X correspond to current and voltage, respectively as seen in Fig. 3, which shows the bond graph for diffusion of a single component. Using the concept of parallel (zero) and series (one) junctions, we may derive the equations showing the dynamics of the transport process

$$\frac{dJ_o}{dt} = \frac{J_o}{C_o R_1} - \frac{J_1}{C_1 R_1} \tag{23}$$

$$\frac{dJ_1}{dt} = \frac{J_o}{C_o R_1} - \frac{J_1}{C_1 R_1} - \frac{J_1}{C_1 R_2} + \frac{J_2}{C_2 R_2} \tag{24}$$

$$\frac{dJ_n}{dt} = \frac{J_{n-1}}{C_{n-1} R_n} - \frac{J_n}{C_n R_n} - \frac{J_n}{C_n R_{n+1}} + \frac{J_{n+1}}{C_{n+1} R_{n+1}} \tag{25}$$

$$\frac{dJ_{n+1}}{dt} = \frac{J_n}{C_n R_{n+1}} - \frac{J_{n+1}}{C_{n+1} R_{n+1}} \tag{26}$$

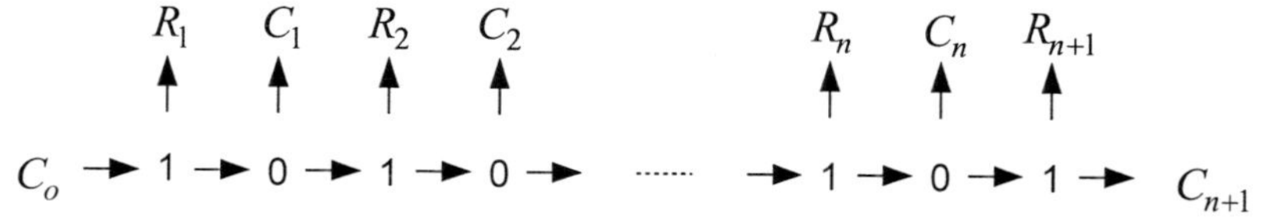

Fig. 3. Network thermodynamic model with bond graph for a single component flow system.

There are n equations for the n lumps of the membrane system, and two equations for the adjacent reservoirs (0 and n+1). All the lump equations have the following common form

$$\frac{dJ_i}{dt} = l_i J_{i-1} - m_i J_i + r_i J_{i+1} \tag{27}$$

where the indices l, m, and r simply mean left, middle and right respectively. The first and the last equations incorporate the boundary conditions. If there are additional flows of $J_{i\text{-}1}$ and J_{n+2}, then corresponding coefficients l_o and r_{n+1} vanish, and we obtain

$$l_o = 0,\ m_o = -\frac{1}{C_o R_1},\ r_o = -\frac{1}{C_1 R_1},\ i = 0 \tag{28}$$

$$l_i = \frac{1}{C_{i-1} R_1},\ m_i = \frac{1}{C_i R_i} + \frac{1}{C_i R_{i+1}},\ r_i = \frac{1}{C_{i+1} R_{i+1}},\quad 1 \le i \le n \tag{29}$$

$$l_{n+1} = \frac{1}{C_n R_{n+1}},\ m_{n+1} = \frac{1}{C_{n+1} R_{n+1}},\ r_{n+1} = 0,\ \ i = n+1 \tag{30}$$

With these assumptions, Eqs. (23)-(26) may be written in a compact form, and Eq. (27) and Eqs. (28)-(30) can be solved numerically.

When there is a two-component flow with coupling, then we have two flows $J_{i,1}$ and $J_{i,2}$ for each lump i, and a matrix of R_{ijk} coefficients. The bond graph is modified additively to accommodate the two-coupled flows, and the two-component coefficients l_{ijk}, m_{ijk}, and r_{ijk} are expressed in terms of R_{ijk} (Fig. 4)

$$l_{ojk} = 0,\ m_{ojk} = -\frac{1}{C_o R_{1,jk}},\ r_{ojk} = -\frac{1}{C_1 R_{1,jk}},\ i = 0 \tag{31}$$

$$r_{ijk} = \frac{1}{C_{i+1} R_{i+1,jk}},\quad 1 \le i \le n, \tag{32}$$

$$l_{n+1,jk} = \frac{1}{C_n R_{n+1,jk}},\ m_{n+1,jk} = \frac{1}{C_{n+1} R_{n+1,jk}},\ r_{n+1,nk} = 0,\ \ i = n+1 \tag{33}$$

The formulation of network thermodynamics bond graph can be used in modeling of coupled nonlinear diffusion in a two-component transport through a membrane. The linear nonequilibrium thermodynamics formulation is used in the network approach to describe the coupled diffusion of water and the cryoprotectant additive in cryopreservation of a living multicellular tissue during the cell freezing [16,2], and in pancreatic islets [23]. Standard membrane transport parameters and interstitial diffusion transport properties have been calculated for the transport of water and cryoprotective agent in pancreatic islets [23]. Assuming that the living tissue is a porous medium, Darcy's law with temperature dependent viscosity is used to model the flows of water and cryopreservation agent. The three independent phenomenological coefficients are expressed in terms of the water and solute permeability, and the reflection coefficient. The network thermodynamics model is able to account for interstitial diffusion and storage, transient osmotic behavior of cells and interstitium, and chemical potential transients in the tissue compartments.

The bond-graph method of network thermodynamics is widely used in the homogeneous and heterogeneous membrane transport [18-20]. Electroosmosis and volume changes within the compartments are the critical properties in the mechanism of cell membrane transport, and the properties can be predicted by the bond-graph method of network thermodynamics [18]. In another study, a network thermodynamics model has been developed to describe the role of epithelial ion transport. The model has four membranes with series and parallel pathways and three transported ions, and simulates the system at both steady state and transient transepithelial electrical measurements [19].

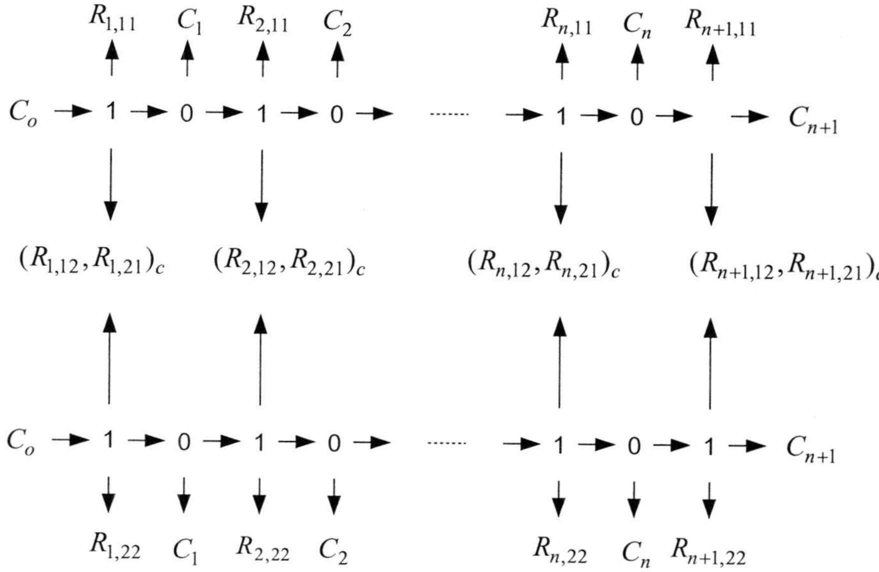

Fig. 4. Network thermodynamic model with bond graph for a two-component flow system.

Network thermodynamics has also been applied for nonstationary diffusion through heterogeneous membranes; concentration profiles in the composite membrane and change of the osmotic pressure have been calculated with the changing boundary and experimental conditions [20,21].

1.2. Chemical reaction processes

Chemical reactions are dissipative processes, and can be easily adapted to a network structure. When there is no diffusion in the system dN_i/dt is related to the flow of reaction J_r, which is measured by the time derivative of the advancement of reaction $d\xi/dt$, and for a reaction

$$_A\mathrm{A} \Leftrightarrow \nu_B \mathrm{B} \tag{34}$$

we have

$$\frac{d\xi}{dt} = \frac{1}{\nu_A}\frac{dN_A}{dt} = \frac{1}{\nu_B}\frac{dN_B}{dt} \tag{35}$$

The rate of flows in terms of components A and B can be expressed based on the forward k_f and backward k_b rate constants

$$J_A = \frac{dN_A}{dt} = -\nu_A(k_f c_A^{\nu A} - k_b c_B^{\nu B}) = -\nu_A J_r \tag{36}$$

$$J_B = \frac{dN_B}{dt} = \nu_B(k_f c_A^{\nu A} - k_b c_B^{\nu B}) = \nu_B J_r \tag{37}$$

The transformation of the capacitive to the resistive flow is expressed by

$$J_i = \nu_i J_r \tag{38}$$

Eq. (38) shows that in bond graph structure all flow contributions will center on a 1-junction

The driving force of chemical reaction is the affinity A

$$A = -\sum \nu_i \mu_i \tag{39}$$

The resistor function of the reaction is given by

$$R_r = \frac{\partial A}{\partial J_r} \tag{40}$$

The resistor function is mostly nonlinear and approaches a constant value only in the vicinity of equilibrium.

The constitutive relation for the capacitative element C_i, given in Eq. (5), and Eq. (38) yields

$$\frac{d\mu_i}{dt} = \frac{\nu_i}{C_i} J_r \tag{41}$$

After multiplying Eq. (41) by $(-\nu_i)$, and summing over i, we obtain

$$-\sum_i \nu_i \frac{d\mu_i}{dt} = -\left(\sum_i \frac{\nu_i^2}{C_i}\right) J_r = \frac{dA}{dt} \tag{42}$$

The change in affinity is related to the progress of the reaction as follows

$$\frac{dA}{dt} = \frac{dA}{dJ_r}\frac{dJ_r}{dt} \tag{43}$$

Combining Eq. (43) with Eq. (40), we have

$$R_r \frac{dJ_r}{dt} = -\left(\sum_i \frac{\nu_i^2}{C_i}\right) J_r \tag{44}$$

From Eq. (44) we may express a typical relaxation time τ

$$\frac{dJ_r}{dt} = -\frac{J_r}{\tau_r} \tag{45}$$

where the relaxation time is expressed by

$$\tau = \left(\sum_i \frac{\nu_i^2}{C_i}\right)^{-1} R_r \tag{46}$$

As Eq. (46) shows that the network structure relates the relaxation time to capacity C and resistance R, which is similar to electrical circuits, and also provides information on any reaction far from equilibrium..

The bond graph structure can be extended for multiple coupled reactions. For example, the change of ith substance in the kth chemical reaction is expressed by

$$\left(\frac{dN_i}{dt}\right)_k = \nu_{ik} J_{r,k} \tag{47}$$

The total chemical transformation of the *i*th component is expressed by

$$\frac{dN_i}{dt} = \sum_k \nu_{ik} J_{r,k} \tag{48}$$

Eq. (48) represents a 0-junction on the capacitor C_i of the component i, since it is a summation of various flows. Such a 0-junction divides the flow for various different chemical reactions, and retains the same chemical potential of this substance in all chemical reactions [5].

A model of a biphasic enzyme membrane reactor for hydrolysis of triglycerides has been formulated according to the bond-graph method of network thermodynamic, and kinetics, permeabilities of fatty acids and glycerides as well as the rates of inhibition of the immobilized enzyme and the concentration of enzyme in a reaction zone are studied [24].

2. MOSAIC NONEQUILIBRIUM THERMODYNAMICS

Another attempt to overcome the phenomenological character of nonequilibrium thermodynamics is called the mosaic nonequilibrium thermodynamics. In the formulation of mosaic nonequilibrium thermodynamics, a complex system is considered as a mosaic of a number of independent building blocks. The species and each process are separately described and hence the biochemical and biophysical structure of the system are included in the description. One of the important feature of the mosaic nonequilibrium thermodynamics model is that it can be expanded to complex physical and biological systems by adding the well-characterized steps. These steps obey the thermodynamic laws and kinetic principles.

The theory of mosaic nonequilibrium thermodynamics has been applied to the following biological free-energy converters [26]:

(i) Bacteriorhodopsin liposomes use the light as the energy source to pump proton beside the passive permeability of protons, K^+ and Cl^-. The flow-force relations for each of the elemental processes are formulated, and by adding the flows of each chemical substance a set of equations are obtained based on the proposed structure of the system. Verification of the mosaic nonequilibrium thermodynamics relations can be used to test the applicability of the proposed structure. If the verification is not realized, then either the formulation is in error or the proposed structure is not appropriate. In testing the formulation experimentally some certain states, such as steady states, are assumed. Effect of addition of ionophores on the rate of light driven proton uptake predictions is experimentally tested; the light-driven pump is inhibited by the electrochemical gradient of protons developed by the system itself.

(ii) The mosaic nonequilibrium thermodynamics formulation of the oxidative phosphorylation uses the chemiosmotic model as a basis, beside assuming that the membrane has certain permeability to protons, and the ATP synthase is a reversible H^+ pump coupled to the hydrolysis of ATP. It is assumed that the reversibility of the reactions allows the coupling of movement of electrons in the respiratory chain for the synthesis of ATP, and the proton gradient across the inner mitochondrial membrane is the main coupling agent. The following flow-force relations are used in the mosaic nonequilibrium thermodynamics formulation

$$J_{\mathrm{H}} = L_{\mathrm{H}} \Delta \tilde{\mu}_{\mathrm{H}} \tag{49}$$

$$J_{\mathrm{O}} = L_{\mathrm{O}} (\Delta G_{\mathrm{O}} + \gamma_{\mathrm{H}} n_{\mathrm{H}} \Delta \tilde{\mu}_{\mathrm{H}}) \tag{50}$$

$$J_{\mathrm{P}} = L_{\mathrm{P}} (\Delta G_{\mathrm{p}} + \gamma_{\mathrm{H}} n_{\mathrm{H}} \Delta \tilde{\mu}_{\mathrm{H}}) \tag{51}$$

The terms L_{H}, L_{O} and L_{P} are the transport coefficients for proton, oxygen and ATP flows, respectively. The γ factors describe the enzyme-catalyzed reactions with the rates having different sensitivities in the free-energy difference for the proton pump and other reactions. This differential sensitivity is a characteristic of the enzyme and is reflected by the mosaic nonequilibrium thermodynamics formulation of the flow-force relationships of that enzyme. The term n_{H} shows the number of protons translocated per ATP hydrolyzed, while J_{H} and J_{O}, and J_{P} indicate the flows of hydrogen, oxygen, and ATP, respectively.

The mosaic nonequilibrium thermodynamics approach was also used for microbial growth [8]. In a simple configuration, aerobic microbial metabolism is considered as combination of three elemental steps that are mutually dependent through the intracellular phosphate potential. First is the catabolism that is the conversion of the growth-supporting energy source with the concurrent generation of ATP. The catabolic substrates are glucose, oxygen, while catabolic products are carbon dioxide and water. Second is the anabolism where the ATP produced in catabolism is utilized for conversion of low-molecular-weight anabolic substrates into biomass. Some of the anabolic substrates are sulphate, phosphate, glucose, and ammonia. Third is the leakage that contains all those processes that utilize ATP without coupling to anabolism. For example, the passive proton flow through the bacterial membrane is a leak. From these three steps the working equations can be derived in terms of free-energy differences in the system [8,26].

In the formulations of mosaic nonequilibrium thermodynamics not all the flows are dependent on all the free energy differences, mainly because only a subset of catalytic components affects each-flow relation. In this respect the models differs from, classical nonequilibrium thermodynamics where all flows are function of all forces.

The mosaic nonequilibrium thermodynamics model uses linear relationships between the rate of anabolism and the rate of catabolism, and the coupling is quantified through the stoichiometric coupling constant. In these linear relations the empirical microbiological constants, such as growth-rate-dependency and grow-rate maintenance, and maximal and theoretical growth can be projected [8]. The mosaic nonequilibrium thermodynamics approach can accommodate biochemically known mechanisms as well as the microbial growth in more complex environments with certain simplifications. For example, the anabolic reactions can be subdivided in various distinctive sections, such as protein and lipids. Also, the effect of dual pathways with different stoichiometry for ATP leading to the same product can be studied. Conductivities, force asymmetry factors, and stoichiometry numbers are treated as real constants. However, in practice composition of the microbial cell is dependent on the environmental conditions and on the growth rate. This, however, increases the mathematical involvement, which is a disadvantage. There are efforts to overcome the shortcomings of the mosaic nonequilibrium thermodynamics in an expanded version of it [27].

One of the important properties of the mosaic nonequilibrium thermodynamics approach is the presumed linearity between flows and forces. Experimental and theoretical studies show that there exists a near-linear region around the inflection point of the flow-force relationships, which is generally sigmodial.

3. RATIONAL THERMODYNAMICS

Rational thermodynamics provides a method for deriving the constitutive equations without assuming the local equilibrium hypothesis. In the formulation, absolute temperature and entropy do not have a precise physical interpretation. It is assumed that the system has a memory, and the behavior of the system at a given time is determined by the characteristic parameters of both the present and the past. However, the general expressions for the balance of mass, momentum and energy are still used [1].

Rational thermodynamics is formulated based on the following hypotheses [31]: (i) absolute temperature and entropy are considered primitive concepts, not limited to near-equilibrium situations, (ii) it is assumed that systems have memory; their behavior at a given instant of time is determined by the past history of the variables, and (iii) the second law of thermodynamics is expressed in mathematical terms by means of the Clausius-Duhem inequality. The balance equations were combined with the Clausius-Duhem inequality by means of arbitrary source terms [32-34], or by an approach based on Lagrange multipliers [35].

The Clausius-Duhem equation is the fundamental inequality for a single component system. The selection of the independent constitutive variables depends on the type of system considered. A process is then described by a solution of the balance equations with the constitutive relations and the Clausius-Duhem inequality.

Studies on thermodynamic restriction on turbulence modeling [36] show that the kinetic energy equation in a turbulent flow is a direct consequence of the first law of thermodynamics, and the turbulent dissipation rate is a thermodynamic internal variable. The principle of entropy generation, expressed in terms of the Clausius-Duhem and the Cluasius-Planck inequalities, imposes restrictions on turbulence modeling On the other hand, the turbulent dissipation rate as a thermodynamic internal variable ensures the mean internal dissipation to be positive and thermodynamic modeling to be meaningful.

Rational thermodynamics is not limited to linear constitutive relations, and when the constitutive equations are expressed in terms of functionals, generally a vast amount of information is necessary [1]. Rational thermodynamics may be useful in the case of memory effects; nonequilibrium processes may approach equilibrium in a longer time than as is generally assumed; as a result nature has a much longer memory of irreversible processes [28]. There are efforts to combine the thermodynamic theories, such as nonequilibrium thermodynamics, rational thermodynamics, and theories using evolution criteria and variational principles into a bracket formalism based on an extension of Hamiltonian mechanics [29]. One result of this bracket approach is a general equation for the nonequilibrium reversible-irreversible coupling (GENERIC) formalism for describing isolated discrete systems of complex fluids [30].

The starting relation of rational thermodynamic is the Clausius-Planck inequality defining the change of entropy between two equilibrium states of 1 and 2

$$\Delta S \geq \int_1^2 \frac{dQ}{T} \tag{52}$$

In the rational thermodynamics formulations Eq. (52) becomes

$$\rho s + \nabla \cdot \frac{\mathbf{q}}{T} - \rho \frac{r}{T} \geq 0 \tag{53}$$

where r is a specific rate of energy supply or energy lost by radiation, and $\mathbf{q}$ is the transport of internal energy due to conduction. Introducing the Helmholtz energy, $a = u\text{-}Ts$, and the following energy balance equation

$$\rho u = -\nabla \cdot \mathbf{q} - \mathbf{P} : \nabla \mathbf{v} + \rho r \tag{54}$$

Eq. (53) becomes

$$-\rho(a + sT) - \mathbf{P} : \mathbf{V} - \frac{1}{T}\mathbf{q} \cdot \nabla T \geq 0 \tag{55}$$

Here **P** is the pressure tensor, and the velocity gradient splits into a symmetric and an antisymmetric part

$$\nabla \mathbf{v} = \mathbf{V} + \mathbf{W} \tag{56}$$

where

$$\mathbf{V} = \frac{1}{2}[\nabla \mathbf{v} + (\nabla \mathbf{v})^T]; \quad \mathbf{W} = \frac{1}{2}[\nabla \mathbf{v} - (\nabla \mathbf{v})^T]$$

Eq. (55) is known as the Clausius-Duhem or the fundamental inequality for a single component system. The selection of the constitutive independent variables is subject to the type of system considered. For example the density, velocity and temperature fields in hydrodynamics are customarily chosen. A process is then described by a solution of the balance equations with the consideration of constitutive relations and Clausius-Duhem inequality.

For simplicity, a set of constitutive equations for a Stokesian fluid, with the absence of memory, are given by

$$\phi = \phi(v, \mathbf{v}, T, \nabla \mathbf{v}, \nabla T) \tag{57}$$

With the absence of memory, the dependence of ϕ is expressed by ordinary functions instead of functionals. With Eq. (57) Clausius-Duhem inequality becomes

$$-\rho\left(\frac{\partial a}{\partial T} + s\right)T - \rho\frac{\partial a}{\partial \mathbf{V}} : \mathbf{V} - \rho\frac{\partial a}{\partial(\nabla T)} \cdot (\nabla T) - \frac{1}{T}\mathbf{q} \cdot \nabla T - \left(\frac{\partial a}{\partial v}\mathbf{U} + \mathbf{P}\right) : \mathbf{V} \geq 0 \tag{58}$$

where the mass conservation is expressed as

$$\rho v = \nabla \cdot \mathbf{v} = \mathbf{V} : \mathbf{U} \tag{59}$$

The rational thermodynamics is not limited to the linear constitutive relations. When the constitutive equations are expressed in terms of functionals with the whole history of the variables, a vast amount of information may be necessary [1].

REFERENCES

[1] D. Jou, J. Casas-Vazquez and G. Lebon, Extended Irreversible Thermodynamics, Springer-Verlag, Berlin, 1993.
[2] R.C. De Freitas, K.R. Diller, C.A. Lachenbruch and F.A. Merchant, In Biotransport: Heat and Mass Transfer in Living Systems, K.R. Diller, ed., The New York Academy of Sciences, New York, 1998.
[3] J.T. Edsall and H. Gutfreund, Biothermodynamics: To study of Biochemical Processes at Equilibrium, Wiley, Chichester, 1983.
[4] D. Jou and J.E. Llebot, Introduction to the Thermodynamics of Biological Processes, Prentice Hall, Englewood Cliffs, NJ, 1990.
[5] G. Oster, A. Perelson and A. Katchalsky, Quart. Rev. Biophys., 1 (1973) 6.
[6] L. Pusher, Studies in Network Thermodynamics, Elsevier, Amsterdam, 1986.
[7] A. M. Simon, P. Doran and R. Paterson, J. Memb. Sci., 109 (1996) 231.
[8] M. Rutgers, K.V. Dam and H. V. Westerhoff, Crypt. Revs, Biotech., 11 (1991) 367.
[9] C. McCallum and R. Paterson, J. Chem. Soc. Faraday Trans., 1. 70 (1974) 2113.
[10] C. McCallum and R. Paterson, J. Chem. Soc. Faraday Trans., 1. 72 (1976) 323.
[11] D.C. Immaculacy, Application of Network Thermodynamics to Problems in Biomedical Engineering, New York University Press, New York, 1994.
[12] R. Paterson, Network thermodynamics, in E.E Bitter, ed., Membrane Structure and Function, Vol. 2, Wiley, New York, 1980.
[13] R. Paterson, Swiss Chem. 10 (1988) 17.
[14] S.J. Paynter, B.J. Fuller and R.W. Shaw, Cryobiology, 39 (1999) 169.
[15] L. Peusner, Studies in Network Thermodynamics, Elsevier, Amsterdam, 1986.
[16] K.R. Diller, J.J. Beaman, J.P. Montoya and P.C. Breedfeld, Trans, ASME J. Heat Trans., 110 (1988) 938
[17] P. Glansdorff and I. Prigogine, Thermodynamic Theory of Structure, Stability and Fluctuations, Wiley, New York 1971.
[18] Y. Imai, J. Memb. Trans., 41 (1989) 3.
[19] S. Mierson and M.I. Fidelman, Math. Comp. Model., 19 (1994) 119.
[20] J. Horno, C.F.G-Fernandez, A. Hayas, and F.G-Cabalero, J. Memb. Sci., 43 (1989) 1.
[21] J. Horno and J Castilla, J. Memb. Sci., 90 (1994) 173.
[22] D.C. Mikulecky, Comp. Chem., 19 (1994) 999.
[23] R.C. de Freitas, K.R. Diller, J.R.T. Lakey and R.V. Rajotte, Cryobiology, 35 (1997) 230.
[24] J. Ceynowa and P. Adamczak, Sep. Purif. Tech., 22-23 (2001) 443.
[25] D.C. Mikulecky, Comp. Chem., 25 (2001) 369.
[26] H.V. Westerhoff and K. V. Dam, Thermodynamics and Control of Biological Free-Energy Transduction, Elsevier, Amsterdam, 1987.
[27] H.V. Westerhoff, J.G. Koster, M.V. Workum and K.E. Rudd, In Control of Metabolic Processes, A. Cornish-Bowden, Ed., Plenum Press. New York, 1990.
[28] D. Kondepudi and I. Prigogine, Modern Thermodynamics, From Heat Engines to Dissipative Structures, Wiley, New York, 1999.
[29] M. Grmela, Phys. Let. A, 102 (1984) 335.
[30] M. Grmela and H.C. Ottinger, Phys. Rev. E, 56 (1997) 6620.
[31[D. Jou and J. Casas-Vazquez, J. Non-Newtonian Fluid Mech., 96 (2001) 77.
[32] B. D. Coleman, Arch. Rational Mech. Anal., 17 (1964) 1.
[33] C. Trusdell and W. Noll, In: Flugge, S., Trusdell, C., eds., Handbook der Physik III/3, Springer, Berlin, 1965.
[34] C. Trusdell, Rational Thermodynamics, 2nd ed., Springer, Berlin, 1984.
[35] I-S, Liu, Arch. Rat. Mech. Anal., 46 (1972) 131.

Chapter 14

Extended nonequilibrium thermodynamics

INTRODUCTION

Care is needed in defining and distinguishing processes with reference to their equilibrium and nonequilibrium states. In equilibrium systems all parts posses the same physical properties, both locally and globally. Nonequilibrium occurs with respect to disturbances in the interior of a system, or between a system and its surroundings; the local stress, strain, temperature, concentration, and energy density vary from one region of a system to another at each time instance leading to an evolution in space and time. A system is constantly disturbed by its surroundings through the boundary. Highly unstable state at the boundary are generally ignored and interpreted as boundary conditions in mechanics or as interfaces across which the stresses are assumed to be continuous. Constantly changing properties cannot be described properly by referring to the system as a whole; some averages of the properties in space and time are necessary. Such averages need to be clearly stated in the utilization and correlation of experimental data, especially when their interpretations are associated with the theories that are valid in equilibrium.

To coordinate components of the generalized flows and the thermodynamic forces can be used to define the trajectories of the evolution of nonequilibriun system in time. A trajectory specifies the curve represented by the flow and force components as function of time in the flow-force space. A useful trajectory can be found and analyzed by a variation principle. In thermodynamics, the variation principles lead to the least energy dissipation and minimum entropy generation at steady states. According to the most general evolutionary criterion open chemical reaction systems are dissipative, and in time they evolve toward an asymptotic state.

Irreversible processes may act as the promoters of disorder at near equilibrium, and as the promoters of order at far from equilibrium. For systems at far from equilibrium there are no general extremum principles to predict the final state of the system. When a system is driven far from equilibrium, the flows are no longer linear functions of the forces. Chemical reactions may reach to nonlinear region easily since the affinities of such systems are in the range of 10-100 kJ/mol. However, transport processes mainly take place in the linear region of thermodynamic branch.

Systems in the nonlinear region can become unstable and evolve to a new organized state. This is caused by the internal fluctuations. The organized states are called the dissipative structures since they constantly need the external energy supply.

1. STABILITY

Stable equilibrium has minimum Gibbs free energy. The necessary (but not sufficient) condition is that the first derivative of the Gibbs free energy G is zero at the possible equilibrium states, and the second derivative of the Gibbs function is positive

$$\delta^2 G = \left(\frac{d^2 G}{dx^2}\right)(\delta x)^2 > 0 \tag{1}$$

here x is a parameter that characterizes the state of the system (for example the concentration of gas). Stability in equilibrium plane in terms of entropy would be stated that the second derivative of the entropy be negative

$$\delta^2 S < 0 \tag{2}$$

Eqs. (1) and (2) represent the energy minimum principle and the entropy maximum principle at equilibrium. Assuming that entropy S is a function of U, V, and N, the general condition for the stability of an equilibrium state to the fluctuations in thermal δT, volume δV, and number of moles δN_i are given by [7]

$$\delta^2 S = -\frac{C_v(\delta T)^2}{T^2} - \frac{1}{T\kappa_T}\frac{(\delta V)^2}{V} - \sum_{i,j}\left(\frac{\delta(\mu_i / T)}{\delta N_j}\right)\delta N_i \delta N_j < 0 \tag{3}$$

where κ_T is the isothermal compressibility

$$\kappa_T = -\left(\frac{1}{V}\right)\left(\frac{\partial V}{\partial P}\right)_T$$

A process is spontaneous, if it obeys the following conditions

$(\Delta S)_{U,V} \geq 0$ at constant U and V

$(\Delta U)_{S,V} \leq 0$ at constant S and V

$(\Delta H)_{S,P} \leq 0$ at constant S and P

$(\Delta A)_{T,V} \leq 0$ at constant T and V

$(\Delta G)_{T,P} \leq 0$ at constant T and P

For a nonequilibrium state the stability condition becomes

$$\frac{d(\delta^2 S)}{dt} = \delta^2\left(\frac{dS}{dt}\right) = \delta^2\Phi > 0 \tag{3}$$

Since a definite function $\delta^2 S$ leads to the stability condition it operates as a Lyapunov function, and assures the stability of the stationary state. As the entropy production is the sum of the products of flows J and forces X, we have

$$\delta^2\Phi = \delta J \, dX \tag{4}$$

where δJ and δX are the perturbations of the flows and forces respectively. For a linear phenomenological law $J = LX$, with $L > 0$, we get from Eq. (4)

$$\delta^2\Phi = L(\delta X)^2 \geq 0 \tag{5}$$

Since $L(\delta X)^2$ is always positive, we consider the stationary states described by the linear phenomenological equations are always thermodynamically stable.

For the systems not far away from equilibrium, the total entropy production reaches a minimum value; this also assures the strability of the stationary state. For the systems far from equilibrium there is no such general criterion to determine the state of the system.

When the phenomenological equations are not linear, the nonlinear stationary states are no longer stable. Starting from these instabilities, the system can display definite structures, sometimes of great biological interest. There are two regions of concern in the thermodynamic branch: the first refers to the states close to equilibrium, without any special order, while the second deals with typical ordering phenomena of states sufficiently far from equilibrium. We may have ordering in time such as appearance of rhythms and ordering in space like morphological structurization. Mainly Prigogine's group in Brussels initiated the studies on such states. There are other schools and trends in this field, such as Nicolis and Prigogine, Haken and Frohlich and Kremer.

If a system is in the thermodynamic equilibrium ($dS/dt \geq 0$), instabilities can occur only at phase transition points, and the new phase may be in a more ordered state (e. g. vapor → liquid), which is a self-sustaining structure. If the stability criterion for a nonequilibrium states ($T\delta^2 S < 0$) is violated a certain class of nonlinear system may appear and is maintained beyond a critical distance from the thermodynamic equilibrium. Such new nonlinear systems may be structured states and can only be maintained on a continuous exchange of energy and matter with the surroundings. Therefore Glansdorff and Prigogine decomposed the change in the entropy production into two parts: one due to the change in forces and the other due to the change in flows, which are expressed as

$$d\Phi = d_x\Phi + d_j\Phi = \sum J_i dX_i + \sum X_i dJ_i \tag{6}$$

They have shown that the first term is negative definite even in cases for which the linear and symmetric phenomenological equations do not hold. By introducing the linear phenomenological equations $J_i = L_{ik} X_k$ with constant coeffcients L_{ik}, we get

$$d_x\Phi = \sum_i J_i dX_i = \sum_i L_{ik} X_k dX_i \tag{7}$$

From the reciprocity relations $L_{ik} = L_{ki}$, we have

$$d_x\Phi = \sum_i X_k (L_{ik} dX_i) = \sum_i X_k dJ_k = d_j\Phi \tag{8}$$

Eq. (8) shows that the contribution of the time change of forces to the entropy production is equal to that of the time change of flows.

In the domain of validity of thermodynamics of irreversible processes, the contribution of the time change of forces to the entropy production is negative or zero.

$$d_x\Phi \le 0 \tag{9}$$

This inequality holds whenever the boundary conditions used are time independent, and it can be extended to include the flow processes as well

$$d\Phi = \int (\sum J_k dX_k) dV \le 0 \tag{10}$$

where the forces X_k and the flows J_k include transport processes, such as convective terms. This inequality is regarded as a general criterion of the evolution in macroscopic physics. However $d\Phi$ is not a total differential, therefore it is not a thermodynamic potential, although it leads to the concept of local potential.

If P is the entropy production in a nonequilibrium stationary state, the change of P due to small changes in the forces δX_i and in the flows δJ_i are

$$\frac{dP}{dt} = \int_V \left(\sum \frac{dX_i}{dt} J_i \right) dV + \int_V \left(\sum X_i \frac{dJ_i}{dt} \right) dV = \frac{d_X P}{dt} + \frac{d_J P}{dt} \tag{11}$$

In the nonlinear regime and for time independent boundary conditions we have

$$\frac{d_X P}{dt} \leq 0 \tag{12}$$

$d_X P$ is not a differential of a state function, so that Eq. (12) does not indicate how the state will evolve, it only indicates that the $d_X P$ can only decrease. So that stability must be determined from the properties of that particular steady state. This leads to the decoupling of evolution and stability in the nonlinear region, and it permits the occurrence of new organized structures beyond a point of instability of a state in nonequilibrium region. The time-independent constraints may lead to the oscillating states in time, such as the well-known Lotka-Volterra interactions where the system rotates irreversibly.

A general criterion for stability of a state is given by the Lyapunov function. A physical system x may be defined by an m dimensional vector with elements of x_i (i = 1,2,..,m) and parameters a_j, and we have

$$\frac{dx_i}{dt} = f_i(x_i, a_j) \tag{13}$$

The stationary states x_{si} are obtained using $dx_i/dt = 0$. We define a small perturbation δx_i and a positive function $L(\delta x)$ called the distance. If this distance between x_i and the perturbed state ($x_{si}+\delta x_i$) steadily decrease in time, the stationary state is stable

$$L(\delta x_i) > 0; \quad \frac{dL(\delta x_i)}{dt} < 0 \tag{14}$$

A function L satisfying Eq. (14) is called a *Lyapunov function*. The second variation of entropy $L = -\delta^2 S$ may be used as a *Lyapunov functional* if the stationary state satisfies $\sum \delta X_i \delta J_i > 0$, hence a nonequilibrium stationary state is stable if

$$\frac{d}{dt}\frac{\delta^2 S}{2} = \sum \delta X_i \delta J_i \tag{15}$$

A functional is a set of functions that are mapped to a real or complex value.

Eq. (15) indicates that the quantity $d(\delta^2 S)/dt$ has the same form for the perturbations from the equilibrium state as well as the nonequilibrium state. In the vicinity of equilibrium the quantity $\sum \delta X_i \delta J_i$ is called the *excess entropy production* [7], which shows the increase in entropy generation. The quantities δJ_i and δX_i denote the deviations of J_i and X_i from the values at the nonequilibrium steady state. The increase in entropy generation for a perturbation from a nonequilibrium state is $\delta P = \delta_X P + \delta_J P$.

Since $\delta^2 S < 0$ under both the equilibrium and nonequilibrium conditions, the stability of a stationary state is accomplished if

$$\frac{d}{dt}\frac{\delta^2 S}{2} = \sum \delta X_i \delta J_i > 0 \tag{15a}$$

We may consider the following autocatalytic reaction, which appears in the reaction scheme of the Brusselator

$$2X + Y \Leftrightarrow 3X \tag{16}$$

We have the forward $r_f = k_1 c_X^2 c_Y$ and backward $r_b = k_2 c_X^3$ reaction rates, respectively. The affinity A and the flow of reaction J_r are given by

$$A = RT \ln \frac{r_f}{r_b} \tag{17}$$

$$J_r = r_f - r_b \tag{18}$$

The excess entropy production is written in terms of $\delta X = \delta A / T$ and δJ_r

$$\frac{1}{2}\frac{d\delta^2 S}{dt} = -R(2k_1 c_{XS} c_{Ys} - 3k_2 c_{Xs}^2)\frac{(\delta c_X)^2}{c_{Xs}} \tag{19}$$

The excess entropy production can become negative, if $k_1 >> k_2$, hence the stationary state may become unstable.

The coupling between chemical kinetics and transport may lead to the dissipative structures which are caused by auto-and cross-catalytic processes with positive and negative feedback, influencing their own rates of reaction. For example, the Belousov-Zhabotinski reaction exhibits a wide variety of characteristic nonlinear phenomena. In the nonlinear region, the possible instabilities in biological systems are (i) multi-steady states, (ii) homogeneous chemical oscillations, and (iii) complex oscillatory phenomena. The thermodynamic buffer enzymes may represent a bioenergetics regulatory principle for the maintenance of a far from equilibrium state.

2. ORDERING IN PHYSICAL STRUCTURES

There are two types of macroscopic structures: equilibrium and dissipative ones. A perfect crystal, for example, represents an equilibrium structure, which is stable and without matter and energy exchange with the environment. On the other hand, the *dissipative structures* maintain their state by exchanging energy and matter constantly with environment. This continuous interaction enables the system to establish an ordered structure with lower entropy than that of equilibrium structure. It is usually assumed that thermodynamics prohibited the appearance of dissipative structures, such as spontaneous rhythms. However thermodynamics can describe the possible state of a structure through the study of instabilities in nonequilibrium stationary states.

2.1. Ordering in convection

If a system is far away from equilibrium, then a dissipative structure associated with the initiation of macroscopic organization such as a motion can appear. The kinetic energy of the motion

$$\Delta S = \frac{-\Delta E_k}{T} \tag{20}$$

accounts for the lower entropy of the system relative to the equilibrium value. One of the best-known physical ordering phenomena is the Benards cells. It is related to heating a fluid held between two parallel horizontal plates separated by a small distance. The lower plate is heated, and the temperature is controlled. The upper plate is kept at a constant temperature. When the temperature difference between the two plates reaches a certain critical value, the elevating effect of expansion predominates, and the fluid starts to move in a structured way; the fluid is divided into horizontal cylindrical convection cells, in which the fluid rotates in a vertical plane. At the lower hot plate the hot fluid rises; later it is cooled at the upper plate, and its density increases again; this induces a movement downward, as seen in Fig. 1. The Benards cells are one of the best-known physical examples of spontaneous structurization as a result of being sufficiently far away from equilibrium that is the large temperature difference between the plates. This structure needs continuous supply of energy, and disappears as soon as the heating stops. For Bernard's cell type of dissipative structure, the relation

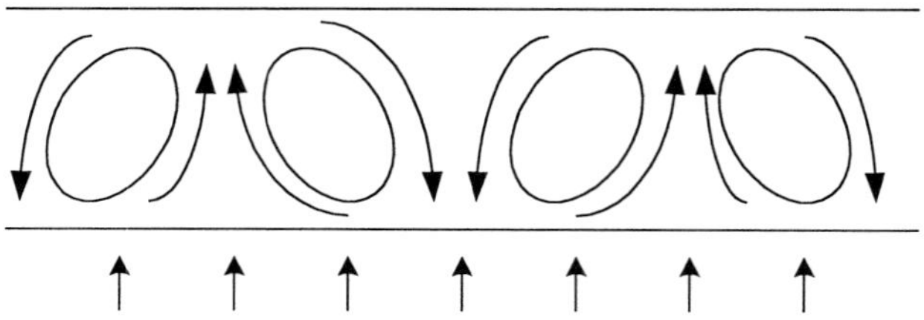

Thermal flow

Heat

Fig. 1. Thermal flow in Benard's cells.

$$d\Phi = \int (\sum_i J_i dX_i) dV \le 0 \tag{21}$$

is introduced, $d\Phi$ is an ordinary differential and Eq (21) can be integrated to determine the system by the maximum value of a certain potential.

The critical temperature difference $(T_2 - T_1)_{\text{crt}}$ can be determined from the dimensionless Rayleigh number Ra

$$Ra = \frac{gH^3\alpha}{\eta k}(T_2 - T_1)_{\text{crt}} \tag{22}$$

where g is the acceleration of gravity, η and k are the viscosity and the thermal conductivity of the fluid, respectively, α is the thermal expansion coefficient, and H is the distance between the plates. The critical value of Rayleigh number is $Ra_c = 1707$. Therefore the critical temperature difference is given as

$$(T_1 - T_2)_{\text{crt}} = 1707\eta k / (gH^3\alpha) \tag{23}$$

Here the dissipative coefficients η or k and the expansion coefficient represent two opposing forces in Benard's phenomenon.

According to the hydrodynamics analysis, the approximate velocity distribution in the Benards cells is given by

$$v(x) = [(Ra - Ra_c)/C]^{1/2} \cos(2\pi x/\lambda) \tag{24}$$

where λ is the repetition length of the horizontal cells, and C is a constant. If we plot the velocity versus the Rayleigh number, we have the bifurcation phenomena as seen in Fig. 2. When the temperature difference is above a critical level, the

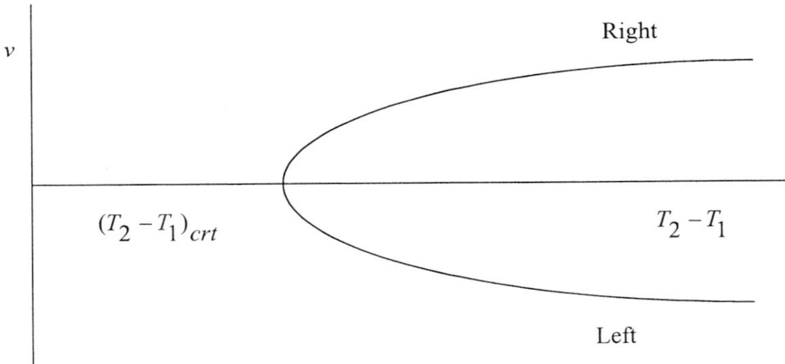

Fig.2. Bifurcation phenomena in Benard's cells

resting fluid becomes unstable and it rotates in two structural states: one rotating toward the right and the other rotating toward the left.

2.2. Ordering in chemical reactions

One theoretical model in ordering in chemical reactions is the Brusselator scheme, which considers the following reactions

$$\mathrm{A} \xrightarrow{k_1} \mathrm{X} \tag{25}$$

$$\mathrm{B} + \mathrm{X} \xrightarrow{k_2} \mathrm{Y} + \mathrm{C} \tag{26}$$

$$2\mathrm{X} + \mathrm{Y} \xrightarrow{k_3} 3\mathrm{X} \tag{27}$$

$$\mathrm{X} \xrightarrow{k_4} \mathrm{E} \tag{28}$$

We assume that the components A and B are supplied continuously so that the concentrations remain constant. The products C and E are withdrawn from the system; the compounds X and Y can diffuse into the system. Assuming that all the rate constants are unity, according to the law of mass action, the change in the concentrations of X and Y are expressed by

$$\frac{d\mathrm{X}}{dt} = \mathrm{A} - (\mathrm{B}+1)\mathrm{X} + \mathrm{X}^2\mathrm{Y} + D_{\mathrm{X}} \frac{\partial^2 \mathrm{X}}{\partial z^2} \tag{29}$$

$$\frac{dY}{dt} = BX - X^2Y + D_Y \frac{\partial^2 Y}{\partial z^2} \tag{30}$$

where D_X and D_Y are the respective diffusion coefficients and z is the distance to the region. This system has a uniform steady-state solution given by X_o = A and Y_o = (B/A). The system is stable if concentration of B is less than the critical value B_c given by

$$B_c = A^2 + 1 + (m\pi / L)^2 (D_X + D_Y) \tag{31}$$

where L is the length of the system, and m is an integer. When B = B_c, the system transforms to a nonuniform state and the concentration of X becomes the function f of the position

$$X = A \pm f[\cos(m\pi z / L)] \tag{32}$$

For example if $m = 6$, a spatial distribution of X appear, and the system would undergo a self-organization showing a banded distribution of concentrations. The Belousov-Zhabotinski reaction also organizes itself into bands. The Belousov-Zhabotinski reaction is a chemical oscillatory reaction, and originally consisted of a one-electron redox catalyst, an organic substance that is easily brominated and oxidized, and a bromate ion is dissolved in acid. The typical catalyst ferroin, in its oxidized state, has a blue color, while in its reduced state ferroin is red. As the Belousov-Zhabotinski reaction alternates between the oxidized state and reduced state, the solution changes its color (Fig. 3).

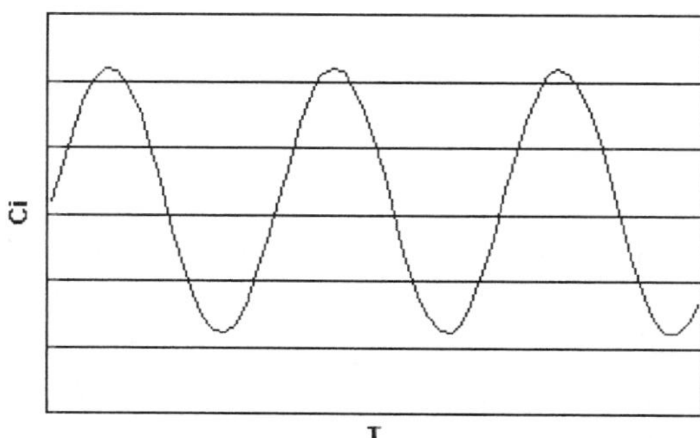

Fig.3. Spatial distribution.

Earlier, concentration oscillations were believed to be not consistent with thermodynamics. Oscillation of the extent of reaction about its equilibrium value is not consistent with the second law. However, oscillation of concentrations about a nonequilibrium value of the extent of reaction is consistent with the second law. It is highly important to investigate the oscillations leading to the dissipative structures in chemical systems.

3. ORDERING IN BIOLOGICAL STRUCTURES

Living objects are self–reproducing and are capable of creating information that influences their evolution and self-reproduction. The ability to create new information gives a meaning to ordering, which leads to the qualitative information needed. We see the ordering in living objects at all level of organization. Entropy is a measure of the degree of ordering. Schrodinger in his book called " *What Is Life*?" stated that living objects feed on negative entropy. The antientropic character of biological evolution is the common trend of discussion among physicists and biologists. Instability and fluctuations to organized states in nonequilibrium systems result the evolution in biological systems.

3.1. Ordering in time: Biological clocks

When the system is far away from equilibrium, ordering in time or spontaneous rhythmic behavior occurs. The *Lotka-Volterra model* of the predator-prey interactions is a simple example of the rhythmic behavior. The interactions are described by

$$J_1 = \frac{dX_1}{dt} = k_1 X_1 - k_2 X_1 X_2 \qquad (33)$$

$$J_2 = \frac{dX_2}{dt} = k_3 X_1 X_2 - k_4 X_2 \qquad (34)$$

where the terms X_1 and X_2 represent the number of individuals of species 1 and 2 respectively, and k_1 and k_4 are the biotic potentials, which are the difference between the birth and death rates respectively. The terms k_2 and k_3 are the interactions between the both populations. The flows shown by J_1 and J_2 have two stationary solutions; first of these is where $X_1 = X_2 = 0$, and the second is that the stationary values of X_{1s} and X_{2s}, which are given by

$$X_{1s} = k_4 / k_3 \qquad (35)$$

$$X_{2s} = k_1 / k_2 \tag{36}$$

To see whether this state is stable or not, we add small perturbations of δX_1 and δX_2 to X_{1s} and X_{2s}, so that

$$X_1 = X_{1s} + \delta X_1 \tag{37}$$

$$X_2 = X_{2s} + \delta X_2 \tag{38}$$

We can introduce these expressions into Eqs. (33) and (34) and neglect the terms containing products of δX_1 and δX_2, and we obtain

$$\frac{d(\delta X_1)}{dt} = k_1 \delta X_1 - k_2 (X_{1s} \delta X_2 + X_{2s} \delta X_1) \tag{39}$$

$$\frac{d(\delta X_2)}{dt} = k_3 (X_{1s} \delta X_2 + X_{2s} \delta X_1) - k_4 \delta X_2 \tag{40}$$

With Eqs. (35) and (36), we can rewrite Eqs. (39) and (40) as follows

$$\frac{d(\delta X_1)}{dt} = -k_2 X_{1s} \delta X_2 \tag{41}$$

$$\frac{d(\delta X_2)}{dt} = k_3 X_{2s} \delta X_1 \tag{42}$$

We now differentiate Eq. (41) with respect to time and after using Eq. (40), we have

$$\frac{d^2(\delta X_1)}{dt^2} = -k_2 k_3 X_{1s} X_{2s} \delta X_1 = -k_1 k_4 \delta X_1 \tag{43}$$

This equation has the form for the motion of a harmonic oscillator, and the solution yields a harmonic oscillation

$$\delta X_1 = \delta X_1(0) \cos(2\pi f t) \tag{44}$$

where f is the frequency given by

$$f = (1/2\pi)(k_2 k_3 X_{1s} X_{2s})^{1/2} = (1/2\pi)(k_1 k_4)^{1/2} \tag{45}$$

Therefore the stationary states given by Eqs. (35) and (36) are not stable. This means that after the small perturbations system does not return to the original state; instead it oscillates. This oscillatory behavior can be explained with the following example: As X_1 increases, species of 2 have more food and tend to increase their population. As X_2 increases, the species of 1 consumed by those of species 2 also increase. Therefore X_1 begins to decrease, and hence the amount of food available for species 2 decreases; this leads to decrease of X_2. As the number of predators X_2 decreases, the population of prey X_1 recovers, and this cause the start of a new cycle.

The amplitude and period of oscillation depend on the initial state. In the dissipative systems, we frequently encounter the behavior called the limit cycle. In this case, regardless of the initial conditions, the system tends toward a certain orbit or an oscillation.

An example of the limit cycle is presented by the set of equations called the Brusselator. From Eqs. (29) and (30), we have

$$\frac{dX}{dt} = k_1 A - k_2 BX + k_3 X^2 Y - k_4 X \qquad (46)$$

$$\frac{dY}{dt} = k_2 BX - k_3 X^2 Y \qquad (47)$$

The stationary solutions are obtained as

$$X_s = A\frac{k_1}{k_4} \qquad (48)$$

$$Y_s = \frac{B}{A}\frac{k_2 k_4}{k_1 k_3} \qquad (49)$$

If we assume the following perturbations

$$X = X_s + \alpha \exp(\lambda t) \qquad (50)$$

$$Y = Y_s + \beta \exp(\lambda t) \qquad (51)$$

and introduce these perturbations into equations (46) and (47), we obtain

$$\alpha\lambda = [B(k_2/k_1) - 1]\alpha + A^2 (k_1/k_4)^2 (k_3/k_4)\beta \qquad (52)$$

$$\beta\lambda = -\mathrm{B}(k_2 / k_4)\alpha - \mathrm{A}^2 (k_1 / k_2)^2 (k_3 / k_4)\beta \tag{53}$$

where we have neglected the terms quadratic in α and β. For the nonzero initial perturbations, the determinant of the Eqs. (52) and (53) should be equal to zero, which leads to the relation

$$\lambda^2 + (a^2 + 1 - b) + a^2 = 0 \tag{54}$$

where

$$a = \mathrm{A}(k_1 / k_4)(k_3 / k_4)^{1/2}$$

$$b = \mathrm{B}(k_2 / k_4)$$

If the solutions for Eq. (54) have a real and positive part, then the perturbations grow and make the system unstable. If the real parts of λ_1 and λ_2 are negative, then the perturbations decrease, and the system becomes stable. The solutions of λ_1 and λ_2 are given by

$$\lambda_{1,2} = -(1/2)(a^2 + 1 - b) \pm [(a^2 + 1 - b)^2 - 4a^2]^{1/2} \tag{55}$$

The system will be stable if $b < a^2+1$, and it will be unstable if $b > a^2+1$. When $b = a^2+1$, and λ_1 and λ_2 are purely imaginary, we have an undamped oscillation of the perturbations similar to Lotka-Volterra model. If $a^2+2a+1 > b > a^2-2a+1$, and λ_1 and λ_2 have a nonzero imaginary part, then we have the oscillatory behavior. These oscillations damp out in the stable zone ($a^2+2a+1 > b > a^2+1$). When $b > a^2+2a+1$, the limit cycle oscillations occur, which are independent of the initial values of the perturbations in X and Y.

One of the best-known oscillatory reactions is the Belousov-Zhabotinskii reaction, which contains a set of oxidation-reduction steps; the oxidizing agent is bromate (BrO_3^-), the reducing agent is malonic acid [$H_2C(COOH)_2$], and cesium ions are used as the catalyst. The concentrations of Ce^{+3} and Ce^{+4} vary periodically with a frequency on the order of 0.01 Hz. The overall reaction is complex, however its oscillatory effects can be understood by studying the following reaction steps

$$Ce^{+3} \xrightarrow{BrO_3^-, H^+} Ce^{+4} \qquad \text{(oxidation)} \tag{56}$$

$$Ce^{+4} \xrightarrow{\text{malonic acid}} Ce^{+3} \qquad \text{(reduction)} \tag{57}$$

At the start, the cycle begins with a certain amount of Ce^{+4} ions. The second reaction provides Br^- ions, which inhibits the first reaction. This leads to increase of concentration of Ce^{+3}. After reaching a certain amount of Ce^{+3}, oxidation reaction starts since little Ce^{+4} remains. The system can no longer produce sufficient Br^- to inhibit the reaction, and Ce^{+3} decreases rapidly, producing Ce^{+4} until the cycle is completed.

It is argued that the biological systems should be oscillatory, which may be the direct result of biological evolution on earth. Earth moves around the sun and rotates about its own axis. These periodicities induce rhythms in the changes of temperature, light, humidity, and have reflected effects in the physiology of living systems. The periodicity of day and night is characteristics of the living systems. Related to these rhythmic changes, the concept of biological clocks has been introduced, along with the circadian rhythms for oscillations with a time span of approximately 24 hours. The circadian rhythms are generated internally, since their periods are practically independent of environmental factors. Periodic self-oscillatory processes are characteristics of the processes of glycolsis (anaerobic catabolism of glucose) and in the conversion of ADP to ATP. Some reactions display oscillatory behavior in the transport of substrates across the membranes, such as the facilitated transport.

4. BIFURCATION

When it is sufficiently far from equilibrium, the system arrives at a *bifurcation* of states of ordered structures in time and in space. Transitions between different modes of dynamic organizations are called bifurcation. If the system continues to move away from equilibrium, the structures become more complex leading to a situation called chaotic in the macroscopic sense. Recently some regularities have been discovered in such chaotic behavior. To study the behavior of the systems far away from the equilibrium a new interdisciplinary field called the *synergetic* is developed. Synergetic is concerned with the cooperation of individual parts of the system that produces macroscopic spatial and temporal structures, which are mainly dissipative.

If we define the distance from equilibrium by some parameter β (for example, a temperature or concentration gradient) a value of β_1 is reached, and the system display ordering characterized by a certain frequency or a wavelength. Fig. 2 shows the bifurcation in the velocity in Bernard's cells. If the parameter β is increased further, we may reach a value of β_2 at which the system is characterized by two frequencies. This behavior can be repeated for the values of β_3, β_4, leading to increasingly complicated states.

In 1975 Mitchell Feigenbaum described a general character among the critical values of β_i given as

$$Lim_{\text{large } n}\frac{\beta_n-\beta_{n-1}}{\beta_{n+1}-\beta_n}\rightarrow 4.669201... \tag{58}$$

and

$$Lim_{\text{large } n}\frac{\varepsilon_n}{\varepsilon_{n+1}}\rightarrow 2.5029.... \tag{59}$$

where ε_l is the separation between the two branches coming from the (i-1)th bifurcation. Therefore the scheme of the bifurcation displays a series of universal characteristics independent of the specific system. These universal numbers were observed for example, in the transition of laminar flow to turbulent flow, and in chemical and electrical systems. These Feigenbaum numbers characterize the order-chaos transition, and reveal certain regularity in chaos or in one of the series of processes that leads to chaos.

Bifurcation and chaos can be expressed in finite difference equations, which have a general form

$$x_{i+1}=f(x_i) \tag{60}$$

If we assume that the function f is a quadratic function, we have

$$x_{i+1}=ax_i(1-x_i) \qquad (0<a<4) \tag{61}$$

As parameters change in finite difference equations, bifurcation (for example changes in qualitative dynamics) is found. A steady state x^* is a value for which $x^*=f(x^*)$. For Eq. (54) there are two steady states: $x^*=0$ and $x^*=(a\text{-}1)/a$.

A cycle of period m is defined by

$$x_{i+n}=x_i \quad \text{and} \quad x_{i=j}\neq x_i \qquad \text{for } j=1,..,m\text{-}1 \tag{62}$$

Stability of steady states and cycles implies restoration of steady states and cycles, respectively, following a small perturbation. One type of bifurcation is period-doubling bifurcation, in which a stable cycle of period n becomes unstable and a new stable cycle of period $2n$ is generated with a new parameter. Eq. (61), for example produces successive period-doubling as a increases. For $3<a<3.57$, stable cycles of lengths 1, 2, 4, 8, 16, 32, 64 are generated.

Bifurcation, instability, multiple solutions and symmetry breaking are all related to each other. Chemical systems of life show an asymmetry. The bifurcation of a solution indicates that a solution that loses stability. This is a general property of the solution of nonlinear equations.

5. EXTENDED NONEQUILIBRIUM THERMODYNAMICS

One of the problems of nonequilibrium thermodynamics is the definition of entropy in nonequilibrium systems. Nonequilibrium thermodynamics is based on the local-equilibrium hypothesis, which allows the classical thermodynamics formulations to be valid at local state. The formulations include the Gibbs equation, thermodynamics potentials, stability conditions, and equations of state. The local thermodynamic equilibrium hypothesis has provided useful formalism with a unifying power, and led to useful predictions and interpretations of transport coefficients.

Typical nonequilibrium variables, such as flows, gradients of intensive properties may contribute to the rate of entropy generation. When the relaxation time of these variables is different from the observation time they act as constant parameters. The phenomenon becomes complex when the observation time and the relaxation time are of the same order, and the description of system requires additional variables.

A system at steady state with minimum entropy production is the continuous extension of the thermodynamic equilibrium state into the nonequilibrium state. After a critical distance from the thermodynamic equilibrium, where the forces are no longer linear functions of the flows, a system may become unstable, and have more than one possible steady state. Chemical reaction systems are in the linear region if the affinities A are small compared to RT ($A \ll 2.5$ kJ/mol at $T = 300$ K). The affinities are mostly in the range of 10-100 kJ/mol, and hence chemical reactions are in the nonlinear region. The stability of an existing structure is constantly tested by internal fluctuations. If the fluctuations are large enough, the system may become unstable and bifurcate to a new space-time structure. As the Belousov-Zhabotinski reaction shows that, these complex structures can arise as the solution of deterministic differential equations. The process and the boundary conditions do not uniquely specify the new nonequilibrium state, which can be highly structured state. *Extended nonequilibrium thermodynamics* is concerned with the nonlinear region and deriving the evolution equations with the dissipative flows as independent variables, beside the usual conserved variables. Some of the keyword and concepts are stability, excess entropy production, and dissipative structures.

Extended nonequilibrium thermodynamics is not based on the local-equilibrium hypothesis, and uses the conserved variables and nonconserved dissipative fluxes as the independent variables to establish evolution equations for the dissipative fluxes satisfying the second law of thermodynamics. For hydrodynamic systems, the independent variables related to conservation laws are the mass density, momentum density, and specific internal energy, while the nonconserved variables are the heat flux, shear and bulk viscous pressure, diffusion flux, and electrical flux. For the generalized entropy with the properties of additivity and convex function [15] considered, extended nonequilibrium thermodynamics provides a more complete formulation of nonequilibrium thermodynamics and nonequilibrium equations of state; gradients are also included in the formulation; it can relate microscopic phenomena to a macroscopic thermodynamic interpretation by deriving the generalized transport laws expressed in terms of the generalized frequency and wave-vector dependent transport coefficients.

In extended nonequilibrium thermodynamics, the generalized Gibbs equation for a fluid characterized by internal energy U and viscous pressure $\mathbf{P}_v$ is expressed by

$$dS = \frac{1}{T}dU - \frac{\tau V}{2\eta T}\mathbf{P}_v : d\mathbf{P}_v \qquad (1)$$

where S is the entropy, V is the volume, T is the absolute temperature, η is the shear viscosity, and τ is the viscoelastic relaxation time for the viscous pressure tensor [21]. When the viscous pressure tensor is a sum of contributions with different relaxation times τ_i, and different viscosities η_i, Eq. (1) is generalized by

$$dS = \frac{1}{T}dU - \sum_i \frac{\tau_i V}{2\eta_i T}\mathbf{P}_{vi} : d\mathbf{P}_{vi} \qquad (2)$$

The extended nonequilibrium thermodynamics theory and the general equation for the nonequilibrium reversible-irreversible coupling (GENERIC) are widely applied for diffusion in polymers, flows through porous media, and description of behavior of helium [16,17,19-21].Diffusion in polymers involves the coupling between viscous stresses and diffusion, and the classical diffusion laws of Flick and Fourier are not applicable. The extended nonequilibrium thermodynamics is also used in expressing a transport equation of the dual-phase-lag model of heat transfer [18].

Polymer solutions may have the memory effects observed in viscoelastic phenomena. This requires additional relaxation terms in the constitutive equations for the viscous pressure tensor, which may be affected by the changes in the velocity gradient. Beside that the orientation and stretching of the macromolecules may have an influence on the flow [16].

Using the extended nonequilibrium thermodynamics for a binary liquid mixture, the viscous pressure tensor and the diffusion flux are considered as additional independent variables; the corresponding extended Gibbs equation is given by

$$ds = \frac{1}{T}du + \frac{1}{T}Pdv - \frac{1}{T}\Delta\mu\, dw_1 - v\alpha_1 \mathbf{J}\cdot d\mathbf{J} - v\alpha_2 \mathbf{P}_v : d\mathbf{P}_v \qquad (3)$$

where T, P, u and v are the temperature, pressure, internal energy, and volume, respectively; w_1 is the mass fraction of the solute, $\Delta\mu = \mu_1 - \mu_2$ is the difference between the specific chemical potentials of the solute and the solvent, α_1 and α_2 are the coefficients, and J is the diffusion flux. In Eq. (3), T, P and $\Delta\mu$ depend on the fluxes J and $\mathbf{P}_v$. Various formalisms on rheology indicating the connection between thermodynamics and dynamics are presented by Jou and Casa-Vazquez [16].

Fluctuation theory has been the basis for the derivation of the Onsager reciprocal relations. This theory also plays important role in extended nonequilibrium thermodynamics in the following stages: (i) describing the theory and the coefficients appearing in the generalized entropy, (ii) relating several parameters to each other, and hence reducing the number of independent parameters in the theory, and (iii) providing a link between extended nonequilibrium thermodynamics and nonequilibrium statistical mechanics with the fluctuation domain.

The following relations are from Casas-Vazquez and Jou [15]. In the classical approach, the state of single component system is described by the velocity **v**, and Two other thermodynamic variables, such as the specific volume v and the specific internal energy u, the time evolution of these variables is described by the general balance equations of mass, momentum, and internal energy, which are expressed by

$$\rho \frac{\partial \mathbf{v}}{\partial t} = \nabla \cdot \mathbf{P} + \rho \mathbf{F} \tag{1}$$

$$\rho \frac{\partial v}{\partial t} = \nabla \cdot \mathbf{v} \tag{2}$$

$$\rho \frac{\partial u}{\partial t} = -\nabla \cdot \mathbf{q} - \mathbf{P} : \nabla \mathbf{v} \tag{3}$$

where **P** is the pressure tensor, **q** is the heat flux, and **F** is the body force per unit mass. In the following sections, it is assumed that **F** vanishes.

The pressure tensor is usually split into the thermodynamic pressure p and a viscous part $\mathbf{P}_v$, which is further split into a scalar bulk viscous pressure p_v and a traceless shear viscous tensor $\mathbf{P}_{vo}$

$$\mathbf{P} = p\mathbf{U} + p_v \mathbf{U} + \mathbf{P}_{vo} \tag{4}$$

where **U** is the identity tensor.

Similarly the velocity gradient is split into a scalar v, a traceless symmetric $\mathbf{v}_o$, and an antisymmetric part $\mathbf{w}$

$$\nabla\mathbf{v} = (1/3)(\nabla\cdot\mathbf{v})\mathbf{U} + \mathbf{v}_o + \mathbf{w} \tag{5}$$

Due to the local equilibrium hypothesis, The Gibbs equation is given as

$$ds = \frac{du}{T} + \frac{pdv}{T} \tag{6}$$

The time derivative of the entropy is given by

$$\rho\frac{ds}{dt} = \rho\frac{1}{T}\frac{du}{dt} + \rho\frac{1}{T}\frac{pdv}{dt} \tag{7}$$

After introducing Eq. (2) and (3) into Eq. (7), we have

$$\rho\frac{ds}{dt} = -\nabla\cdot\left(\frac{\mathbf{q}}{T}\right) - \frac{\mathbf{q}}{T^2}\cdot\nabla T - \frac{1}{T}\mathbf{P}_v\nabla\cdot\mathbf{v} - \frac{1}{T}\mathbf{P}_{vo}:\mathbf{v}_o \tag{8}$$

The most general form of the entropy balance is expressed by

$$\rho\frac{ds}{dt} + \nabla\cdot\mathbf{J}_s = \Phi \tag{9}$$

where Φ is the entropy production, Comparison of Eqs. (8) and (9) yields

$$\mathbf{J}_s = \frac{\mathbf{q}}{T} \tag{10}$$

and we obtain

$$\Phi = -\frac{\mathbf{q}}{T^2}\cdot\nabla T - \frac{1}{T}\mathbf{P}_v\nabla\cdot\mathbf{v} - \frac{1}{T}\mathbf{P}_{vo}:\mathbf{v}_o \tag{11}$$

Eq. (11) shows that the entropy generation Φ is a bilinear form in the thermodynamic fluxes $\mathbf{q}$, $\mathbf{P}_v$, and $\mathbf{P}_{vo}$ and their respective conjugate forces. According to the second law of thermodynamics, the entropy generation is positive and related phenomenological relations are given by

$$\mathbf{q} = -s_1 \frac{1}{T^2} \nabla T \tag{12}$$

$$\mathbf{P}_v = -s_o \frac{1}{T} \nabla \cdot \mathbf{v} \tag{13}$$

$$\mathbf{P}_{vo} = s_2 \frac{1}{T} \mathbf{v}_o \tag{14}$$

The positive coefficients s_1, s_0 and s_2 are related to the thermal conductivity *k*, bulk viscosity μ and shear viscosity η, respectively

$$s_1 = kT^2\,,\ s_o = \mu T\,,\ \ s_2 = 2\eta T \tag{15}$$

These classical transport laws are not satisfactory at high frequencies, or in systems with long relaxation times, such as viscoelastic liquids, and heat conduction at low temperatures.

Jou and Casas-Vazquez elaborated the theory of rational extended theory [16]; the space of independent variables is modified; beside the classical state variables, the response of the system is described by the evolution differential equations for the additional independent variables rather than in terms of the constitutive functional in rational thermodynamics. The entropy inequality is combined with the evolution equations by means of Lagrange multipliers.

REFERENCES

[1] S.R. Caplan and A. Essig, Bioenergetics and Linear Nonequilibrium Thermodynamics, The Steady State, Harvard University Press, Cambridge, 1983.

[2] J.T. Edsall and H. Gutfreund, Biothermodynamics: To study of Biochemical Processes at Equilibrium, Wiley, Chichester, 1983.

[3] P. Glansdorff and I. Prigogine, Thermodynamic Theory of Structure, Stability and Fluctuations, Wiley, New York 1971.

[4] H. Haken, Synergetics: An Introduction to Nonequilibrium Phase Transitions and Self-Organizations in Physics, Chemistry and Biology, 2nd ed., Springer, Berlin, 1978.

[5] D. Jou and J.E. Llebot, Introduction to the Thermodynamics of Biological Processes, Prentice Hall, Englewood Cliffs, NJ, 1990.

[6] A. Katchalsky and P.F. Curran, Nonequilibrium Thermodynamics in Biophysics, Harvard University Press, Cambridge, 1967.

[7] D. Kondepudi and I. Prigogine, Modern Thermodynamics, From Heat Engines to Dissipative Structures, Wiley, New York, 1999.

[8] G. Nicolis and I. Prigogine, Self-Organization in Nonequilibrium Systems, Wiley, New York, 1977.

[9] S. Sieniutycz and P. Salamon, (eds.), Nonequilibrium Theory and Extremum Principles, Taylor & Francis, New York, 1990.
[10] S. Sieniutycz and P. Salamon, (eds.), Extended Thermodynamic Systems, Advances in Thermodynamics , Vol. 7, Taylor & Francis, New York, 1992.
[11] A.I. Zotin, Thermodynamic Bases in Biological Processes, Walter de Gruyter, Berlin, 1990.
[12] D. Jou, Extended Irreversible Thermodynamics, Springer-Verlag, New York, 1996.
[13] J. Nedelman and S.I. Rubinov, J. Math. Biol., 12 (1981) 73.
[14] J. Nedelman and S. I. Rubinow, J. Mathematical Biology, 12 (1981) 73.
[15] J. Casas-Vasquez and D. Jou, In Extended Thermodynamic Systems, Advances in Thermodynamics, Vol. 7. Ed. by S. Sieniutcyz and P. Salamon, Taylor & Francis, New York, 1992.
[16] D. Jou and J. Casas-Vazquez, J. Non-Newtonian Fluid Mech., 96 (2001) 77.
[17] N. Despireux and G. Lebon, J. Non-Newtonian Fluid Mech., 96 (2001) 105.
[18] S.I. Serdyukov, Phys. Lett. A, 281 (2001) 16.
[19] M. Grmela, D. Jou and J. Casas-Vazquez, J. Chem. Phys., 108 (1998)7937.
[20] D. Jou, J. Casas-Vazquez and M. Criado-Sancho, Physica A, 262 (1999) 69.
[21] A.N. Beris and S.J. Edwards, Thermodynamics of flowing Fluids with Internal Microstructure, Oxford University Press, New York, 1994.

Appendix A

Table A1. Coefficients in the smoothing equation $k(W/mK)=\sum_{i=0}^{4} a_i x_1^i$ for the thermal conductivity of alkanes in chloroform.

Solute (comp. 1)	a_o	a_1	a_2	a_3	a_4	% E*
n-hexane	0.11076	-0.03777	0.04917	0.01362	-0.02051	0.22
n-heptane	0.11079	-0.04462	0.11050	-0.07782	0.02156	0.05
n-octane	0.11081	-0.04457	0.14569	-0.14027	0.05211	0.16
3-methylpentane	0.11079	-0.06830	0.10737	-0.04171	-0.00121	0.11
2,3-dimethylpentane	0.11080	-0.04081	0.04925	-0.00950	-0.00453	0.17
2,2,4-trimethylpentane	0.11080	-0.04350	0.01113	0.05298	-0.03542	0.08

* $E=\frac{1}{N}\sum_{i=1}^{N}\left|(f_{i,calc}-f_{i,\exp})/f_{i,\exp}\right|$

Table A2. Coefficients in the smoothing equation $D(10^{-9}m^2/s)=\sum_{i=0}^{4} a_i x_1^i$ for the mutual diffusion coefficients of alkanes in chloroform.

Solute (comp. 1)	a_o	a_1	a_2	a_3	a_4	% E
n-hexane	2.4394	-1.0243	3.9930	-0.1670	-0.7710	0.14
n-heptane	2.2368	-0.7534	2.3321	2.7348	-3.0324	0.08
n-octane	2.0367	-1.0923	3.2118	0.6468	-1.7149	0.06
3-methylpentane	2.2851	-0.2589	2.7260	-1.3769	1.2063	0.08
2,3-dimethylpentane	2.0936	-1.1928	8.4556	-11.7748	6.0184	0.08
2,2,4-trimethylpentane	1.9529	-0.4573	1.2118	1.4582	-1.0782	0.00

Table A3. Coefficients in the smoothing equation $-Q_1^{**}(kJ/kg) = \sum_{i=0}^{3} a_i x_1^i$ for the heats of transport of alkanes in chloroform.

Solute (comp.1)	a_o	a_1	a_2	a_3	% E
n-hexane	69.7765	-89.6873	127.5539	-78.3904	0.21
n-heptane	63.8333	-71.3086	68.3994	-31.5108	0.21
n-octane	61.3523	-79.0694	110.1027	-73.4022	0.10
3-methylpentane	66.8969	-75.9630	90.5502	-47.8497	0.08
2,3-dimethylpentane	62.2974	-71.8306	80.2944	-47.6851	0.14
2,2,4-trimethylpentane	59.7674	-76.4213	99.8976	-66.2341	0.45

Table A4. Coefficients in the smoothing equation $k(W/mK) = \sum_{i=0}^{4} a_i x_1^i$ for the thermal conductivity of alkanes in carbon tetrachloride.

Solute (comp. 1)	a_o	a_1	a_2	a_3	A_4	% E
n-heptane	0.09977	-0.01821	-0.02666	0.15875	-0.09150	0.05
n-octane	0.09979	-0.04871	0.16253	-0.14561	0.05612	0.01
3-methylpentane	0.09980	-0.06310	0.15139	-0.12861	0.04580	0.01
2,3-dimethylpentane	0.09969	-0.03883	0.07447	-0.02811	0.0	0.11
2,2,4-trimethylpentane	0.09981	-0.07541	0.18953	-0.21637	0.10198	0.02

Table A5. Coefficients in the smoothing equation $D(10^{-9} m^2 / s) = \sum_{i=0}^{4} a_i x_1^i$ for the mutual diffusion coefficients of alkanes in carbon tetrachloride.

Solute (comp. 1)	a_o	a_1	a_2	a_3	A_4	% E
n-hexane	1.5886	1.9757	-2.7016	6.6258	-3.6101	0.08
n-heptane	1.5027	1.4798	-0.5978	1.9214	-1.1210	0.08
n-octane	1.4127	0.3903	2.3090	-1.0357	-0.5235	0.15
3-methylpentane	1.4940	1.7198	-1.3199	4.6769	-2.5842	0.06
2,3-dimethylpentane	1.3964	0.9422	1.1393	-0.7047	0.4628	0.02
2,2,4-trimethylpentane	1.2280	0.4974	2.3803	-1.9014	0.4919	0.01

Table A6. Coefficients in the smoothing equation $-Q_1^{''*}(kJ / kg) = \sum_{i=0}^{3} a_i x_1^i$ for the heats of transport of alkanes in carbon tetrachloride.

Solute (comp. 1)	a_o	a_1	a_2	a_3	% E
n-hexane	68.5474	-90.6118	118.5839	-63.0289	0.01
n-heptane	65.8806	-85.9901	111.5736	-69.6799	0.01
n-octane	63. 0431	-80.5948	83.2332	-50.3668	0.13
3-methylpentane	70.7849	-122.0631	221.6195	-138.4118	0.07
2,3-dimethylpentane	61.3236	-75.3408	88.1979	-55.9357	0.40
2,2,4-trimethylpentane	58.4977	-106.5203	179.0495	-122.3974	0.01

Appendix B

Tensors

Scalars are specified by a single numerical value. Vectors come with directions as well as numerical values. In a three-dimensional space, a tensor of rank n is determined by 3^n elements. A scalar is a tensor with the rank zero, hence $3^0 = 1$. A vector is tensor with the rank 1, hence $3^1 = 3$. For a tensor with n = 2, we have $3^2 = 9$ elements.

Differentiation of a tensor with respect to a scalar does not change its rank. The spatial differentiation of a tensor raises its rank by unity, and identical to multiplication by the vector ∇, called del or Hamiltonian operator or the nabla

$$\nabla = \frac{\partial}{\partial \mathbf{x}}$$

The gradient of a scalar field *a* is a vector

$$\text{grad}a = \frac{\partial a}{\partial \mathbf{x}} = \nabla a$$

The derivative of a scalar a with respect to a vector is a vector. The gradient of a vector field ***v*** is a tensor of rank two

$$\text{Grad}\mathbf{v} = \frac{\partial \mathbf{v}}{\partial \mathbf{x}} = \nabla \mathbf{v}$$

When contraction is performed once (summation over repeated indices) the divergence is obtained instead of the gradient. The divergence of a vector field ***v*** is a scalar

$$\text{div}\mathbf{v} = \frac{\partial}{\partial \mathbf{x}} \cdot \mathbf{v} = \nabla \cdot \mathbf{v}$$

The divergence of a tensor field **T** is a vector

$$\text{Div}\mathbf{T} = \frac{\partial}{\partial \mathbf{x}} \cdot T = \nabla \cdot \mathbf{T}$$

The Laplace operator or Laplacian is a scalar

$$\nabla^2 = \nabla \cdot \nabla = \text{div grad} = \frac{\partial}{\partial \mathbf{x}} \cdot \frac{\partial}{\partial \mathbf{x}}$$

Nonequilibrium thermodynamics often uses the Gauss-Ostrogradsky theorem, which states that the flux of a vector through a surface a is equal to the volume integral of the divergence of the vector v for the space of volume V bounded by that surface

$$\int_a \mathbf{v} \cdot d\mathbf{a} = \int_V \text{div } \mathbf{v} dV = \int_V \nabla \cdot \mathbf{v} dV$$

LARS ONSAGER

Lars Onsager was born in Oslo, Norway, November 27, 1903 to parents Erling Onsager, Barrister of the Supreme Court of Norway, and Ingrid, née Kirkeby. In 1933 he married Margarethe Arledter, daughter of a well-known pioneer in the art of paper making, in Cologne, Germany. They have sons Erling Frederick, Hans Tanberg, and Christian Carl, and a daughter Inger Marie, married to Kenneth Roy Oldham.

After three years with the experienced educators Inga and Anna Platou in Oslo, one year at a deteriorating private school in the country and a few months of his mother's tutoring, he entered Frogner School as the family returned to Oslo. There he was soon invited to jump a grade, so that he was able to graduate in 1920.

Admitted to Norges tekniske høgskole in the fall of that year as a student of chemical engineering, he entered a stimulating environment; the department had attracted outstanding students over a period of years. Among the professors particularly O.O. Collenberg and B. Holtsmark encouraged his efforts in theory and helped him in the evaluation of background knowledge.

After graduation in 1925 he accompanied Holtsmark on a trip to Denmark and Germany, then proceeded to Zurich, where he remained for a couple of months with Debye and Hückel and returned the following spring, for a stay of nearly two years. There he organized his results in the theory of electrolytes for publication, broadened his knowledge of physics and became acquainted with a good many leading physicists.

The great importance of irreversible thermodynamics becomes apparent if we realize that almost all common processes are irreversible and cannot by themselves go backwards. As examples can be mentioned conduction of heat from a hot to a cold body and mixing or diffusion. When we dissolve a cold lump of sugar in a cup of hot tea these processes take place simultaneously.

Onsager's great contribution was that he could prove that if the equations governing the flows are written in an appropriate form, then there exist certain simple connections between the coefficients in these equations. These connections - the reciprocal relations - make possible a complete theoretical description of irreversible processes.

The proof of the reciprocal relations was brilliant. Onsager started from a statistical mechanical calculation of the fluctuations in a system, which could be directly based on the simple laws of motion, which are symmetrical with regard to time. Furthermore he made the independent assumption that the return of a fluctuation to equilibrium in the mean occurs according to the transport equations mentioned earlier. By means of this combination of macroscopic and microscopic concepts in conjunction with an extremely skilful mathematical analysis he obtained those relationships which are now called Onsager s Reciprocal Relations.

ILYA PRIGOGINE

I was born in Moscow, on the 25th of January, 1917 - a few months before the revolution. My family had a difficult relationship with the new regime, and so we left Russia as early as 1921. For some years (until 1929), we lived as migrants in Germany, before we stayed for good in Belgium. It was at Brussels that I attended secondary school and university. I acquired Belgian nationality in 1949.

My father, Roman Prigogine, who died in 1974, was a chemical engineer from the Moscow Polytechnic. My brother Alexander, who was born four years before me, followed, as I did myself, the curriculum of chemistry at the Universite Libre de Bruxelles. I remember how much I hesitated before choosing this direction; as I left the classical (Greco-Latin) section of Ixelles Athenaeum, my interest was more focused on history and archaeology, not to mention music, especially piano. According to my mother, I was able to read musical scores before I read printed words. And, today, my favorite pastime is still piano playing, although my free time for practice is becoming more and more restricted.

"The more deeply we study the nature of time, the better we understand that duration means invention, creation of forms, continuous elaboration of the absolutely new." Fortunate coincidences made the choice for my studies at the university. Indeed, they led me to an almost opposite direction, towards chemistry and physics. And so, in 1941, I was conferred my first doctoral degree. Very soon, two of my teachers were to exert an enduring influence on the orientation of my future work.

I would first mention Théophile De Donder (1873-1957).[2] What an amiable character he was! Born the son of an elementary school teacher, he began his career in the same way, and was (in 1896) conferred the degree of Doctor of Physical Science, without having ever followed any teaching at the university.

It was only in 1918 - he was then 45 years old - that De Donder could devote his time to superior teaching, after he was for some years appointed as a secondary school teacher. He was then promoted to professor at the Department of Applied Science, and began without delay the writing of a course on theoretical thermodynamics for engineers.

In order to understand fully the originality of De Donder's approach, I have to recall that since the fundamental work by Clausius, the second principle of thermodynamics has been formulated as an inequality: "uncompensated heat" is positive - or, in more recent terms, entropy production is positive. This inequality refers, of course, to phenomena that are *irreversible*, as are any natural processes. In those times, these latter were poorly understood. They appeared to engineers and physico-chemists as "parasitic" phenomena, which could only hinder something: here the productivity of a process, there the regular growth of a crystal, without presenting any intrinsic interest. So, the usual approach was to limit the study of thermodynamics to the understanding of equilibrium laws, for which entropy production is zero.

Given my interest in the concept of time, it was only natural that my attention was focused on the second principle, as I felt from the start that it would introduce a new, unexpected element into the description of physical world evolution. No doubt it was the same impression illustrious physicists such as Boltzmann and Planck would have felt before me. A huge part of my scientific career would then be devoted to the elucidation of macroscopic as well as microscopic aspects of the second principle, in order to extend its validity to new situations, and to the other fundamental approaches of theoretical physics, such as classical and quantum dynamics.

In this way, I was confronted with the precise application of thermodynamical methods, and I could understand their usefulness. In the following years, I devoted much time to the theoretical approach of such problems, which called for the use of thermodynamical methods; I mean the solutions theory, the theory of corresponding states and of isotopic effects in the condensed phase. A collective research with V. Mathot, A. Bellemans and N. Trappeniers has led to the prediction of new effects such as the isotopic demixtion of helium He^3+ He^4, which matched in a perfect way the results of later research. This part of my work is summed up in a book written in collaboration with V. Mathot and A. Bellemans, *The Molecular Theory of Solutions*.[8] From the very beginning, I knew that the minimum entropy production was valid only for the linear branch of irreversible phenomena, the one to which the famous reciprocity relations of Onsager are applicable.[13] And, thus, the question was: What about the stationary states far from equilibrium, for which Onsager relations are not valid, but which are still in the scope of macroscopic description? Linear relations are very good approximations for the study of transport phenomena (thermical conductivity, thermodiffusion, etc.), but are generally not valid for the conditions of chemical kinetics. Indeed, chemical equilibrium is ensured through the compensation of two antagonistic processes, while in chemical kinetics - far from equilibrium, out of the linear branch - one is usually confronted with the opposite situation, where one of the processes is negligible.

Symbols

A_j	chemical affinity of reaction j, Helmholtz free energy
Be	Bejan number
B^*	mobility of component
c_i	molar concentration of component i
c_p	heat capacity
D	diffusion coefficient, diameter
D'	thermal diffusion coefficient
D''	Dufour coefficient
e	total specific energy
e_p	specific energy potential
f	function, friction coefficient
F	Faraday constant, frictional forces
$\mathbf{F}$	mass force
G	Gibbs free energy
h	specific enthalpy, heat transfer coefficient
h_i	partial specific enthalpy of component I
H	enthalpy
I	charge
$\mathbf{j}_i$	diffusion flux of component i
$\mathbf{J}_i$	flux of component i
J_q	heat flux in the energy balance equation
J_q	heat flux in the entropy balance equation
J_s	entropy flux
J_u	internal energy flux
k	Boltzmann's constant, thermal conductivity
K_c	chemical equilibrium constant of a reaction
K_{ik}	phenomenological coefficients (resistance)
L_{ik}	phenomenological coefficient (conductance)
m	mass, number of independent chemical reactions
M	molar mass
N	number of moles
Nu	Nusselt number
P	pressure, total entropy generation
Pr	Prandtl number
q	degree of coupling
Q	heat, volume flow
Q^*	heat of transport
$\dot{Q}$	heat flow
r	number of chemical reactions,
R	universal gas constant
J_r	rate of reaction
s	specific entropy
s_i	partial specific entropy of component i
s_T	Soret coefficient
S	entropy
S^*	entropy transport

t time
t_i transference numbers
T absolute temperature
u specific internal energy, mobility
U internal energy
U^* internal energy transport
v specific volume, centre-of-mass or barycentric velocity
V volume
z stoichiometric parameter, valance
w_i mass fraction of component i
W work
X ratio of forces
X_i thermodynamic force

Greek Symbols
α thermal diffusivity
β electroosmotic permeability
δ unit tensor
ε local field intensity
ϕ irreversibility distribution coefficient
γ activity coefficient
η ratio of dissipations, ratio of flows
κ electrical conductance of solution
λ ratio of forces
λ_{eq} equivalence conductance
μ_i chemical potential of component i
ρ density
σ stress tensor, reflection coefficient
τ viscosity part of stress tensor, transport number
ν stoichiometric coefficient
ω permeability
ξ extent of reaction
ψ electrostatic potential
Φ entropy generation
Γ thermodynamic factor
Π osmotic pressure
Θ viscous dissipation function
Ψ dissipation function

Subscripts
e effective
gen generation
i,j,k components
max maximum
min minimum
q heat
r chemical reaction

Index